W9-DGW-421

3 –

The
Nearly
Men

The
Nearly
Men

A CHRONICLE OF
SCIENTIFIC FAILURE

MIKE GREEN

TEMPUS

This book is dedicated to Rhiannon

Frontispiece: Alan Turing's statue, Manchester.

First published 2007

Tempus Publishing
Cirencester Road, Chalford
Stroud, Gloucestershire, GL6 8PE
www.tempus–publishing.com

Tempus Publishing is an imprint of NPI Media Group

© Mike Green, 2007

The right of Mike Green to be identified as the Author
of this work has been asserted in accordance with the
Copyrights, Designs and Patents Act 1988.

All rights reserved. No part of this book may be reprinted
or reproduced or utilised in any form or by any electronic,
mechanical or other means, now known or hereafter invented,
including photocopying and recording, or in any information
storage or retrieval system, without the permission in writing
from the Publishers.

British Library Cataloguing in Publication Data.
A catalogue record for this book is available from the British Library.

ISBN 978 0 7524 4232 7

Typesetting and origination by NPI Media Group
Printed and bound in Great Britain

Contents

Acknowledgements

I would like to take this opportunity to express my gratitude to all those involved in assisting me in completing this project, and stating that I take full responsibility for making such a mess of it.

Firstly, thank you to the staff at the British Museum Library, the Royal Society, the New York City Library, the Institute of Electrical Engineers, the Boston Science Museum, the Garibaldi-Meucci Museum, the Edison Museum, the Tesla Museum in Belgrade, and the Musee Nationale d'Histoire Naturelle in Paris. To all the librarians and archivists who I pestered while putting together my research, please forgive me. I promise not to bother you again in the future (well not for a while at least).

My unreserved appreciation also goes out to Texas Instruments, Fairchild Semiconductor, Ericsonn-Marconi, Bletchley Park, and the Royal Greenwich Observatory, for helping me to get interviews, supplying photographs, and just generally being nice people. I am indebted to Emily Pearce, Campbell McCutcheon and Leafy Robinson for helping to get this book into print.

Finally thanks to all the individuals (friends, family, colleagues, etc.) that in some way supported my misguided efforts, and countless other poor unfortunates who happen to be lumbered with knowing me.

About the Author

Mike Green was born in Cardiff, in 1972. He studied at Swansea University, in his native Wales, and the University of Ghent, in Belgium. Since then he has lived in London, Antwerp and Paris, and written articles which have been published in the UK, France, Germany, Israel, China, Japan, Russia and the United States for a variety of different magazines, including *Electronic News*, *Movers & Shakers*, *Electronic Business*, *Electronics Weekly*, *Electronic Component News*, *Composants Industrielle Electronique* and *Telecom Plus*. He is the Managing Editor of *Electronics Product News*, and is on the editorial board of *Who's Who in Semiconductors*.

The Nearly Men is his first book, and was mainly written while residing in France.

Introduction

How do you attempt to begin your first book? After the best part of two years working on this project, the thought has never been far from my mind. Through the almost continuous strain and endless cycles of optimistic elation followed morbid self-doubt, a constant worry which has always been located back in the rarely visited recesses of my mind – what the hell I was going to say to get the whole thing started. To be honest, even as I tap away at my bruised and battered laptop right now, I am still unsure.

'All the world's a stage' – definitely been done.

'It was the best of times. It was the worst of times' – too pretentious.

'A long time ago, in a galaxy far, far away' – hmm, seen it somewhere before.

Of course the underlying reason that I want these first paragraphs to capture everyone's attention is that I would like this venture, which I have invested so much time and effort into, to some extent at least, to be a success. Every one of us in our own small way would like to make some sort of mark after all, nothing much, just something to say we did not completely waste our 'three score and ten'. I guess this is my little attempt at doing that, so try to be gentle with me.

The fact is that at some point we will all look back and ask what we have done with our lives and question whether we made the most of the time that as allotted to us. So let's hear it then, what did you do? Come on! Did you compose a symphony, create a work of art, discover a continent, or in someway make the world a more tolerable place for its populace?

'Hang on a second,' you are probably thinking, 'I bought this book to have a bit of easy reading, you know, nothing too taxing. I certainly wasn't expecting the 'third degree'. It's alright, relax, let's face it there are very few people in the world who really make what could be considered a lasting contribution to our civilisation. On the whole we are just too busy trying to enjoy the frustratingly short lives that we have been given to bother getting involved in anything that serious.

Sure, if somebody has the energy and the inclination to strive for something more, that is fine with us, no problemo. They do so with our blessing and hopefully get the rewards befitting the additional effort put in. But therein lies the central theme of this book (which you may already regret having picked up – next time stick with the Andy McNab, you're on safer ground there). What of those who made these sacrifices, but found that fate didn't hold up its end of the bargain? How would you feel if you had tried to do something exceptional, but you still got ranked along with the rest of us worthless layabouts? It would be pretty annoying don't you think?

Sadly history is a bit like this, it does not always get things right. Now and again it chooses to forget about some people who did go that extra mile and aim for greatness, but ended up coming a little way short. Thus the back shelves of the annals of time are strewn with little-read tales of those destined to remain unsung heroes. Each deprived of their place in the who's who of science and technology. The money, the kudos, the girls, all passed these poor, unfortunate souls by. As a result each has become part of a very exclusive club, though its membership may not be a highly coveted one.

The name of this collection of the disgraced and discarded is the 'Nearly Men'. Okay, not the most respectful, or tactile, of nomenclatures it must be said, but I think it sums up the nature of their association quite well.

For the record, the 'Nearly Men' are:

- Nineteenth-century Italian mechanical engineer and theatre technician, Antonio Meucci
- Twentieth-century English mathematician and code breaker, Alan Turing
- Eighteenth-century French botanist, Jean-Baptiste Lamarck
- Victorian naturalist and social reformer, Alfred Russel Wallace
- Seventeenth-century English geometry professor and surveyor, Robert Hooke

• Nineteenth-century Serbian electrical engineer, Nikola Tesla
• Twentieth-century American electronic engineer and Nobel Laureate, Jack Kilby
• Victorian chemist and photography expert, Sir Joseph Swan

Some of you may have heard of one or two of these individuals, but the majority of them will be a mystery to most of you for sure. Even for those that you are familiar with, it is unlikely that you realise the full extent of their involvement to the progression of humankind.

None of them ever met, they were born in different parts of the world, in different eras, and lived markedly different lives. So what is it that actually connects these people? What commonality do they share? Well, the thing that binds them all, in one way or another, is that they each managed to miss out on one of the most important inventions or greatest scientific discoveries since civilisation began, and all the subsequent admiration that came along with it. Yes, but all of us could say that, couldn't we? Sure, that is very true, but these guys were close, painfully close, and in some cases they were just robbed. Almost all of them died in states of complete poverty, neglect, or even ignominy, overlooked by a society that did not respect or understand their particular brand of genius. By contrast the people they lost out to are virtually household names: Isaac Newton, Charles Darwin, Alexander Graham Bell and Guglielmo Marconi among them. Each of these received a string of awards, widespread public acclaim, plus riches beyond measure, gaining credit for such innovations as the radio, the telephone and the light bulb. While our chaps, the C-list celebrities if you like, got zip.

What follows is a series of tales filled with treachery, deceit and sheer bad luck, spanning a period of over 300 years. There are a number of reasons why fame and fortune did not choose to look favourably upon these guys, and we will try to analyse each in turn. For some it was a lack of time, resources, or business acumen. Others were hindered by a language barrier, a tendency to be too trusting, or simply they could not convince people that their ideas were worth listening to. One or possibly more of these proved to be fatal flaws in their character, and thus meant they would forever be assigned a second-tier rating in the chronicles of human endeavour. In the following chapters we will look at their lives and try to uncover what it was that lead to their disappointment, the 'Achilles heal' that each of them was blighted with, exploring the good, bad and ugly aspects of each of their personalities.

Articles and books have been written about just about all of these people and the altercations that will be covered in this treatise have been discussed at some point before. But most if not all of these other publications are, at best, partisan in their outlook. They are championing the cause of a particular individual, and effectively only telling one side of the story. To get any sort of accurate depiction of how the events that shaped their lives were to unfold, and how they failed to realise the potential that they possessed, it is important to look at their accounts in context of the people who managed to reach the particular goal they had also desired. I personally have no axe to grind, for me the objective is simply finding out the truth. For that reason, in each case we will look at the person that beat them to it.

But wait a second, is all this really necessary? Is there any serious doubt about who was actually behind these important discoveries, or is it all just ill-informed speculation? After all, type the words 'conspiracy theory' or 'priority claim' into a search engine, and you are certain to get bombarded with all manner of outlandish and bizarre proposals. In a time where there seems to be a half-baked supposition about virtually everything, it is important to ensure that no rash conclusions are jumped to. The Internet Age has meant that any idiot can fabricate a semi-plausible story, so it seems prudent to tread carefully when it comes to building a body of research. I would therefore like to thank the various libraries and museums across Europe and North America for access to their countless volumes that helped me to prepare the reams of notes I scribbled down, in order to condense them in to the following piece.

Our journey will take us through every branch of the scientific arena; from biology to physics, from telecommunication to electronics, from mathematics to computer technology, and as your trusted guide I will try to convey as accurate a re-enactment as possible of the events that took place in each case. The last couple years has seen me partake in a world tour which has included stops in Florence, Paris, London, Cambridge, New York, Belgrade, San Jose, Stockholm, Dallas, Boston and Munich. It has given me the chance to visit the places where these people once lived, the sites of their great triumphs, and their bitterest defeats.

To cover all these people in the same depth as a biography devoted specifically to one individual would be impossible, but by looking at them in this particular form it will enable us to make comparisons, and note the similarities in their make-up that somewhere down the line led them down the wrong path. The aim of this work is to raise people's awareness, and encourage readers, like your good self, to look further into the stories that are briefly described within these pages. Hopefully this will act as a catalyst for some of you to do further investigations of your own. It should not be judged as a comprehensive or definitive reference, but more as a starting point for more widespread reading on any of the particular topics covered that might prove of particular interest, or on the lives of those involved that the reader may feel they identify with especially.

As no other book has looked at all these people together, hopefully this collection will offer some insight into what it takes to be a true innovator, and how to avoid the six-lane expressway towards error. So those planning on rising above the rest of us freeloaders and actually making some sort of dent on society might want to start taking notes. The basic lessons apply now, just as they did back then.

Maybe it has always been this way, for all we know the incidents that will be described have been going on for millennia. When the Neanderthals first migrated into Europe out of Africa, and settled in the plush evergreen woodlands they came across, perhaps they were corrupted by the same motives that would plague future generations. Who knows, maybe the bright spark (well the one with the slightly less sunken forehead than the rest) that managed to first create fire, soon afterwards found himself on the receiving end of a jawbone, wielded by another member of the tribe who thought that such an idea would secure him a larger mud hut with a nice view of the swamp.

There is something about failure that has always fascinated us. Just like novelists tend to concentrate on unrequited love, and tragedies have been what theatre-going audiences bayed for since the art form was first developed. The world has been, is, and will probably always stay a nasty and devious place, and in truth, most of us like nothing better than hearing about it. So with that in mind let's begin.

Ladies and gentlemen, for one night only, I give you … the 'Nearly Men'.

CHAPTER 1

Crossed Lines

'The telephone that I invented, ... was stolen from me'[1]
Antonio Meucci

The importance of the rights to the invention of the telephone can never be underestimated. In one US Congress report it was described as 'the most valuable single patent ever granted'.[2] There are now close to two billion lines in service across the globe.[3] Without this remarkable innovation the speed with which we live our lives today, and the ability to encompass any part of the world, no matter how remote, into the transom of our experience, would be unthinkable. In short, modern society would simply not function without it.

The huge significance of this tool in determining the path along which civilisation has progressed over the last hundred years, and the financial rewards that would come with it, goes a long way to explaining why there is such a prominent element of controversy about who was actually responsible for developing it. The story of the creation of the telephone is a highly convoluted one, filled with contradiction, catastrophe and betrayal. This chapter will examine the lives of its principal players, in particular the often-overlooked Antonio Meucci, and the more widely revered (though possibly not so deserving) Alexander Graham Bell.

I first heard the story of Meucci while working for a publishing company in Antwerp. An Italian colleague told me how history had been misled, and despite what the majority of people assume, it was not Bell who invented the telephone, but it had in fact been the work of an obscure theatre technician from Tuscany. Sceptically I looked into the matter, and began to discover that there could be some truth in what he had said.

Antonio Santi Giuseppe Meucci was born in Florence on 8 April 1808. He was the son of Amatis Meucci, a government official, and Domenica

Pepi. In November 1821, at the age of fourteen, he started studying mechanical engineering at the Academia di Bella Arti, remaining there for six years. Here he first learnt of fellow Italian Luigi Galvani's theories on electricity, and absorbed as much information as he could on the subject.

The inhabitants of Florence were an oppressed people at this time. Tuscany was captured by the French some eight years before Meucci's birth, and had been under the rule of Napoleon since his ascendancy in 1804. The Florentine population suffered great hardships at the hands of their subjugators.

After completing his schooling Meucci gained employ with the City Guard, working as a customs inspector checking the documentation and baggage of people entering the city. He was forced to leave this job after managing to get himself into serious trouble. Ironically the young customs officer had something he had neglected to declare. Unbeknownst to his commander, he had been carrying on with a married woman while he should have been working. He was reported, and soon afterwards found himself in prison, guilty of dereliction of duty (although certain books imply Meucci was imprisoned for his political activities and not this far less honourable reason). The punishment for this transgression was confinement for a period of one month.

In early 1830, Meucci started work as a stagehand at the Pergola Theatre, Florence's main operatic venue. On 7 August 1835, he married Ester Mochi, a seamstress by trade, who worked in the theatre's wardrobe department. The couple remained in Florence until that autumn, when circumstances forced them to leave their homeland, never to return. In early October they embarked upon a journey across the ocean, taking their places on a boat bound for the Caribbean.

The reason for Meucci's eagerness to leave was that he had been implicated in a plot to oust the puppet government the French had installed in the region. His suspected involvement in this rebellion meant he was under threat of further incarceration. He was living on borrowed time, and needed a way out. When Spanish impresario Don Francisco Marti Y Torrens came to Italy looking to recruit opera singers for his company, he met Meucci and offered him the position of chief engineer at his establishment in Havana. Ester was also given a post, as head of costuming. It presented Antonio with an opportunity to escape, and gave him and his wife the prospect of beginning a new life for themselves.

The Meuccis arrived in the Cuban capital on 17 December, taking up their posts at the Tacon Theatre just a few days afterwards. The theatre had been built in the late 1830s, in memory of Cuba's former governor, Miquel Tacon. (It still stands today, but since the revolution in the 1950s it has been known as the Theatre Federico Garcia Lorca.) This enormous building could seat close to 3,000 people. Antonio and Ester rented a small apartment next to it. Once fully settled in Havana, the newlyweds began to make plans to start a family, but their wishes for a child were not quickly contented. It was not until early 1844 that Ester finally became pregnant. On 16 December she gave birth to a daughter, but sadly the girl died in infancy.

To supplement his income from the theatre, Antonio made use of his training in electrical engineering. He set up a small electroplating business, the first outside Europe. Here he would galvanise equipment to give it greater ruggedness. Deals with the local government to electroplate military hardware brought him some considerable wealth. In addition, he worked on electro-therapeutic techniques to help alleviate various medical conditions. It is likely that Ester's growing arthritis was his principle reason for choosing this area of research, but it also proved successful as another commercial sideline, and Meucci built a considerably large clientele.

He would later explain how he constructed 'an electrical machine for the purpose of using it on people who were sick.'[4] Meucci's apparatus was simple enough; wires were placed on the body of the patient, and a small electric current applied for a short period. The instrument used, as he later described it, had 'a copper plate at its end, which I placed in the patient's mouth, or other part of the body.'[5]

It was in Spring 1849, while administering treatment to a patient suffering from acute migraines, that Meucci made a discovery that would affect, in no small way, the future of civilisation. It would truly change the world in which we live, even though it had no connection with what he was actually trying to do at the time. Meucci had just attached the electrodes of his machine to the patient, and started to administer his remedy, when something happened. He realised that despite the thick wall that separated them, he could hear the patient in the next room, as if they were standing right beside him. He would later state, 'When the man cried out from the effect of the shock, I thought I heard this sound more distinctly than natural. Then I put the copper of my instrument to my ear, and heard the sound of his voice through the wire.'[6] He had discovered that sound could be transferred via electrical currents, and thus in principle it was possible to transmit speech over distances by use of wire.

Meucci began the slow and arduous process of trying to transform this discovery into something that had practical use. He soon found that by placing a magnetised iron coil at the mid-point of the wire he could increase the transmission range considerably.

With the onset of old age, Don Francisco decided to withdraw for the theatre business, informing the Meuccis that he would not be able to renew their contracts again. They came to the decision that perhaps it was time to move on, and became attracted to the idea of starting anew in the nearby United States, as opposed to heading back to the Mediterranean. Antonio and his wife left Havana on 23 April 1850, reaching New York on 1 May. They soon found a small but comfortable house on Forest Street, in the Staten Island town of Clifton. Antonio continued work on his invention, improving the design and trying to increase the clarity of the sound transmission, as well as expanding the range over which it could operate.

Meucci always stayed true to his revolutionary principles. Although he was far away from the drama taking place in his country of birth, this did not dampen his fervour to see the creation of a unified Italy. He continued to aid the struggle, helping to fund the campaign led by Giuseppe Garibaldi, and trying to garner support from New York's substantial Italian immigrant community. Garibaldi and his red shirts' valiant defence of Rome against the French failed, and the city was sacked. Following his defeat, he came to North America and for a while stayed at the Meuccis' home. The revolutionary stayed with them from October 1850 through to the following April before returning to Europe. Antonio went on to name his house after his illustrious compatriot, following his death in 1882.

He filed a petition to become a US citizen in March 1854, and by the end of that year seemed to have made considerable headway on his 'Telletrofono'. He set up a telephone link between the house and his workshop. Ester was by this stage so troubled with arthritis that she could barely leave her bedroom.

He found that it did not need a large electrical current for successfully communication to take place, the number of battery cells required was not as great as needed in his electro-therapy sessions, in fact he wrote that 'the current not being so strong gave better results in the transmission of sound.'[7] In early 1865, after fifteen years of continuous trial and error, Meucci finished constructing what he described as his 'Best Telephone'. Though his research looked to be progressing well, this good fortune was not to last. Meucci suffered a series of unrelated calamities that although singularly would not have prevented him reaching his objective, when compounded together formed an insurmountable barrier that eventually led to his ruin.

On 30 July 1871, there was an explosion on one of the ships used to transport commuters between Staten Island and Manhattan, while Meucci was
aboard. The whole front page of the *New York Times* was devoted to the disaster. It stated that the city had 'never been afflicted with tragedy so appalling
as that which burst upon it at 1 o'clock yesterday afternoon, when the boiler
of Staten Island ferry-boat *Westfield* exploded while the vessel lay in her
slip'[8] and that 'in an instant, hundreds of human beings were killed, maimed
or scalded.'[9] *Harper's Weekly* reported that 'over four hundred souls were on
board, lured by the delightful weather, from their crowded homes'[10] and that
the crew were about 'to unhook the chains when suddenly there came a
terrible crash, and in an instant the steamer was a wreck. The huge boiler
exploded.'[11] The blast caused the main deck of the ferry to be 'forced upwards
for a considerable distance; the beams and planks were torn into fragments.'[12]
The magazine told how 'scores of men, women and children who escaped
the full force of the explosion were enveloped in a cloud of scalding steam.'[13]

Over 120 people were killed. Meucci escaped, but was badly burnt nonetheless. He was infirmed for many months, unable to continue his work.
Antonio and Ester relied on charitable donations from friends to support
them over the ensuing period. The incident was to hold back the progress
of his communication system for some time, and meant that much of the
momentum he had built up was lost.

To add to the scars and mental trauma he had already endured, Meucci
was to find out, once he had finally recovered, that his wife had been given
no option but to sell many of his models to help save them from destitution. This set the Italian's research back even further.

In December 1871, Meucci, who was still convalescing after his accident,
and three associates formed the Telttrofono Company, to fund the filing
of patents for his telephone design, as well as lawyers' fees and so forth.
Antonio Tremeschin, Angelo Zilio Grandi and Sereno Bregiglia all trumped
up $20 to this end, but that was far less than Meucci had hoped for. With so
little money forthcoming, the kitty at Meucci's disposal was not adequately
stocked to file a patent application. It meant he had little choice but to
settle for just taking out a caveat, which was basically just an intention to
patent in the future. Essentially this prevented anyone else gaining priority
for an invention. If another person tried to file a patent containing material
deemed to interfere with the incumbent caveat, then its holder would have
three months to pay for a full patent to counter the application and assure
their precedence. This cost only $20 to purchase, as opposed to the princely

sum of $250 required for a complete patent; on the downside, however, it was not permanent, and had to be restored every year. It was unfortunately Meucci's decision to go for this option that proved to be his undoing. If he could have somehow found the money to take out the full patent, the history of telecommunication would have been completely different, and I would have been saved the need to tell you this story.

With Meucci's notes, patent lawyer Thomas Stetson was asked to draw up a caveat application. Caveat 3335 for Meucci's 'Teletrophone' or 'Sound Telegraph' was filed on 28 December 1871. In it he stated that by employing 'the well-known conducting effect of continuous metallic conductors as a medium for sound, and increase the effect by electrically insulating both conductors and parties who are communicating' the apparatus described allowed the creation of a 'speaking telegraph.'[14] He explained how 'each of the persons holds to his mouth an instrument analogous to a speaking trumpet, in which the word may easily be pronounced, and sound concentrated upon the wire'[15] then another instrument would be 'applied to the ears, in order to receive the voice of the opposite party.'[16] He continued by stating that the 'ear utensils being of a convex form, like a clock glass, enclose the whole exterior part of the ear, and make it easy and comfortable for the operation.'[17] The mouth and earpieces were made of metal helixes with magnets at the end, connected to an electric current, and beneath this lay the diaphragm, which would convert the electrical energy into sound.

An illustration depicting the operation of his telephone system, drawn by his friend Nestore Corradi, was unfortunately omitted from the caveat by the rather lacklustre Stetson. Alterations that Meucci asked to be made to the first draft were also left uncompleted. The lawyer was not getting paid much for this assignment, and so only put in the bare minimum of effort to get the job done. Meucci would be left to live with the consequences of these oversights. The rather rushed description in the caveat, and absence of any visual aid, left the document somewhat ambiguous.

Meucci had at least gained a little security for his invention. The caveat meant that for twelve months his work was protected by US law. The next stage was clearly to gather commercial backing to start producing this equipment. This had already proved to be a stumbling block though. For more than a decade he had been trying to find a way to bring the fruit of his labour to the public's attention, but to no avail.

L'Eco d'Italia, a magazine for the Italian community in New York, had published an article on Meucci's work on using electricity as a means of communication back in 1860, however it failed to rouse much interest. He eventually abandoned hope of fellow immigrants helping him, and tried to encourage people in his country of origin instead. The Genoan local newspaper, *Il Commercio di Genova*, mentioned Meucci's experiments in an editorial published in December 1865, but again it did not raise any real enthusiasm. Meucci had sent details of how his 'Telletrofono' operated, and that 'with the help of electricity and a small metal wire, one could transmit the exact word'[18] so 'two people could be put in direct communication.'[19]

The real question is why he did not let the English-speaking population hear about it too. Surely it would have made it easier to find funding to get this venture off the ground? The majority of Italians inhabiting the United States at this time lived in poverty, and so this was never going to be a conducive environment in which to generate venture capital. It appears clear that Meucci was not too trusting of people that were not of Italian origin, and as we will see had good reason to be. When he did begin to broaden the scope of people whom he would deal with on this matter, he ended up getting his fingers burnt.

Though he spoke perfect Italian and French, and had become fluent in Spanish while in Cuba, Meucci never mastered English. He moved in the circles of New York's Italian immigrant community, not really integrating into American society. His grasp of the principal language of his adopted home never got beyond an elementary level. This proved to be a costly omission on his part. On the occasions when he did have to deal with people that weren't of Latin persuasion, he relied on his friend Angelo Bertolini to interpret for him.

The obvious path that lay ahead, if his creation was to aspire to anything, was for Meucci to talk with the mighty telegraph giant Western Union. The company was established back in 1856, and by the 1870s had almost total control of the telegraphic communication market – its reach was huge. If Meucci could convince them that there was some notable merit to his findings, he was a made man.

Accompanied by Bertolini, he met with Edward Grant, Vice President of American District Telegraph (an affiliated laboratory of Western Union), in Spring 1872. He gave Grant descriptions and prototypes of his telephone equipment in an attempt to get the company to test it on its lines. George Durant, Grant's assistant, was given the responsibility of taking care of these artefacts and seeing the trials were undertaken. Time went by, and Meucci heard nothing. When he contacted them to enquire if there was any news, he was just fobbed off.

The months rolled on and still nothing seemed to happen. Meucci began to get impatient. He had been waiting for almost a year and a half by this stage, and though he chased them constantly, he got little in the way of response. Eventually furious at the company's lack of interest, he demanded his materials be returned, but was mortified to hear that they had apparently been 'lost'. The models were most likely moved to another Western Union laboratory, but no trace could be found of them. Meucci struggled to believe that so much misfortune could befall one man. His plight was comparable to that of the Norse goddess Frigga; granted the gift of seeing into the future, but unable to tell any one about it. Meucci had managed to witness something of equal wonder, and though he could inform other people, nobody actually wanted to listen.

By 1873, two of the three original backers had lost interest in the project. Regardless, Meucci managed to scrape together enough money to get another caveat renewal that year, but by 1874 funds had dried up completely. From this point on, he allowed the payments needed to retain the rights to his invention to lapse.

There is not a single one of us who can say that they never missed out on an opportunity; whether it was the job interview where you said the wrong thing, the knockout girl that you were really getting on great with until you spilt your drink down her top, the last-minute shot on goal that would have made you the hero of your Sunday league club had it reached the back of the net rather than being ballooned over the crossbar, or the driving test that was going so well until you inadvertently mounted the pavement and ploughed into that bus queue of old aged pensioners. Meucci's was far more costly than the everyday chances that we did not pick up on though. He had been given a period of three decades to make something of the discovery he had stumbled upon, but it proved to be insufficient. One way or another, partially down to circumstance but to some extent due to his own inadequacies, he squandered one of the biggest breaks in history. During the mid-1870s another contender entered the arena, and though he was little more than a babe in arms when Meucci first commenced his research, he would achieve, in just a handful of years, what his rival had not in over thirty.

Alexander Bell was born in Edinburgh on 3 March 1847. He was the second of three sons sired by Melville Bell. His father met an English girl, Eliza Symonds, in late 1848 and they married the next summer. In the mid-1830s his grandfather, also known by the name Alexander Bell, had established a practice specialising in the treatment of speech impediments. Melville had followed in his father's footsteps, offering speech training, and developing his own practice in Edinburgh, as well as lecturing at the university. Eliza, who was herself afflicted with deafness, decided to tutor the children at home, so they might receive greater nurturing.

Alexander took an interest in his lessons, though he appeared to have a relatively short attention span. On his eleventh birthday he adopted the name Alexander 'Graham' Bell in admiration of family friend Alexander Graham, who studied under Melville, and had become something of a role model for the young boy.

In Autumn 1862, Alexander Jnr arrived in London. He spent the best part of a year there, under the charge of his grandfather and namesake, in order to broaden the scope of his education. His grandparents' influence proved to be a positive one. The boy gained a new-found industriousness while in England. During his time there he was afforded the opportunity to attend lectures on physiology at University College. Returning to Edinburgh in the closing months of 1863, he took up a post at Weston House, a private school in Elgin, teaching elocution and music.

Melville was seen as a bit of a trailblazer in the field of speech therapy. His main claim to fame was something called 'Visible Speech', which he developed in the mid-1860s. The system employed graphical means as a guide, so deaf people would be able to pronounce specific vowels and consonants with more accuracy. It would show the orientation the tongue should take to produce a particular sound. A u-shape would inform whether it should be placed upwards or downwards, other symbols signified the size of aperture the mouth should form, and duration of the sound. By utilisation of this method it was possible for those afflicted with deafness to communicate better.

When Alexander Snr died in Spring 1865, Melville decided to take up his higher profile, and potentially more lucrative, practice in the Empire's capital. Meanwhile his son continued to teach in Elgin, though he also began to do his own work on the concept of 'Visible Speech'. It was at this stage, or so he would later assert, that Bell originally heard of Herman von Helmholtz' work on electro-acoustics. The German scientist had used an electric current to make a tuning fork vibrate, thus turning electrical energy into sound.

Bell took a teaching post in Bath, at Somerset College, in 1864, so he could be a little closer to his family. Then in Spring 1868, he was presented with the opportunity to work at a school for the deaf in the wealthy London suburb of Kensington.

Back in late 1866, Bell's younger brother Edward had contracted tuberculosis, and was taken by the disease early the following year. He was just eighteen at the time. His other brother Melville suffered exactly the same fate, developing the condition in late 1868. He had only married a year before, settling in Edinburgh with his wife and newborn child. His consumption finally took him in May 1870.

The Bell family became worried when their only remaining child also fell ill. Alexander began complaining of headaches and dizziness. His father decided it was no longer sensible for his loved ones to stay in Britain. In July 1870, the remaining members of the clan left Europe, taking an ocean liner from Liverpool to Canada, and settling in Brantford, Ontario. The healthier environment proved beneficial; away from the smog-filled air of industrial England Alexander slowly regained his strength.

In 1871, once fully recovered, he gave a series of lectures on 'Visible Speech' in Boston, which received considerable acclaim. Alexander took up the mantle of his father, and became the leading campaigner for the acceptance of this system. Given the success of the tour, and warm welcome obtained throughout New England, plus the fact there were bound to be greater possibilities of receiving financial backing for the project there, he decided to move to Boston. As the summer of 1872 drew to a close, he set up home there, hiring a couple of rooms on West Newton Street. At first he taught at the Clarke School for the Deaf, in the Massachusetts town of Northampton. Then toward the end of that year, he established his own institute to train teachers of deaf children.

By September 1873, he had taken a post at Boston University, becoming Professor of Vocal Physiology at its Oratory School, using private tuition to supplement his income. Bell thought how new developments of electrical science might facilitate the teaching of his pupils. He reasoned that use of electrically generated vibrations could help distinguish between letters in the alphabet that deaf people tended to mix up while lip reading (for example 'b' and 'p'). He began to consider how Helmholtz' work might be applied to this end.

To begin with this was just a minor project that Bell made a little time for when not too busy with his many other concerns, but as things progressed it became his primary interest. At this time the telegraph companies were keen to find a way to send more information down their lines, in order to increase throughput. Bell could see that the research he had been doing on tools to aid the deaf might have far wider implications.

He began work on the development of what he would refer to as a 'Harmonic Telegraph'. He planned to take the studies that Helmholtz had done a stage further. If electricity could be used to make sound, he reasoned it would in principle be possible to use several different sounds to send multiple signals down a telegraph line at once, and this would mean more messages could be transported across the network. Bell felt it would be more prudent to use a number of different pitched reeds rather than a set of tuning forks. The reeds would denote different sounds so multiple messages could be formed.

In 1874, top-flight Boston lawyer Gardiner Greene Hubbard approached Bell with a proposal to help fund his investigation. He came from blue-chip stock, and had all the makings of a powerful ally. Hubbard ran a profitable law firm, and had his fingers in many different pies within the local business community. Bell first met him after delivering a lecture at the Clarke School, and soon afterwards started giving voice coaching to his daughter Mabel. The girl had lost her hearing after contracting scarlet fever as a toddler. Back in October 1872, Bell had started to teach Georgie Sanders, who had been deaf since birth. He was the young son of Thomas Sanders, a wealthy merchant from the Massachusetts town of Haverhill, who made his fortune in the leather trade. Sanders was likewise intrigued by the precocious young Scot's talents, and also offered Bell support.

For the next three years Bell lodged with the Sanders family, at their house in Salem, thus allowing him to forego any additional expense so all financial backing he received could be directed towards his research. Thus he did not have to worry about doing additional tuition work to stretch his income further, as he had in the past, although he still continued to lecture on 'Visible Speech' whenever the opportunity arose.

He worked on the 'Harmonic Telegraph' in the basement of Sanders' home, helped by his assistant George Hamilton. Often they worked into the early hours of the morning. Bell also managed to secure use of the facilities at the Massachusetts Institute of Technology (MIT) in Autumn 1874, which further aided the project.

Bell's basic design consisted of a pair of electromagnetic inductors, each with a set of staves. A wire connecting the devices was wrapped around the inductor cores at each end. The principle was that when one stave was caused to vibrate it would create a change in the magnetic field. Now as we will look into in slightly more detail in chapter four, when there is a change in the strength of a magnetic field an electric current is produced.

So the current that resulted in this particular case would pass through the wire to the second coil, where it would cause a change in its magnetic field, thus the second set of staves would begin to vibrate, being attracted and repelled by the magnetic fluctuations, and thereby create sound. Thus electrical energy would be turned into mechanical energy, and this would in succession form sound energy.

In February 1875, Hubbard and Sanders signed a formal agreement to provide further financing to Bell's work. It was around this time that Thomas Watson replaced Hamilton as Bell's assistant. Watson was only in his early twenties, but had a considerable mastery of electrical engineering, and this addition to the team proved most valuable. By this stage Bell was no longer boarding with Sanders, but had taken an apartment on Exeter Place in central Boston.

In the early spring he was in New York, using the Western Union facilities to test the operability of his invention. It must be noted that he did not start looking specifically at the transmission of speech until after he returned from New York. The first records of his contemplation of the idea begin to appear in April and May, before that it had only concerned telegraphic signals. This could suggest that he had unearthed something at the Western Union laboratories that would help him to go down this particular route. As we know, Meucci had left his prototypes at the company's offices in New York three years before, in order that trials could be embarked upon, and that sometime after this equipment disappeared.

On 6 April 1875, Bell was granted a patent for the 'Harmonic Telegraph' (US Patent 161,739). Throughout the summer he worked on honing his device. Then on 2 June, or so the story goes, a watershed moment arrived. While experimenting, one of the reeds on the apparatus with which Bell was trying to pass messages to Watson in an adjoining room became jammed. When he managed, with the use of some force, to release it, the reed gave out a loud twang. Bell, with his trained ear for such things, could hear that it was not just one tone being emitted, as he would normally have expected, but several different harmonics too. This meant that the system would, in principle, be able to carry complex sounds, not just separate notes. He realised that if this was the case then it could be possible to send vocal speech across the wires. That autumn, after a somewhat tumultuous courtship, Alexander and Mabel were engaged to be married, their betrothal being formally announced on 25 November. Bell was eleven years her senior.

Bell filed his patent for the telephone on 14 February 1876, under the title 'Improvements in Telegraphy'. It was granted on 3 March (US Patent 174,461). Luckily the Patent Office had dispensed with the need for a working model some five years before, because at this stage Bell's work was incapable of achieving what it claimed.

It was slightly after this that the final breakthrough took place. While Bell was hard at work one morning, a minor accident happened to make him call out, 'Watson come here. I want you'. Watson, who was in the next room, rushed in and informed Bell that his voice had been carried through the telephone wires. Until this point they had struggled to get anything close to intelligible voice out of the apparatus.

On 25 June 1876, at the Centennial Exhibition in Philadelphia, Bell's telephone was first shown to the public. The Bell Telephone Company was formed later that summer, Hubbard taking the role of president, and the company's headquarters being established at Milk Street, Boston. On 1 August, the first shares were issued, split mainly between Hubbard, Sanders, Bell and Watson. Initial attempts at long-distance telephone conversation took place later that month. On 3 August, Bell placed a call from his family home in Brantford to nearby Mount Pleasant, covering a distance of 8km, and six days later he was successful in conversing with Watson in another Ontario town, called Paris, some 15km away.

A further patent (US Patent 186,787) for 'Improvements in Electric Telephony' was filed on 15 January 1877 and granted two weeks later. By the end of the summer, over a thousand telephones were in service.

When Meucci discovered someone had started to promote an invention with similar function to the product of research he had been doing for all those years, he became suspicious. Then upon learning more about it, he came to the awful realisation that this was to all intents and purposes the same device he had taken out the caveat for in 1871. The problem now was that he was in no position to fight against it; a serious legal challenge required money, and Meucci's coffers were empty.

Sadly the telephone was not the first time he had been duped out of the rights to one of his inventions. There were other examples of when unscrupulous investors had succeeded in gaining ownership of his patents and making a lot of money, while leaving him with nothing. He had been diddled out of capital from paper, candle and kerosene lamp businesses he had established.

Though Meucci had been a moderately wealthy man when he arrived in America, his badly judged ventures failed, and meant his affluence was short-lived. He had always been a generous soul, willing to give financial support to his friends and countrymen, to the point where Ester became so despairing of his philanthropy that, in order to save them from bankruptcy, she took charge of the family finances. In late 1879, to make ends meet, the Meuccis were forced to sell their home to the Meyer & Bachmann Brewery Company, though they would be allowed to stay there till they died.

As is commonplace when the possibility to capitalise on a priority dispute arises, opportunists look around to see if they can find someone who offers a good case against the patent owner. This is exactly what a group of Philadelphia businessmen did after learning of Meucci's plight. They set up Globe Telephone, and negotiated a deal with the Italian, acquiring his telephone rights and taking up the battle for his precedence. In October 1885, Globe petitioned the US Government to bring charges against Bell, regarding irregularities in his patents. The first hearing in this case commenced during the second week of November. On 15 December, Meucci himself took the stand.

Bell's legal team was made up of well-trained engineers and hotshot lawyers, who showed no mercy to the old man. James Storrow, an eminent patent lawyer with many years of experience, spearheaded the attack. He was flanked by technical support from Charles Cross, an MIT professor in electrical engineering that Bell had known for over a decade. Between them they managed to hunt out inconsistencies in Meucci's story, and implied the caveat was too ambiguous to claim priority over Bell's patent. It was not hard for them to make the account of an aged inventor with precious little understanding of English look spurious, and his case looked to be on a weak footing from the start.

Nevertheless, fifty witnesses stated under oath that they could uphold the Italian's priority, including Stetson, the patent lawyer who had prepared the caveat. The *New York World* brought attention to the throng of people which 'testified that Meucci's invention was in use and a good "speaking telephone" – five years before Bell'[20] filed his first patent.

Stetson admitted during his cross-examination, which took place on 8 January 1886, that he had not really understood what Meucci's patent was trying to do. He went onto acknowledge that he had not added the additional comments Meucci had requested after the first draft, and failed to include the illustration provided.

Fighting in Bell's corner, Storrow proved himself to be an indispensable advocate. He managed to convince the court (quite unbelievably I feel) that his invention was not electrical but mechanical in operation; that effectively it was a glorified paper cup and string telephone, like those you might have made at school. Sadly the drawing meant to accompany the application, but left out, clearly showed batteries. This would have made the argument that Meucci's invention was not electrical simply ludicrous. However the hazy wording of the caveat and lack of a diagram, made Meucci's case against Bell that much harder to contest.

There is little doubt that many people involved in the trial had vested interests in its outcome. Nearly everyone had a hidden agenda. As a result it is hard to see that justice was correctly apportioned. On 19 July 1887, despite scores of affidavits supporting Meucci's claim, the court finally decided in Bell's favour. Judge William Wallace's opinion was that the caveat was too vague to be certain that it described an electric telephone, and there was no reason to think this was not the best description that could have been managed. Suspicions have been raised in the time since the case was closed about his validity to adjudicate, as it is believed that Wallace may have personally had shares in Bell Telephone. However this is only speculation.

Globe did not give up. It filed an appeal in August 1887 and continued to contest Bell's priority. However, there were a series of delays in it coming to trial. The case was continually postponed. Bell's camp orchestrated a number of hold-ups in order to buy extra time. They clearly hoped that Meucci, who was very frail and fatigued by this stage, would not live long enough to see it through to its conclusion. On the morning of 18 October 1889, their prayers were answered; Antonio Meucci died, aged eighty-one, at his home in Staten Island.

The *New York Herald* said in its obituary that Meucci 'died in the full belief of the priority of his claim as inventor of the telephone.'[21] Bell's stalling tactics had worked; the case was finally abandoned shortly after Meucci's death. As Globe's chances of progressing any further were at best slim, the partnership decided to cut their losses.

In a letter to Garibaldi's daughter, Teresta, two years earlier, Meucci had mentioned 'the telephone which I invented and first made known, as you know, was stolen from me.'[22] He continued, 'I would like to see my rights restored to me before my death.' His wishes were not to be fulfilled.

The release of the first telephone brought a great deal of uproar with it. Over 600 lawsuits followed. It was to be a golden age for lawyers. Many small firms tried to take advantage of this new media by bringing out imitations of the equipment. Bell Telephone was quick to act in such instances.

Copies of the Bell patent were distributed to every branch office in the company's rapidly expanding network. Each of the area managers was told to keep an eye out for possible piracy. However, it was not all one-way traffic. A number of cases were brought before the Supreme Court, each claiming that Bell stole the idea from them.

Johann Philipp Reis, a German schoolteacher and amateur electrical engineer, had developed a device for transmitting basic sounds across electric wires back in 1860. The system could not carry true speech though. Its transmitter mechanism worked on the principle of a diaphragm moving in and out as sound waves hit it. This had a needle connected to it which made contact with an electrical circuit as the diaphragm became distended (i.e. when strong enough sounds were incident). However, this was effectively an 'all or nothing' arrangement, only differentiating between the existence of sound and its complete absence. At best this could make a rough approximation of some single letters, but it certainly was not capable of dealing with a variation in the magnitude of the sound while it was being produced.

What was needed was something that could create an analogue representation of the different undulations present in human speech. Reis's 'Telephon', as he called it, was suitable for transmitting musical notes or perhaps separate vocal sounds in isolation, but nothing more than that. Even Reis himself did not foresee any form of commercial potential for the device. He had only used it as a toy, making no attempt to patent his work. After a few public demonstrations in the early 1860s, he lost interest in the project, nothing more being said about it until the opportunity to capitalise on Bell's success arose. Unfortunately for him, Reis died long before the case made it to trial, though it would be cited by others trying to avoid prosecution by Bell. Amos Dolbear, another inventor who resided in Boston and was a professor at Tufts College, had developed a magnetic transmitter and claimed that Bell had assimilated this work and patented it as his own. However, the courts were not convinced, and he got nothing.

One contender who managed to put up more of a fight was Elisha Gray, a Quaker from Barnesville, Ohio. Gray had been successful in making substantial improvements to the telegraph relay, and had a string of patents to his name. In March 1875, he had read a paper entitled 'Transmission of Musical Tones' at the American Electrical Society. He had created a system in which a number of reeds would act as a means of representing different notes. This 'Musical Telegraph' was very similar in conception to Bell's 'Harmonic Telegraph'. Both men's research had led them in similar directions, and before long Gray came to the same conclusion as Bell; if multiple signals could be sent down a wire, then in principle the same could be done for a human voice.

At the end of the year Gray wrote the caveat 'Instruments for Transmitting and Receiving Vocal Sounds Telegraphically'. It stated that the objective of this invention was 'to transmit the tones of the human voice through a telegraphic circuit, and reproduce them at the receiving end of the line so that actual conversations can be carried on by persons at long distances.'[23] His apparatus consisted of a hollow chamber, across the outer end of which was stretched a parchment diaphragm 'capable of responding to all of the vibrations of the human voice.'[24] Attached to the diaphragm was a light metal rod through which the electrical current would pass. The vessel, which was filled with a high-resistance liquid (such as acid) would be in close proximity with the conductor rod. Any changes in the rod's position, due to sound vibrations, would induce a subsequent change in the liquid's resistance 'and consequently, in the potential of the current passing through'[25] a rod placed in an identical set of apparatus at the other side of the line. The key to Gray's design was this 'liquid transmitter'. Basically it offered a way of carrying a signal that was effectively analogous to the original sound. When the diaphragm vibrated the connected rod moved up and down close to the diluted acid, thus altering the current. This distinguished it from the on/off-type circuit Reis had utilised, and meant that human voice could be accurately transferred into electrical signals. However the caveat was not actually submitted until just after Bell's patent application had been received.

Though Bell had got past the post first, at the time his equipment was still not operable. He had yet to successfully transmit human voice. Gray's application seemed to hold some of the answers. With the help of a rather shady patent clerk, by the name of Zenas Fisk Wilber, it seems that the Scot managed to get his hands on Gray's dossier and he proceeded to glean information from it. A total of four paragraphs were added to the application Bell had already submitted. In them he replaced his previous transmitter technique with one that closely mimicked Gray's.

Western Union had finally woken up to the threat posed to its existing business by then, and with Gray's intellectual property, and the aid of a number of hard-working and eager to please telegraph engineers, most notably the young Thomas Edison, as well as Dolbear, it set about trying to make a better telephone system. With the emergence of these competing devices on the market, Bell sued for patent infringement in September 1878.

The Supreme Court heard how Gray's work had been 'interpolated into Bell's specification' and that there were differences between the original filed on 14 February 1876, and the one he re-submitted some five days later. Wilber would admit that Bell had bribed him in order to get information on what his rival had submitted. He went on to say that he had shown Bell drawings of Gray's apparatus and explained how the mechanism worked in detail. Unfortunately for Gray, the court ignored it all. Wilber was seen as far too unreliable for it to base any concrete evidence, for or against, on his deposition. Bell had sworn under oath that he had not paid Wilber, and he was perceived as far more credible, under the circumstances.

However, it is worth noting that the whole 'Watson come here' incident, which was the first practical use of Bell's system, is based on him working on a transmitter that incorporated a cup of acid, and it was the spilling of this on his clothing that had caused him to call out. Until then there had been no suggestion of using a liquid transmitter. All this happened only a couple of days after his patent was granted, and clearly shows he was trying to utilise what he had obtained to improve his own design.

The dispute between Bell Telephone and Western Union would finally come to a close in late 1879. On 10 November, the two companies struck a deal. Rather than having to fight to the death, an out of court settlement gave Western Union a fifth share in Bell Telephone's revenues from phone rentals, for the remaining seventeen years of the patent.

In the middle of all the turmoil surrounding the telephone's introduction, Bell somehow managed to find the time to marry Mabel. The ceremony took place on 11 July 1877, in Cambridge, Massachusetts, at the Hubbards' family home. The couple travelled to Great Britain soon after, where they stayed for many months. In January 1878, Bell demonstrated his handiwork to Queen Victoria, at Osborne House, her Isle of Wight residence. Soon after this, Bell Telephone set up the world's first telephone exchange, in New Haven, Connecticut. Within the first six months of that year, the company declared it had over 10,000 telephones in operation across the Eastern Seaboard, and this figure almost trebled by early 1879. It was at this point that phone numbers started to be assigned, to facilitate easier organisation of the growing quantity of users.

Mabel and Alexander had several children together. Elsie was born in May 1878, and in February 1880 a second daughter, Marian, arrived. Sadly both their sons, Edward and Robert, died within only months of their birth. Following the birth of his second child, Bell took a back seat in the running of Bell Telephone. He relied on his royalties to keep him and his growing family in the manner to which they were now accustomed. It also allowed him the time needed to concentrate on other areas of research. Bell moved to Washington DC in 1881. By this stage the company had 130,000 telephones in service. Within three years, long-distance connections reached between Boston and New York.

In July 1881, President James Garfield was shot at close range by Charles Guiteau, a rather peeved lunatic who wanted a cabinet post, and when this had not been forthcoming started looking for some payback. Bell was asked to try to locate the exact position of the bullet in the president's body, as it appeared to be lodged close to his spine, and this would make operating incredibly risky. Using induction coils he created a probe, which when brought into close contact with a metal object induced a change in the magnetic field around the coils, giving off an electric signal. On several occasions Bell attempted to find the lead slug, but it was to no avail. The president's condition deteriorated quickly, and he died later that summer.

In November 1882, Bell was finally granted American citizenship. When he was awarded the Volta Prize soon after, which came with a handsome sum of F50,000, he decided to set up a laboratory in his new home of Washington. It was in the 'Volta Laboratory' that he developed the photophone, which used selenium crystals to take sound transferred into an electrical signal and turn them it into light, allowing different shaped symbols to be generated. He also began to look seriously into the field of voice recording. Edison had invented the phonograph some four years earlier, but Bell thought there must be a better method to reproduce sounds.

In a letter to Hubbard, in March 1878, he stated that he believed 'that the phonograph will be enormously improved' and that he had 'apparatus that will give still better results than those produced by Edison.'[26] He continued by describing a mechanism that would 'drag a slip of tinfoil rapidly under a vertical style which caused to press upon the tin foil with uniform force by means of a weight or spring.'[27] Since the 'style' (or stylus) only moved in the vertical plane (i.e. up and down), and the recording plate rotated horizontally, there was less chance of it being dragged and thus recordings were far less prone to distortion. The medium used on Bell's recordings proved much easier to score with the stylus needle, giving it higher sensitivity, and resulting in a more accurate replication of the original sound. Flat disks were also lighter to carry, and more convenient to store. He worked on his 'Graphophone' through the summer of 1882, along with his cousin, Chichester Bell. The apparatus was completed early the following year.

In mid-1886, Bell bought a few hundred acres of land on Cape Breton Island, Nova Scotia, and had a summerhouse built there, which he named Beinn Breagh, or 'Beautiful Mountain' when translated from Gaelic. A long-distance line between New York and Chicago was set up in 1892. By this time nearly 250,000 telephones were in use.

The Bell name had quite some pedigree by now, and both his daughters were seen as quarry for the most eligible bachelors. In October 1900, Elsie married Gilbert Grovenor, editor of *National Geographic*. Then in April 1905, Marian married the botanist and explorer David Fairchild.

In January 1915, the first transcontinental telephone line was set up, connecting San Francisco with New York, and crossing some 7,000km. Bell took part in the opening ceremony; he was based on the East Coast talking with his old colleague Watson located out on the West Coast.

Along with Casey Baldwin, Bell developed a hydrofoil boat, and in September 1919 it set the World Marine Speed Record. He also took a keen interest in the development of aviation, and helped design a number of early aircraft. The plane *Red Wing* that he and fellow enthusiasts built, took to the air in March 1908 with Baldwin at the controls. It was one of the first public demonstrations of manned flight, though the flying machine crashed during the trial and was deemed beyond repair.

Bell still carried on improving his inventions right into his old age, but his health worsened considerably as he entered into his seventies. He suffered from both diabetes and anaemia. He finally died on the morning of 2 August 1922, in the arms of his beloved Mabel, and was buried at Beinn Breagh two days later. Out of respect for the man who had given life to the cables that ensnared the entire continent, exchanges across North America enforced a minute silence on their lines. Mabel was diagnosed with cancer before the end of that year, and told it was inoperable. She died in January 1923, and was buried alongside her husband.

As there are so many complexities to this story, and such a large number of possible versions of what took place, it maybe worth trying to recap and briefly go through some of the key points once again. Firstly, there is a striking resemblance between the design of Meucci's 'Best Telephone' of 1864 and the device Bell and Watson produced in 1875. It is hard to believe that their construction was totally unrelated. Then there is the fact that Bell was known to have been doing experiments at Western Union's labs in New York (where Meucci's models were being stored) in early 1875. What is more, George Bartlett Prescott of Western Union, who was overseeing

Bell's trials at this time, was most definitely aware of Meucci's work. Also it is worthy of note that Durant, Grant's second in command at American District Telegraph, who was put in charge of dealing with the Meucci affair, would later take up an important post with Bell Telephone.

Meucci was widely held to be an honourable man, and generous to a fault. It was said of him that he was the last person who would take anything that did not belong to him. Bell on the other hand seems to have perjured himself on more than one occasion, so by that proviso, it is hard to believe wholeheartedly in anything else he did.

It should not go without mention that Bell would employ similar underhand tactics to those used during the telephone's development with his other inventions; compensating for his limited technical skills by using the research of others as a staging post of much of his work. The phonoautograph he created in 1874 bore a great deal of resemblance to a device that Leon Scott developed, his photophone borrowed heavily from the work of Rudolf Koenig, and the graphophone used many of the techniques that Edison had publicised with his phonograph.

Bell later said that he would 'never have invented the telephone if I had been an electrician. What electrician would have been so foolish as to try such a thing? The advantage I had was that sound had been the study of my life – the study of vibrations.'[28] It is true that he had no training in electrical engineering, nor any telegraph experience. Often this is used as an argument for why Bell succeeded and the others failed, and that it was his understanding of acoustics and human hearing that gave him the edge over his rivals. However, this seems just a bit too fanciful to be the real reason for his triumph. Gray and Edison were both die-hard telegraph men, while Reis, Dolbear and Meucci were accomplished electrical engineers. It is hard to believe that some novice could beat all of them to it, by sheer luck.

Also, what he was trying to achieve with his original goals still emanate an impression of being very badly defined. Initially he was just experimenting on means for improving speech for the deaf, but then at some undetermined moment he completely switched to the field of telegraphic communication, where he had no prior knowledge. This just does not appear to make any real sense. His sudden change of direction, and his interest in telegraphy, come out of nowhere, and no matter how Bell, or the long line of biographers that have written about him, tried to justify this unexpected detour in the focus of his research, there does not seem to be any way to explain why he would jump into a whole new area he knew nothing about.

Even though he felt that a better understanding of electricity had not been necessary, there were clearly big gaps in his knowledge. His patent of 1876 included the use of brass as a conductive component. This really shows Bell's ineptness as an electrical engineer, as brass does not actually conduct a current. Little things like this suggest that though he had a good perception of audiology and vibrational motion, he knew precious little about the nature of electricity.

Only having been familiar with one side of this particular equation does not appear to be enough. It makes it highly unlikely that he could have created the telephone on his own, and this is probably why he hired Watson, who was already considered something of a whizz-kid. This would also explain why his supposed 'assistant' took on the capacity of almost being an equal partner in the research process, and was rewarded with such a large portion of the Bell Telephone stock. Most inventors would have not been so willing to give such high regard to someone who was just supporting their work. It suggests that Watson played a more important part than Bell would openly admit, or alternatively, that he knew something more than he should have about how the invention had actually been arrived at.

It is of course important to keep a sense of perspective here. Once Bell had his patent it was effectively open season for anyone who could find a remotely viable inventor able to put up a good case against him. Any chink in his amour would be under attack. People like Dolbear, Reis, and countless others were trawled up by opportunist companies looking to try to break Bell's patent rights and snatch a piece of the action.

Though to some extent Meucci's tale seems plausible, it must be said there are a couple of things that just don't quite add up. To begin with, Ester is supposed to have sold his prototypes for a measly $6 while her husband recovered from his accident. Now this appears to be a little bit suspicious. Meucci may have been unwell, but he was still able to communicate. If Ester had known that had he spent twenty years perfecting this invention, she would not have been so quick to part with the fruits of his labour, despite their financial plight, and it seems absurd to think she would not have at least consulted him first.

Secondly, those who champion Meucci are always keen to show what a fertile and productive mind he possessed, and as a result are often over-zealous to state the large number of patents he took out during his life, and the wide spectrum of subjects these covered. However, this inadvertently reveals a major anomaly in Meucci's argument; namely that he simply could not afford the $10 required to renew his caveat. In 1871, Meucci acquired another patent (US Patent 122,478) for an 'Effervescent Drink' that was rich in vitamins, which he had taken during his convalescence. If he was so poor that he could not afford to take out a full patent for the telephone, and truly saw the value of his discovery, then why the hell would he not prioritise this over some health supplement? In 1873, around the same time as the telephone caveat ran out, he managed to find the money to file another patent. His patent for a 'Sauce for Food' (US Patent 142,071) again seems to be a strange endeavour to be involved with if he realised the worth of his work on the telephone. Meucci took out another two full patents in the time between his telephone caveat expiring and when Bell applied for his patent. So suddenly the story that he had not got the funds to take his invention further seems to lose its credibility. In this period he filed a total of four patents, amounting to $1,000 worth of investment, but still he could not find ten lousy bucks to save the crowning glory of his inventive powers. This suggests that he did not feel there was enough interest in the 'teletrophone' project to warrant further investment. However, that seems equally unlikely.

It is possible, and to be fair quite probable, that he got other people to raise the money for these later inventions, but was unable, however hard he tried, to rally enough support for his telephone. This rings true (sorry for the pun), as he does not seem to have been able to pitch this to anybody (Western Union, his fellow Italian-Americans, or the press). It is here perhaps where the difference lies. Bell could sell the idea, and whatever else he did (or did not do) towards developing the telephone, he was the man who transformed it into something commercially viable. It still sounds ridiculous that Meucci could get people to invest hundreds of dollars in a food sauce, but could not find a single person to lend him a far smaller sum to help create something that would enable the entire world to talk with one another.

In conclusion, it is almost certain that Bell took much of the work Meucci, Gray and others had done to perfect his telephone. Though, in his favour, he was the only one of the characters discussed who managed to really convince anyone of the value of this innovation. Reis had fobbed it off as an amusing plaything, Meucci had failed to even persuade his own friends to back the project, and his articles published throughout the Italian-speaking world gained little interest. Although his integrity is certainly doubtful, Bell did succeed in enthusing others of the telephone's potential. Even if it is possible for Meucci to claim rightful ownership of the telephone's creation, it was still Bell who brought it to the world.

It must be considered truly unfortunate that Meucci, who had abetted so many people over the years, could not get any help from those around him during his time of need. But though he seemed sadly lacking in supporters while he was alive, there have been no shortage of people willing to fight for his honour since. In September 1923, a memorial bust was erected in Clifton to commemorate Meucci, at the site of the cottage where he lived. Upon the centenary of his death, a monument was placed in Meucci Square, Brooklyn by the Italian Historical Society. Rudolph Guiliani, then Major of New York City, proclaimed 1 May 2000 'Antonio Meucci Day', in recognition, as he described it, of 'this important New Yorker'. Then in 2003, the Italian Postal Service issued a stamp to honour his life.

The greatest boost to Meucci's claim to the invention of the telephone came in September 2001, 113 years after his death, when US Congress officially recognised his contribution to its development. Some of his supporters have chosen to see this as a complete validation of his precedence over Bell, but in reality it is a rather nebulous statement.

107th Congress Resolution 269

....

....

That it is the sense of the House of Representatives to honour the life and achievements of Nineteenth Century Italian American inventor Antonio Meucci, and his work in the invention of the telephone.

Meucci was a great Italian inventor, and had a career that was both extraordinary and tragic.

Upon immigrating to New York, Meucci continued to work with ceaseless vigor on a project he had begun in Havana, Cuba, an invention he later called the 'teletrofono', involving electronic communications;

Whereas, having exhausted most of his life's savings in pursuing his work, Meucci was unable to commercialize his invention, though he demonstrated his invention in 1860 and had a description of it published in New York's Italian language newspaper;

Whereas Meucci was unable to raise sufficient funds to pay his way through the patent application process, and thus had to settle for a caveat, a one year

renewable notice of an impending patent, which was first filed
on December 28 1871;
Whereas Meucci later learned that the Western Union affiliate laboratory
reportedly lost his working models, and Meucci, who at this point was living
on public assistance, was unable to renew the caveat after 1874;
Whereas in March 1876 Alexander Graham Bell, who conducted experiments in the same laboratory where Meucci's materials had been stored, was
granted a patent and was thereafter credited with inventing the telephone;
Whereas on January 13 1887 the US Government moved to annul the patent
issued to Bell on the grounds of fraud and misrepresentation, a case that the
Supreme Court found viable and remanded for trial.

As you can see it does not really state anything, other than the blatantly obvious. All it confirms is that Bell could not have patented the telephone had Meucci been able to raise the money to maintain his caveat. Anyone who has read a little about the Meucci–Bell affair would be able to reach that conclusion without any great show of mental agility.

Talk is cheap, and government statements, which nobody will ever read, do not realign the course of history. As far as the world as a whole is concerned, Bell was the man behind the telephone. There needs to be a lot more work done if Meucci's integral part in the defining of contemporary society is to be fully recognised, and the appreciation of the masses correctly ascribed.

In the words of John La Conte, President of the Italian Historical Society of America, 'We can only credit Mr Bell with the commercialisation of the invention of Meucci. In the tradition of fair play and honesty, let Meucci have the honour to be recognised as the "Father of the Telephone". Let Bell have the money.'[29] Today at his house on Staten Island, you will find a small museum celebrating Meucci's life and works; its staff continue to fight to achieve recognition for the man they feel was the true inventor of the telephone.

The number of telephone lines installed passed the one billion[30] mark in 2004, and there are now just over a billion mobile phones in service with this figure still rapidly increasing. Though what has happened in telephony during the 120 years since Bell's patent was granted is well documented, there is still no certainty as to how he arrived at that point, and whether it was his ingenuity that proved crucial to its conception. Unfortunately the debate may continue for another six score years or more, but the matter is unlikely to ever be resolved.

CHAPTER 2

There was a Crooked Man

While each of us who inhabits the Earth follows our own little life, trying to convince ourselves that it might have some sort of significance, way above us something is occurring that until recent times we still knew precious little about. It hammers home just how unimportant everything going on down here really is. Which team wins the Stanley Cup, what is the hottest band in the charts, who is elected Pope, or what Posh and Becks are up to – it all means zip.

When judged from the celestial perspective anything that happens on this pointless little world is not even worth thinking about. In the meantime huge gaseous giants thunder through the darkness, each with a litter of satellites pirouetting around them. Cold desolate rocks and raging spheres, with tumultuous volcanoes interspersed over their surface, arc across the vast empty void unable to escape from the binds that tether them. A blazing globe measuring a million kilometres across lies at the apex of this scene, and has an influence so great that bodies more than five billion kilometres away from it are held tightly within its grasp. The Greeks and Egyptians observed it in the sky and thought of it as a god travelling across the heavens in his chariot. Both the Aztecs and the Maya made human sacrifices to appease the fiery object.

This spectacle could not be described as a transitory one. It has a sort of permanence to it that humans can barely comprehend. The whole thing has been going on for at least four and a half billion years, and since the very day that our species first gained the ability to reason we have looked upwards in wonder and asked what it all meant? Every culture has had a different view on how to explain what they saw. According to Hindu lore the Earth was borne of the backs of four gigantic elephants being carried through space by a monstrous turtle (a concept used more recently by Terry

Pratchet in his Discworld novels), while early Christian astronomers had thought that we maintained a fixed position and the stars were painted onto an enormous revolving shell that encapsulated the unique creation they judged our planet to be. With time, and more comprehensive investigation, such ideas would be marginalised.

The cosmic ballet by which the Moon, Earth, Sun and planets were all kept in perfect order has been a theme employed by countless poets and composers through the ages. Scientists and theologians have pondered over it, not just for centuries, but for entire millennia. Whether the sacred orchestration of some divine creator or chance occurrence had brought this arrangement together, there had to be something keeping it all in place. What stopped these immense bodies crashing into one another, or alternatively, prevented them from spinning off into the nothingness? An unseen hand guided the component parts of the cosmos, but nobody could say categorically what its nature was.

Virtually the first lesson of physics taught to school kids concerns Hooke's Law. This basically shows how the extension of an elastic material is proportional to the stretching force applied, i.e. the more weight hung from it, the more it stretches. It is not the most exciting of phenomenon, but it is well suited to serve as pupils' initial excursion into the scientific world, since it is simple to follow and allows them an early opportunity to perform a practical experiment, taking measurements and producing graphs with the collated data. The law was formulated by Robert Hooke, however it does not really seem fair that a person who had such a great influence on the development of scientific knowledge should be solely remembered in connection with springs and rubber bands.

During his life Hooke played many roles: engineer, inventor, architect, anatomist, cartographer, microscopist, astronomer, not to mention hypochondriac, junkie, skinflint and possibly even sexual deviant (these last comments may be a little harsh, you can make your own judgement on them later). The seventeenth-century biographer, and his close personal friend, John Aubrey, summarised Hooke's accomplishments thus:

Robert Hooke – Experimental philosopher. Curator of Experiments at the Royal Society (1662). Fellow (1663) and Secretary (1677-1682) of the

Royal Society. Gresham College Professor of Geometry (1665). Designed Bethlehem Hospital, Montague House and the College of Physicians. He helped Newton by hints in Optics, and his anticipation of the Law of Inverse Squares was admitted by Newton. He pointed out the real nature of combustion (1665); proposed to measure the force of gravity by the swinging of a pendulum (1666); discovered the fifth star in Orion (1664); inferred the rotation of Jupiter; first observed a star by daylight; and made the earliest attempts at telescopic determination of the parallax of a fixed star. He also first applied the spiral spring to regulate watches; expounded the true theory of elasticity and kinetic hypothesis of gases (1678); constructed the first Gregorian telescope (1674); first asserted the true principle of the arch; and invented the marine barometer.[1]

An impressive catalogue of achievement for any man. Another of his acquaintances, who took it upon himself to get some of Hooke's unpublished works put into print after his death, was Richard Waller, a fellow member of the Royal Society and eminent scientist of his day. He described him as 'one of the greatest promoters of experimental natural knowledge.'[2]

So how did Hooke fall into almost total anonymity after his death? One reason is clearly that he made enemies of several high-standing and powerful people. Among these was none other than Sir Isaac Newton. Their prolonged animosity would prove particularly detrimental to Hooke's reputation, with many of his contributions to the advancement of civilisation being not fully appreciated. Also it must be considered that because of his incredible versatility, he would become involved in so many different projects that he never managed to devote adequate time to truly complete many of them. As we will see, unlike Meucci, or the others discussed in this book, Robert Hooke did not just miss the boat, he missed a whole flotilla.

Hooke was destined to live through possibly the most tumultuous and eventful period in British history. He would witness civil war, regicide, religious upheaval, epidemic disease, destruction of the capital, foreign invasion and his nation's eventual ascendancy to the position of the most powerful on the globe (a mantle it would not relinquish for the best part of three centuries). In particular Hooke was born into an era in which England first displayed its scientific prowess. London, the city which he inhabited his entire adult life, would become the core of activity for the scientific world, and for over thirty years it would be his thin and disfigured frame that carried a hefty proportion of this load.

Robert Hooke came into the world at around noon on 18 July 1635. He was the son of the Revd John Hooke, the curate of Freshwater parish church, and his wife Catherine. He was the couple's third child; their daughter Katherine had been born in May 1628, and their first son John arrived two summers after. Robert's schooling while on the Isle of Wight was minimal. He had been a sickly child and not expected to live long, so his parents saw little point in wasting any formal tuition on him. As a result the boy was left pretty much to his own devises. He spent most of his time drawing wildlife, making models, and so forth.

In 1647, his elder brother left home to take up an apprenticeship with a grocer in Newport, the island's main town, but no decision seems to have been made at that stage as to what career Robert might go into. His parents were probably just glad that he had made it through infancy. When his father died in Autumn 1648 after a prolonged illness (probably liver related, as he was said to be heavily jaundiced at the end), young Hooke was left a small inheritance of £100(according to Aubrey and Waller, though other sources state that it was less than this amount, possibly £60). He was sent to London to serve as an apprentice to portrait painter Peter Lely. Lely had studied under the Flemish masters, and gained widespread renown for his likenesses of royalty. Originally the training period was to have taken seven years, but this was cut short. After just a little while Robert decided this was not likely to prove itself a satisfying way of life for him. It is also believed that the paint fumes he was exposed to did not agree with his less than hearty constitution.

Hooke decided to use the legacy left by his father to secure himself a good education. He enrolled at Westminster School, and was taken under the wing of the headmaster, Dr Richard Busby. Hooke does not seem to have attended many actual lessons during his stint at the school, but studied by himself and probably received private tutoring from Busby.

The power struggle between Parliament and the king had been going on through most of Robert's childhood, but by this stage was finally coming to an end. Charles I had been incarcerated, and was put on trial for treason in January 1649. He was found guilty and beheaded at the end of that month, with Parliament abolishing the monarchy soon after. Nevertheless, as far as the inhabitants of Westminster School were concerned, there was no swaying of allegiance, Busby remained loyal to the Crown throughout Cromwell's rule, and Hooke seems to have been greatly affected by this (he is believed to have made a point of fasting on each anniversary of the king's execution).

In 1653, he began his studies at Christ Church College, Oxford. He had managed to gain a chorister's place, which helped cover his living expenses. In addition, he acted as an assistant to Dr Thomas Willis, his chemistry lecturer, and also aided Dr John Wilkins, the Warden of Wadham College. He learnt about the art of astronomy from Dr Seth Ward, the University's Savillian Professor at this time.

Even in his youth he had the demeanour of an old man. Waller described him as 'very crooked' and 'low of stature, though by his limbs he should have been moderately tall.'[3] Some have conjectured that it was the many hours that he spent hunched over a lathe, fashioning different types of paraphernalia, which caused him to develop a pronounced stoop. He concluded his studies in 1656, though he never actually graduated. It would later be said that 'Oxford had given him more than a thousand degrees could match.'[4]

Wilkins recommended Hooke to Robert Boyle, believing he was more than adequately capable of supporting him in his scientific research. Hooke became his assistant soon after. Boyle was the son of the Earl of Cork, and hailed from County Waterford. The Irish aristocrat was interested in investigating the nature of gases and had been looking for someone to aid him in his experiments. Hooke was charged with the task of completing the design and construction of a more efficient air pump with which to undertake these experiments. Although Hooke's position was quite clearly that of an underling, Boyle was well aware of his young assistant's abilities when it came to building instrumentation. It is certain that Hooke both created the means and also helped perform the experimental research used to formulate Boyle's Law (which states the volume of a gas is inversely proportional to its pressure). The pump's effectiveness seems a little modest when compared to what this sort of equipment can do today, but back in the seventeenth century it was a revelation. It was far more efficient than any contraption previously devised. By use of a ratchet, air was pushed into a cylinder, the seal was then closed and the cylinder withdrawn, the whole procedure was subsequently repeated many times. The evacuation process would normally take several minutes to complete. Today a plaque is to be found on Oxford High Street, next to the entrance to University College, which states that this was the site of Boyle and Hooke's laboratory, and gives them equal accreditation.

In 1661, still in his mid-twenties, Hooke's first paper was published. It discussed an occurrence that he had happened to stumble upon, namely that of capillary action. What basically happens when this phenomenon take place is this: if you put a thin tube into a dish of water, the water entering into the tube will rise to a higher level than in the dish. The basic reason for this is liquids tend to be more tightly bonded at their surface than elsewhere, and thus a film covers them (incidentally this is how certain insects are able to walk across the surface of ponds). So when a tube is placed into a liquid, it will have a stronger adhesion to the tube's surface, and rise up it.

Back in September 1658, Oliver Cromwell had died, his son Richard taking on the responsibility of Lord Protector. His leadership proved ineffectual however, and by spring the following year he was ousted, throwing the whole of England into turmoil once more. The Royalist faction had taken control of the country by early 1660, and the way was now clear for the remnants of the Stuart family to return home. Charles II took the throne in August, following years of exile in Holland. Soon after his ascendancy, he started to mete out reprisals on those men still living which had played a part in what he saw as his father's murder, and even the dead did not escape his wrath. Cromwell's grave was exhumed, and the remains of his body hung from the gallows at Tyburn, symbolically executed.

Following the Restoration, the scientific community that had built up in Oxford started to migrate back to London, and Gresham College in particular became the new focal point of research into what was then known as natural philosophy. The leading lights of England's academia, such as Christopher Wren and William Petty, took up professorships there, and in time the likes of Boyle and Wilkins also relocated to the capital.

According to Thomas Birch's historical accounts of the time, the 'greatest part of the Oxford Society coming to London about the year 1659, they usually met at Gresham College at Wednesday's lecture upon astronomy by Mr Christopher Wren.'[5] It was out of this maelstrom of scientific activity that the Royal Society was formed. This entity came into existence in July 1662, drawing together England's greatest minds. Robert Moray, George Ent, William Hoskins, Gilbert Talbot and John Evelyn were among the names that blessed its membership roll, not to mention Wren and Boyle. Lord Brouckner took the presidency, and German immigrant Henry Oldenburg was given the role of secretary. Hooke was installed as the organisation's Curator of Experiments on 5 November that year. The workload involved was to prove monumental, and this had to be added to his far from insubstantial obligations to Boyle. The Society would expect Hooke to prepare three to four new experiments for each weekly meeting.

The group looked to capture the spirit of the Elizabethan scientific virtuoso and former Lord Chancellor Sir Francis Bacon, who believed that through methodical investigation (it was as a direct result of one such experimental investigation that was to take his life), a greater understanding of the world could be achieved. In Bacon's eyes it was not unsubstantiated theorising, but rather continued compilation of data from scientific observation that would unlock the secrets of the universe. Nothing could be assumed, all had to be questioned and interrogated.

Eventually Boyle agreed to allow Hooke to leave his service so he could concentrate on his other duties. Although Boyle had been forced to sacrifice the full-time attention of his most valuable servant in order that he could meet the demands of his function within the Society, they still worked together closely, and Hooke would always make the experimental requirements of his former master a higher priority than those of other members.

His position in the organisation would give Hooke the opportunity to dip into many different areas of investigation, and secured his reputation as one of the most versatile minds in history, but it also proved to be his downfall, as it would mean he was never furnished with enough time to make a real mark in any one field. In many cases he would just end up as a sounding board for others, who would eventually take all the glory.

He was elected a Fellow of the Royal Society on 20 May 1663. The council meeting of 3 June confirmed his position, and taking into account that it had still not been able to pay him for his experimental work, his membership fees were overlooked. The minutes stated that 'Mr Hook [sic] was elected a Fellow of the Society by the Council, and exempt from all charges.'[6] At the meeting of 19 October, he was given further responsibility. Taking charge of the Society's rapidly growing collection of scientific curiosities, it was 'ordered that Mr Hook have the keeping of the repository of the Society.'[7]

One of the most famous, and regularly quoted, descriptions of Robert Hooke is the one given by the diarist Samuel Pepys, whose works helped form the basis of much that we know about the goings on in seventeenth-century England. Upon his visit to the Society on 15 February 1665, he noted that, 'Mr Boyle today was at the meeting, and above him Mr Hooke, who is the most, and promises the least, of any man in the world that I ever saw. Excellent discourses till 10 at night.'[8] Hooke would become the Society's stalwart, his experiments enlivening the early years of this institution.

In mid-1664, a man stepped into the realm of the Royal Society who could offer the means to finally give its curator some form of salary, albeit a meagre one. Until then it had still not managed to pay him a bean for his far from insignificant duties; it is likely he had continued to receive some sort of remuneration from Boyle until this time. Transcripts state: 'The Society did yesterday chuse Sir John Cutler as Honorary Member, and ordered, he having declared his resolution to settle upon Mr Hook [sic], during his life an annual stipend of 50 pounds, and to refer to the Society ye direction of ye kind of imployment ye stipend shall be putt upon, should have solemne thanks returned to him from this singular favour.'[9] Cutler was a London-born merchant, and one-time member of parliament, who had been knighted four years earlier. He agreed to pay the Society £50, out of his considerable assets, towards the £80 needed for Hooke's salary. In return Hooke was obliged to give regular lectures on the 'history of trades'.

Hooke entered Gresham College in September 1664, and became its Professor of Geometry early the following year. The college, on Bishopsgate Street, would be his home for close to forty years. By this stage Hooke's talent for designing apparatus, and his experimental skills, were already highly respected, but now he needed to show he also had true intellect. In the early 1660s, Wren had started a series of drawings using a magnifying glass that he planned to present to the king as a gift from the Society. Unfortunately, other more pressing matters meant he had to abandon the undertaking. He turned to Hooke, and asked him to continue the work. Hooke was more than happy to oblige, and, by the time he was finished, had amassed a portfolio of more than 200 drawings. Hooke used three lenses placed in a pasteboard tube to form a rudimentary microscope. A large glass globe filled with water was put close to the bottom of the tube, used to condense sunlight and centre it on the sample, thus helping to illuminate it. The collection he produced, accompanied by details of the techniques and apparatus utilised, was named Micrographia, and published in January 1665. The book proved to be a huge success, and was the first occasion that the general public were given insight into what went on within the Royal Society. Pepys was one of those to be impressed by Hooke's creation, describing it as 'the most ingenious book that I ever read.'[10]

The Society heaped praise upon Hooke for this work. It gave the organisation its first notable achievement and raised its profile considerably. The following review was given in its Philosophical Transactions: 'The ingenious and knowing author of this treatise. Mr Robert Hook, considering with himself of what importance a faithful history of nature is to the establishing of a solid system of Natural Philosophy, and what advantage by experimental and mechanical knowledge hath over the philosophy of discourse and disputation, hath lately published a specimen of his abilities in the kind of study.'[11] Its members were beginning to realise the increasing significance of the talents possessed by their previously undervalued curator.

In the book he proclaimed that with the 'help of microscopes there is nothing so small as to escape our inquiry'[12] and 'in every particle of matter we may behold as great a variety of creatures as we were able before to reckon in the whole universe itself.'[13] Probably the most well-known illustration from the publication is that of the flea. The detail shown in this piece bears witness to Hooke's incredible powers of observation, and allowed seventeenth-century readers to see an aspect of the world they lived in that had previously been unexplored and alien to them. He described how 'the microscope manifests it to be all adorn'd with a curiously polished armour, neatly jointed and beset with a multitude of sharp pins.'[14] The drawing of the head of a drone fly showed 'two large protoberant bunches' the surface of each being 'shaped into a multitude of small hemispheres, placed in a triangular order.'[15] It is the earliest known occasion of anyone describing the compound eyes of an insect, nobody seems to have seen this before him.

It was also in Micrographia that Hooke used the word 'cell' to outline what he saw when studying a slice of cork. He wrote that he could 'exceeding plainly perceive it to be all perforated and porous, much like a honeycomb.'[16] He observed that 'air is perfectly enclosed in little boxes or cells distinct from one another.'[17] He chose this word as he felt that they resembled monastery cells. It was the first time in history that this term was applied to biological specimens, and is a nomenclature that has been utilised ever since to describe the basic functional unit of all living things.

Micrographia had put him on the map; he had proved that his technical abilities in performing experiments and designing apparatus were matched by his capacity to create original and thought-provoking research. Over the following years Hooke demonstrated the true depth of his aptitude for scientific investigation. He showed masterly understanding across more branches of the scientific arts than any other man could hope to achieve.

Hooke performed a series of experiments on the nature of combustion. He noted that substances cannot burn unless in the presence of air. He went further by discovering that if combustion took place inside an enclosed container, the pressure inside it would be reduced up until a certain point. From this he deduced that it was some specific component in the air which caused combustion, and when it was used up the flame would be unable to continue (it would later be affirmed that this is in fact the action of oxygen, but until this stage nobody had even suggested that air might be made up of different gases).

He observed several other important phenomena, though usually they were not picked up on at the time by any of his fellow scientists. He managed to perceive 'a small spot on the biggest of the three obscurer belts of Jupiter, and that observing it from time to time he had found that within two hours after the said spot had moved East to West about half the length of the diameter.'[18] Based on this, he was able to suggest that the other planets in the solar system revolved, just like the Earth does(although Kepler and Galileo had also considered these possibilities some years before).

Pepys recalled that in February 1665, while at Gresham College, Hooke read a 'very curious lecture about the late comett, among other things, proving very probably that this is the very same comett that appeared before in the year 1618, and that in such a time probably it will appear again – which is a very new opinion.'[19] Up till then comets were not thought to orbit the sun like planets, they were just assumed to travel in straight lines, occasionally happening to pass through our solar system.

Hooke was the first scientist to suggest that the Earth would not have a perfectly spherical shape, but would actually be more like a slightly squashed tennis ball. Its diameter around the equator being greater than it was across its poles. Once more he was proved right. The reason for this being that its constant rotation distorts its shape. He would also be the first to question 'whether the axis be fixt in the Earth, or not: and among other queries, whether the vast sandy deserts of Africa and Arabia owe not their origin to the sea?'[20] Again no other scientist had previously speculated as to whether our planet's poles have always stayed the same. Today we know that they have moved, and as Hooke had correctly proposed, this has had an effect on the Earth's geographic structure.

In his 'Discourses on Earthquakes & Subterraneous Eruptions' Hooke put forth his opinion of how fossils were formed. He conjectured that these strangely 'shaped stones, which the most curious naturalists most admire, are nothing but the impressions made by some real shell in a matter that at first was yielding enough, but which had grown harder with time.'[21] Until this stage fossils had just been seen as 'tricks of nature'. Hooke was the first to ascertain their true origin, but his ideas fell on deaf ears.

Hooke's persistently heavy schedule left little time for other activities, but whenever possible he would try to spend an hour or two in one of his favourite coffee houses. There were somewhere in the region of 150 of these establishments that sprang up in London during this period, and Hooke seems to have frequented more than his fair share of them. Particularly regular haunts were Garraways, Joes's and Jonathan's. Along with other members, he would go to one following the Society meetings and while away the evenings discussing different theories, and catching up on gossip.

In late Summer 1666, an event took place that had a profound effect on Hooke's life. In some ways it made him. For one thing it would solve his financial problems; the Society's stipend not going very far, his Gresham professorship only covering his food and board, and the untrustworthy Cutler proving increasingly hard to receive payment off in return for his lecturing services. It would also improve his standing, making him an important figure in London's social scene, and bringing him in to close contact with the king himself. However, all this did not come without a price, it would consume so much of his already limited time that it prevented him from using his scientific powers to their fullest. On the morning of 2 September, a bakery on Pudding Lane, just off Fleet Street, caught fire, and as it was not dealt with quickly enough soon spread. When that matter was brought to the attention of the Lord Mayor, Sir William Bolton, he underestimated its seriousness. His lack of concern meant that it proceeded to become completely out of control. Pepys witnessed the pandemonium that ensued. He recalled seeing 'the fire rage in every way, and nobody to my sight endeavouring to quench it.'[22]

The blaze continued to tear through the streets of Europe's largest metropolis unabated for three full days, and by the time it was finally extinguished, the heart of the nation's principal city had been razed. From its inception at Thomas Farynor's bakery, in central London, it spread up as far as Holborn to the north, and along the Thames from the Temple all the way to Tower Hill. This 'most horrid malicious and bloody flame'[23] as Pepys described it, destroyed more than 13,000 houses, and close to ninety churches, leaving over 400 acres of land in smouldering ashes.

The manpower needed to deal with the gargantuan task of restoring London was not at the city's disposal, it was going to require many additional people to volunteer. Because of his skills in taking measurements, and his grasp of physics, Hooke was enrolled as a surveyor just a few days after the fire had been brought under control. There was huge pressure to get the rebuilding operation underway as quickly as possible; the 'Great Fire' had left over 60,000 people homeless and devastated the commercial nerve centre of the entire country. Hooke, along with three others, started the extensive and laborious process of staking out the buildings and streets. The surveying job took many months to complete. Meanwhile Wren had wasted no time in petitioning the king with his plans to restructure the city, and after being given Royal approval to commence this colossal assignment, he asked his friend Hooke to act as his deputy. The two had been close since they were at Westminster School (though Wren was slightly older), and Hooke had followed in his footsteps by taking a university place at Oxford. Although Wren seems to have gained all the historical recognition for the construction of the new London, it is more likely that Hooke's role was of almost equal substance. Wren delegated much of the work to his 'second in command' and concentrated on dealing with just the high-profile projects himself. Unfortunately for Hooke, this seems to have been the way of things in many other incidents in his professional life. He was forever consigned to a role of understudy to the leading man. Although on many occasions he would pull off great things and make completely new discoveries, he was never destined to get the praise and admiration he so desperately craved.

Hooke was personally responsible for the design and construction of such buildings as the London College of Physicians, completed in 1679; the Bethlehem Hospital for the Insane – better known as 'Bedlam', finished in 1676; and probably contributed the lion's share to the Monument to the Great Fire (a 200ft column that to this day marks where the whole sorry affair had begun), which saw completion in 1677. He also created a method for accurately drawing large circles (consisting of a wheel attached to taut wire). His invention was capable of marking out much larger diameters than previous solutions, and was far lighter, making it easier to operate. It allowed Wren to map out the dome for St Paul's Cathedral, which was to become the second largest in the whole of Christendom (only St Peter's in Rome being bigger).

The Society, which was so dear to Hooke's heart, was on several occasions lampooned for its bizarre behaviour, and Hooke himself, as one of it leading proponents, would take much of the brunt of this unwanted attention. Jonathon Swift's book *Gulliver's Travels*, first published at the early stages of the eighteenth century, told of an academy on the island of Laputa (which means 'the whore' in Spanish), where scientists performed all sorts of pointless experiments, such as trying to extract the sunlight out of cucumbers. It was clearly aimed as an attack on the Society's worth, but this was far from being the first time it was satirised. In Spring 1676, Thomas Shadwell's play The Virtuoso opened in London. The performance was again meant to poke fun at the scientific elite of the Royal Society, implying that this body was completely superfluous, and had no real value in the progression of civilisation. Hooke first mentions the play in his diary entry for the 25 May. It appears he heard about it while visiting one of his regular coffee house haunts. After learning more about it from his Society colleagues, on 1 June he decided to see for himself what it was all about. The next evening, he and Thomas Tompion, a watchmaker and the creator of much of Hooke's experimental apparatus, went to the theatre. He was not best pleased with what he saw. 'Damned dogs,' he wrote following the ordeal, 'people almost pointed.'[24]

The principal character in the play seemed to be a rather scathing caricature of him, at least in Hooke's eyes. In fact it was probably a fusion of a number of different people, used to embody the Society as a whole, however nobody personified the organisation as well as Hooke did, so it was he who would get the most brutal mocking. To be frank, it is hardly surprising the public found their behaviour more than a little peculiar, because it has to be said that it was. The use of animals, for example, in cruel and seemingly unnecessary experiments, appears to have been commonplace.

Pepys' recollection of a Society meeting in July 1668 tells of an experiment in which a dog was 'tied through the back about the spinal artery, and thereby made void of all motion.'[25] Though it does not mention Hooke by name, it is safe to assume that it was him who performed this procedure. Waller was present on an occasion where Hooke was 'keeping a dog alive, his thorax being laid open, by blowing fresh air into his lungs.'[26] Another experiment investigated how the transfusion of blood between animals might effect their disposition. It tried to show 'whether a fierce Dog, by being often quite new stocked with the blood of a cowardly Dog, may not become more tame'[27] or if it would lose the ability to do tasks that it had been taught previously, such as fetching. In January 1664, Hooke 'proposed an experiment against the next meeting of shutting up an animal and a candle together in a vessel, to see whether they would die at the same time or not.'[28] With such examples to draw upon, it is fairly understandable that the outside world thought of these men, and Hooke in particular, as slightly weird. However, though such things appear horrific by our modern standards of decency, we should not be too quick to condemn the actions of the Society, given that at this time the treatment of animals in everyday life was little better.

And so we come to the man who would become Hooke's nemesis. In the early 1670s, a young Cambridge professor became involved with the goings-on of the Society; his name was Isaac Newton. Hooke would have many confrontations with other great philosophical minds during the course of his career, in fact he managed to pick a fight with just about every prominent figure in science who occupied the same era as him. These heated debates stretched across the full gamut of areas where he was proficient; He would be involved in arguments with England's John Wallis with regard to geology, and Flanders' Anton Van Leeuwenhoek on the subject of microscopy. He had conflagrations with the Dane Johannes Hevelius as well as Englishman John Flamstead on astronomical matters, and Dutchman Christian Huygens concerning watch making. But none of these would compare to the battles that took place between him and Newton. Their treatment of one another would go from thinly veiled contempt right through to pure unadulterated scorn. Neither man would miss an opportunity to reproach their opponent, the mutual detest of each other that they displayed would become legendary. They were to have two major altercations during the spell of their troubled acquaintance, and we will examine both of them.

Upon joining the Society in 1672, Newton put forth the idea that light was not, as had previously been accepted, a wave, but in fact a transmission of particles. Hooke championed the widely accepted wave theory, and gave little credence to Newton's musings on the subject. He said of Newton's prognosis that, 'I confess, I cannot see yet any undeniable argument to convince me of the certainty thereof.'[29] He upheld the belief that, 'light is nothing but a pulse or motion, propagated through an homogenous, uniform and transparent medium, and that colour is nothing more but a disturbance of the light, by the communication of that pulse to other transparent mediums, that is by the refraction thereof.'[30]

Newton's basis for disputing the wave theory was thus: if light had the form of a wave it should be able to go around corners, in the same way sound did, but light appeared to just move in straight lines. Hooke tried to combat Newton's 'corpuscular theory' by showing that when the edge of a razor blade was placed across a beam of light, an echo of illumination would be produced overrunning into the shadow of the blade, thus preserving his belief that light was a wave, not a particle. He also argued that if light was made up of particles, how could it travel with such high speeds as those witnessed?

Hooke was not the only person to pour cold water on Newton's ideas. Many of his colleagues joined in the derision of the young scientist. In fact, what we now know is that light has characteristics of both; it is not a solid particle (like a hard billiard ball) or a wave (like a ripple on water), it cannot be described in terms of everyday objects such as these. It actually consists of packets of oscillating energy (called photons) that display some of the attributes of a wave, and some of a particle, so in fact they were both right. However, the damage was done. Newton, as we will discover later, was a mass of neuroses, taking the criticism Hooke and his followers heaped upon him very much to heart. In the end it would cost Hooke dearly.

> 'Nature and Nature's laws lay hid in night, God said, Let Newton be!
> And all was light.'[31]

Alexander Pope

Let us now turn our attention to the other participant in this clash for a while, and see what information we can gather from his story that may help us to unravel what really happened between them. The life of Newton is incredibly well documented, so for many this may seem a little redundant, but of course it is important to set the scene, and try to make some sense of the sort of man he was, in order to be able to look at his actions in later life. Newton played the irresistible force to Hooke's immovable object. As already mentioned, these scientific heavyweights would clash on two major subjects. First of all on the composition of light, then some years later, and perhaps more decisively, on the laws of gravitation and planetary motion.

Isaac Newton came from a somewhat better background than his adversary. He was born in Woolsthorpe Manor, in the small hamlet of Colsterworth, which is nestled in the heart of the Lincolnshire countryside. He arrived on Christmas Day 1642, though the change to the Gregorian calendar actually makes this 4 January 1643 in our terms. He was born prematurely, and said to be so small that he would easily fit into a 'quart pot'.

His father, also called Isaac, had died before his birth. He had been an illiterate, but affluent, farmer. His mother, Hannah Ayscough, remarried a couple of years after young Isaac was born, to a wealthy landowner by the name of Barnabas Smith, who was considerably older than her. She moved to North Witham with him, and had another three children. In the meantime, Isaac stayed at Colsterworth, in the care of his grandmother, Margery Ayscough. Like Hooke, he showed an early affinity for scientific matters, amusing himself by reading, making sundials, constructing windmills, drawing, and such like.

After Barnabas' death, in 1653, his mother returned to Woolsthorpe. Later that year Isaac started to attend grammar school in the nearby town of Grantham, where he lodged with William Clarke. Clarke was the local apothecary, and introduced him to the work of John Wilkins (the man who had taught Hooke) and probably gave him some instruction on the fundamentals of the scientific arts. It was thanks to this tuition that he first gained an interest in chemistry, a discipline that proved of great use to him in his attempts to unlock the secrets of alchemy, in later life. His schoolmaster, Henry Stokes, gave him a rudimentary understanding of mathematics, but apparently little else.

Once his schooling was completed, Newton was expected to return home and take up his responsibilities looking after the farm. However, he appeared to be sadly inadequate in this role, possibly intentionally trying to make a mess of things. Newton was clearly not a farmer. Stokes, along with Isaac's uncle John, persuaded his mother that a university career was more in keeping with his character.

He went back to school to prepare for the Cambridge entrants exam. Newton managed to gain a place at Trinity College, being admitted in June 1661. Though his family were wealthy enough, his mother did not want to part with too great a sum of money when it came to Newton's higher education. To cover his board he was forced to carry out tasks for the wealthier students, but this seems not to have bothered him too greatly, and he still managed to find time for his studies. He familiarised himself with the works of ancient scientists and philosophers like Aristotle, Ptolemy and Plato, as well as more modern thinkers such as Thomas Hobbes, Rene Descartes and Galileo Galilei. He attended the mathematics lectures of the much-esteemed Cambridge professor Isaac Barrow, and became greatly intrigued by the subject.

In early 1665, he received his bachelor's degree and decided to stay on to take a master's. By now he knew that he wanted to remain in academic circles. He had no intention of returning to the agrarian life he had been bought up in. Nonetheless, later that year he was given little choice in the matter. Newton was forced to flee Cambridge, as risk of contracting the plague became too high. He went back to Lincolnshire in June, and would stay there for eighteen months. It was back home that he did much of his most important work on the topic of gravitational motion and movement of the planets, as well as the propagation of light. It would be in this more relaxed setting that he put together the core of his most seminal work and, almost without dispute, the most important book in the history of science, his *Principia Mathematicae*.

The quiet country lifestyle that he followed for this period provided a conducive environment in which he could philosophise. Here, completely free from all distraction, he worked ceaselessly, with almost superhuman industriousness. It was while back in Woolsthorpe that he first read Hooke's *Micrographia*, and thereby gained an interest in optical physics. Using this as a starting point, he began to investigate the form that light took, and tried to find a way to explain the various unusual characteristics that it displayed.

By the time it was safe for him to go back to Cambridge, Newton was a changed man. Over the next few years he would perfect the hypotheses that he had pondered over, with regard to both optical phenomena and gravitational motion, then eventually publish groundbreaking works on each of these subjects. The seeds from which a new era of scientific discovery would grow had begun to germinate. Newton said of his return to Lincolnshire that, during this particular period, he had been at the 'prime of his age for invention'. His *anni mirables* (years of wonder), as they became known, would have the most profound effect on the future of science.

Though Newton had benefited from a prolonged period of absence from his duties at Cambridge, while living in Woolsthorpe, the fact was that even when he was back at the university his schedule was probably not that harsh, and he had plenty of time to devote to his own work. On the other hand, time was one thing that Hooke never had the dispensation of.

Newton had been made a fellow at Trinity by September 1667, and the following year obtained his master's degree. It was in 1669 that he published the paper 'On Analysis by Infinite Series', with which he made the transformation into one of the country's leading authorities in mathematical research. Over the space of just a few years he had gone from a Lincolnshire yokel, almost completely ignorant of mathematics, to the standard bearing of its new epoch. Later that same year he succeeded Barrow as Lucasian Professor.

As briefly mentioned prior, Newton was of the opinion that the long-accepted belief that light had the form of a wave seemed to have too many contradictions to be truthful. He felt that many of the phenomena that had been observed fitted better with the model that light was in fact made up of millions of tiny particles. As he saw it, the light from the Sun was 'an aggregate of an indefinite variety of homogeneal colours.'[32] Each colour would be represented by a different type of particle, and when all of these were brought together they would create white. Newton supposed that the particles would be 'of various sizes, densities, or tensions do by percussion or other action excite sounds of various tone, and consequently vibrations in Air of various bigness.'[33] He felt that our 'eyes effect in us a Sensation of Light of a white colour, but if by any meanes those of unequall bignesses be separated from one another, the largest beget a sensation of a Red colour, the least or shortest of a deep Violet.'[34] Thus the occurrences like refraction and diffraction could be clarified much more satisfactorily. The medium through which light travelled could have a more profound effect on the weaker rays (blue and violet), than it would on the stronger ones (red).

Newton sent a small, but highly powerful, telescope that he had created to the Royal Society in late 1671. It used a mirror, which allowed the images to be a lot sharper but the distance between the lenses to be smaller, thereby making the device less cumbersome. The instrument was 'composed of two metallic speculum, the one concave (instead of an objective), the other plain, and also a plano-convex eye glass.'[35] He noted that the 'telescopical tube may be shortened without prejudice to the magnifying effect.'[36] His work on the 'reflecting telescope' won him a great deal of admiration, and allowed him to take his place as a Fellow of the Royal Society in January 1672. However, Hooke was by that time used to getting most of the attention there, and did not want some young upstart encroaching on his territory. He was not impressed, and claimed to have constructed a similar device some three years prior, though true to form he had not pursued the project further.

A text describing Newton's work with prisms was read to the Society the month after he was admitted, and created a great deal of controversy. He described how 'having darkened my chamber, and made a small hole in my window shuts, to let in a convenient quantity of sunlight, I placed my prism at this entrance, that might be thereby refracted to the opposite wall. It was at first a very pleasing divertissement.'[37] Using one part of the spectrum produced he passed it through a second prism, only to find the light was not further diffracted but remained the same colour. As it could not be separated into further component parts, it showed, as he had supposed, that white light was in fact made of a combination of all the colours in the spectrum, and its diffraction was not simply an optical effect as Hooke had suggested.

Several members of the Society questioned his experimental technique. Newton responded quite abruptly, clearly upset with their lack of faith in his abilities. Hooke was joined by Giovanni Rizzetti, a wealthy Italian virtuoso who piled condemnation on Newton's experiments, claiming he could not produce the same results. France's Adrian Auzout, and Holland's Christian Huygens, members of the Society's foreign collective, were equally derisive. As was his way, he took their criticism very seriously, but with time the scientific community learnt to show him some respect. He would demonstrate that he was a man of will and quite phenomenal boldness. His commitment to finding the truth was secondary even to his personal safety.

In a lecture that October, he continued to voice his opinion that white light was a fusion of many different colours. He talked of everyday examples of things that backed up his doctrine, such as the 'Rainbows or Coronas such as those Descartes sometimes observed around a candle,'[38] but this was not enough to convince his peers that his ideas had any basis in fact. However, Newton was not someone to give up easily, he was willing to go to any lengths to substantiate his theory. On one occasion, he pushed a bodkin (a blunt needle-like object, usually made of metal or bone) into the corner of his eye, so as to touch the optic nerve. Newton later wrote that he had seen 'severall white darke and coloured circles'[39] when he pushed the implement between the eye and the bone, causing the shape of the eyeball to change. As the coloured circles appeared 'when the eye's shape is deformed by some externally applied force, they necessarily arise from some newly formed curvature or creases in its membrane'[40] he concluded. This meant that alteration in the optical apparatus of his eye was causing the way light was perceived to change, and he was seeing it in its component parts. The pain involved, and the possible risk of doing serious damage to his eyesight, seem not to have fazed him in the least.

In 1675, when he sent his completed work on the nature of light to the Royal Society, Hooke was asked by its council, as resident expert in optical phenomenon, to study it and give a report on its findings. He could not resist the opportunity to stir things up. He was far from kind about it, and, to rub salt into the wound, accused Newton of claiming certain discoveries originally put forward in his Micrographia.

It appears that the Society's secretary, Oldenburg, tried to accentuate the rift between these scientific titans for his own ends. He reported to Newton of Hooke's remarks, but seems to have been somewhat economical with the truth, shall we say. Then when Newton corresponded with the Society, Oldenburg was again selective with what he disclosed, in order to have maximum effect. He implied that Newton held scant regard for Hooke's work, and this probably contributed to the latter's merciless criticism of his counterpart's theorising. This tactic was bound to work, as it played to both Hooke's arrogance and Newton's hypersensitivity.

On 20 January 1676, Hooke wrote a letter to him to try to repair the damage done, stating that he felt Newton had been 'misinformed' of his attitude towards him. Hooke went on to say that he in fact valued Newton's 'excellent disquisitions' and was pleased to see some of the notions that he personally had never had time to complete improved upon.[41]

In response, Newton sent the following lines on 5 February. He stated, in a more respectful and less emotionally charged letter to Hooke, that if he had 'seen further it is by standing on the shoulders of giants.'[42] Though the atmosphere had perhaps calmed a little between them, it is unlikely that Newton really meant those words. It has been suggested that this was just petty snipe at Hooke's diminutive stature. They would continue to keep up a façade of tolerance, but the bridges between them had been permanently burnt, their differences were never to be reconciled. Much of this was down to the interference of others, but it was not helped by their own difficult personalities.

When not locking his intellectual horns with Newton, Hooke could always find other ways of keeping himself occupied. During the mid-1670s, he and Huygens, who was scientific advisor to the 'Sun King' Louis XIV of France, would be involved in a race to create an accurate marine time-keeper that would allow sailors to correctly judge their longitudinal position. On this occasion neither would be able to claim victory. It was several decades before John Harrison successfully produced a pocket watch capable of doing this.

Following Oldenburg's death in September 1667, Hooke became the Society's secretary, but this was clearly not a duty he was suited to. What with his continuing commitments to the experimental activities of the organisation, and rebuilding of the city, he was unable to deal with the huge number of correspondents his predecessor had built up. Several of Europe's most eminent scientists who made up the Society's overseas membership, Huygens and Van Leeuwenhoek among them, complained at the communication breakdown, and Hooke was reprimanded on several occasions. Just a few years after his appointment he was asked to stand down.

Hooke was more of a doer than a writer. He clearly had no aspirations of being another Pepys, his diary was a purely functional affair. It mainly consisted of a brief account of his ailments, complaints of his insomnia, his repasts, his scientific work, and the occasional sexual encounter. He seldom elaborated on anything, just sticking to his basic daily activities, cataloguing them in a concise, rather abrupt manner. For example: '9th November 1672 – Eats roast lamb and chicken broth, with conserve of roses. Slept not well, rose at six,'[43] and '14th January 1673 – Eat beef at noon, eggs at night.'[44] This of course does not make it any easier to get any real impression of what sort of character he was. There is very little information that can be collected from, '18th August 1675 – At Garrways with Fitch and Hayward, Vomit from hewk.'[45] or '7th August 1678 – Poysoned by drinking glass of sour beer,'[46] but this was pretty much all there was to go on.

One of his more interesting little quirks, which has been commented on by many of his biographers, was his habit of denoting acts of sexual congress with the symbol for Pisces 'ƎϹ'. For instance: '25th July 1674 – Dreamt of ƎϹ.'[47] This seems a little strange, but it was actually not that uncommon to use symbols to represent such acts. The Marquis de Sade, among others, employed similar techniques in his diary.

Hooke's sex life has drawn quite some attention. His affairs seem to have been mainly restricted to the series of housemaids who worked in his service over the years. The first of these was Nell Young, with whom he had quite regular intercourse before she left his employ in mid-1673: '12th September 1672 – Played with Nell ƎϹ, hurt small of back'[48] and '28th November 1672 – Nell ƎϹ, not slept well'[49] being examples of this. Doll Lord replaced Nell in fulfilling both his domestic and amorous demands in Autumn 1673, his diary entry of '11th March 1674 – Doll sat up late, ƎϹ'[50] laying testament to this. The next victim of his sexual advances was Beth Orchard. She enrolled as his housekeeper in Summer 1674, and like her predecessors soon found herself occupying Hooke's bed. He wrote that on '26th July 1674 – First saw Betty ƎϹ.'[51] Their liaison was short-lived, and as with both Nell and Doll before, appears to have done little more than cater to his animal urges.

Hooke's one great love, if indeed he ever really had one, was to come from a rather unexpected source. Robert's brother, John, had married Elizabeth Niamer in 1658, and had opened a grocers shop on Newport High Street. The couple had a daughter, Grace. Hooke's niece first came to London in Summer 1672. John Hooke had hopes of her marrying the son of Sir Thomas Bloodworth, the Mayor of London at that time, and asked Robert to look after the arrangements. Hooke acted as intermediary between the two parties, trying to barter a favourable contract. While these negotiations took place, the young girl lodged with her uncle.

The prospect of marrying into the Bloodworth family appears to have faded away in early 1673, but Grace had become accustomed to the pleasures of city life by then, and wanted to make use of the opportunities that it afforded, going to fairs and parties. It appears that she had many suitors during this period.

Hooke was not immune to her charms either, and with time a connection that was more than kindred began to build between them. It was in mid-1676 that he seems to have first consummated his physical relationship with her. '14th June 1676 – Slept with Grace.'[52] She would have been sixteen at the time, and he was in his early forties. Over the months that followed, he and Grace would be regular sexual partners. His diary tells of their union about once a fortnight, though it appears that she continued relations with other men too. This arrangement continued for a little over a year, before she returned back home. Hooke's diary entry for 10 August 1677 tells of 'Cousin Grace into the country.'[53]

John Hooke seems to have struggled to carve out an adequate living as a grocer, requesting for Robert to lend him money in order to help alleviate his fiscal woes, and it looks likely that his brother had been forthcoming. Unfortunately, this did not manage to rectify the situation. Things continued to get worse for him.

On 27 February 1678, John committed suicide. He hung himself. The definite reasons for this act are uncertain. It is known that he was still heavily in debt, despite Robert's alms, but it could also be that he had learned more than he should have about the nature of the relationship between his brother and daughter. Of course this is pure speculation, and there is no real point in pursuing that line of enquiry further, as there is simply no way of knowing what the true cause was.

Left without parental support, Hooke would take care of Grace, making her his ward. She returned to London on 7 June, four months after her father's death. She would remain with Hooke for the rest of her life, serving as his housemaid, and effectively becoming his common-law wife.

By this stage the problem of payment from Sir John Cutler had become an increasing concern. Countless entries in his diary tell of his requests. At first they are relatively good humoured, '29th December 1673 – Spoke again about Sir J. Cutler's money'[54] or '5th January 1674 – Cutler promised money, but I know not when,'[55] but soon they degrade into 'at Garraway's Sir J. Cutler played the fool with me,'[56] and comments like 'cheating rogue'[57] became increasingly common. Throughout much of the 1670s he had tried to get out of not paying Hooke for his lecturing services, and so drastic measures were called for. In the early 1680s, Hooke took his debtor to court, but the legal wrangling would continue for several years, and when Cutler died it was held up even longer. Cutler's family finally paid Hooke the arrears due him in 1693. By then he had been waiting for over two decades.

In July 1679, Grace became seriously ill, his diary tells that on '24th Thursday – Grace sick to stomach.'[58] The following day he describes her as 'desperately sick'. By the weekend Hooke's worst fears were realised when he 'discovered small pocks'.[59] That night he wrote to her mother informing her of the seriousness of Grace's plight. Smallpox was one of the most potent of diseases of this period of history, and the chances of survival were at best slim. Miraculously Grace managed to recover; over the next few months she convalesced, Hooke taking care of her.

He was no stranger to illness either, though it is possible that he inflated the seriousness of his medical problems, and certainly made matters worse for himself by overwork and increasing dependence on various drugs.'20th August, 1679 – At Jonathon's, Pappin, Loddwick, Hunt, Lamb, Hally. Drank chocolat, eat apple py, vomiting'[60] and '1st March 1678 – Drank Senna, very sick with it – slept in clothes'[61] are just a couple of examples of almost ceaseless complaints about the status of his health. Although originally intended to relieve his suffering, the long-term effect of the medication he used would have probably only served to make his condition graver.

Hooke first heard about the beneficial properties of 'hemp' (or cannabis) from one of his coffee house cronies, the sea captain Robert Knox, in the late 1680s. He became fascinated by the subject, and gave a lecture on it at the Society meeting on 18 December 1689. It is almost certain that he never got the opportunity to try it, but he did take other substances to help relieve his discomfort, in particular laudanum (diluted opium). The continued use of this is likely to have had a detrimental effect on his health, and may have played some part in the increasingly erratic characteristics he displayed as he entered old age. The notion of a flawed genius seems to be a recurring theme in literature, and is consistently echoed in real life. The perception that even the greatest heroes have some aspect of shame which makes them more human perhaps explains Sherlock Holmes' addiction to cocaine, the alcoholism of Shakespeare's Falstaff, or JFK's womanising. In some small way Hooke lived up to this good-guy/bad-guy duality.

The question of how the movement of the heavens was orchestrated was something that had perplexed history's greatest scientific minds, and it was a matter that Hooke would contemplate many times through the course of his career. Unfortunately, yet again, he would not be the one to determine the answer to this riddle, it would be Newton who took that honour.

As far back as early 1664, Hooke had looked at how gravity exerted its power on a body. He performed a series of experiments on the forces experienced by weights dropped from different heights. Before the chill of winter had full evaporated, he had constructed an apparatus with which he tried to calculate the force that a falling object would exert. It consisted of a balance, on one side of which was placed a small weight. Upon the other side was a metal pan, and suspended over this a steel ball bearing. It became clear to him that a ball dropped onto the pan could move a heavier weight on the other side of the balance. This showed the ball was being accelerated by an attractive force, the closer it came to the Earth the greater was the gravitational pull, but the experiment was far too basic to give any idea of how this increased attraction could be defined. He later performed this procedure at the Society, showing his 'instrument for making the experiment of the force of falling bodies.'[62] However, without any accurate data about how the value of the force changed with height it was only of marginal interest.

In Summer 1666, he returned to the subject, giving a lecture on 'Gravity & Celestial Motion'. Then, in his Cutler Lecture of 1674 'Attempt to Prove the Motion of the Earth', Hooke finally started to make some headway. He proposed a system based on three suppositions:

1. That all Celestial bodies whatsoever have Attraction or Gravitating power towards their own Centers, whereby they attract not only their own parts, and keep them from flying from them (as we may observe the Earth to do), but also all other Celestial bodies that are within the Sphere of their activity.

2. That all Bodies whatsoever, that are put in direct and simple motion, will so continue to move forward in a Straight line, till they are by some other more effectual power deflected and bent into a motion that describes some Curve line.

3. That these Attractive powers are so much the more powerful in operating, by how much the nearer the body, wrought upon, is to their own Centers.[63]

He knew gravitational force was not directly proportional to the distance between the two bodies, he had shown in his previous experiments that it was far greater at closer distances, but he still did not know the nature, or at least not chose to divulge any details, of the relationship between force and distance.

At the beginning of the seventeenth century, the German Johannes Kepler, Imperial Mathematician to the Holy Roman Emperor, had through astro-

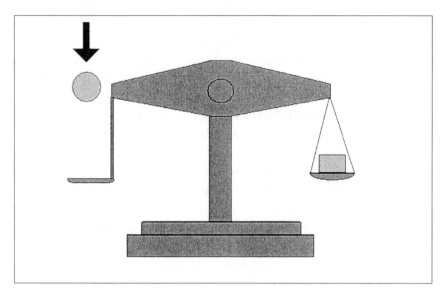

Hooke's apparatus for investigating the force of falling bodies.

nomical observation postulated that planets moved in elliptical orbits (not circular, as Nicholas Copernicus had proposed). He had also surmised that they moved slower in parts of their orbit furthest away from the Sun, and if you took the square of the time required for a planet to complete one orbit, it would be proportional to the average distance that the planet was from the Sun. He also noted that the area within the arc of the ellipse that a planet traversed over a given time, no matter whether it was at its furthest away or at its closest proximity to the Sun, would always be the same figure. Hooke could see that the answer lay somewhere in Kepler's findings, but it still alluded him.

He and Wren discussed the matter in 1677. Hooke's diary mentions that on 20 September they met at Jonathon's coffee house and Wren 'affirmed that if the motion of were reciprocal to the distance, the degree of velocity should always be as the areas.'[64] This meant that if the gravitational forces were directly proportional to the distance between the bodies involved then no acceleration would take place. Now Hooke knew that this was not the case, that there was an acceleration, he had shown that velocity increased as a body came close to the Earth.

In his capacity as the Society's secretary, Hooke wrote to Newton, hoping to continue correspondence between him and the organisation. He also realised that Newton might be able to help him in his quest to unlock the secrets of gravitational motion. In his letter of 24 November 1679, Hooke stated his enthusiasm for Newton to 'continue his former favours on the Society by communicating what shall occur to you that is philosophicall, and in return I shall be sure to acquaint you with what we shall receive considerable from other parts or find out new here.'[65] He mentioned that there have been some who 'have indevour'd to misrepresent me to you'[66] and continued by saying 'for my part I shall take it as a great favour if you should please to communicate by letter your objection against any hypothesis or opinion of mine'[67] and 'particularly if you will let me know your thoughts of that compounding the celestiall motions of the planets.'[68]

In Spring 1679, Newton had been forced to return once more to Lincolnshire. His aged mother's health was deteriorating rapidly, and so he tried to care for her as best he could in her final days. She passed way on 2 June and was buried in Woolsthorpe two days later. He stayed on for several months to tie up outstanding affairs, and returned to Cambridge as winter began to take hold. The end of that year, when he arrived back in Cambridge, Newton must have been surprised to find Hooke looking to initiate a dialogue with him regarding the enigma of planetary motion. It probably did not take Newton long to deduce that he was fishing for clues. Newton had first given serious contemplation to the topic while he was back in Colsterworth during the late 1660s. After intense deliberation he concluded that the force that kept the moon in orbit was the same as we witnessed here on the surface of the Earth. The mechanism that meant an apple would fall from a tree also dictated the motion of planetary bodies.

Newton's reply to Hooke was guarded to say the least, if he did know something more at this stage he was certainly in no rush to divulge it. He stated how 'heartily sorry I am that I am at present unfurnished with matter answerable to your expectations. For I have been this last half year in Lincolnshire, cumbered with concerns amongst my relation.'[69] Keen not to let anything slip, and implying that he was no longer involved in research in this area, he wrote that 'I had for some years past been endeavouring to bend myself from Philosophy to other studies, in so much ye I have long grutched the time spent in yt study, unless perhaps at idle hours something for diversion.'[70] Nevertheless, Newton became drawn into a discussion over how the movement of bodies towards the gravitational centre of the Earth would take place if there was not a solid mass there to obstruct it. On 9 December, Hooke responded to Newton trying to sustain the interplay. Four days later Newton replied agreeing with his supposition on how a falling body would be affected by the Earth's gyratory motion.

At the Society's meeting on 18 December, Hooke gave an account on the 'three trials of the experiment proposed by Mr Newton' in which a weight was dropped from a tall edifice, to see if it would move to the west or the east. As the Earth spins from west to east, it would be expected if an object was dropped from a great height, the planet would have rotated a small distance in the time that it had taken to fall, and it would land slightly to the west. However (and here Hooke and Newton were for perhaps the only time in their lives in complete agreement) they felt this would not be the case. Both men correctly postulated that if the body was being dropped from a distance above the Earth's surface then it must have been travelling faster to start with, as the further away from the planet's centre it was, the quicker it would have to rotate in order to keep up. For instance, on certain fairground rides there is a difference between the speed of the inner sections, and those further out. So the more adventurous tend to place themselves on the outside, while the more squeamish elect to stay closer to the centre. Another example is a slingshot; if the length of the string used to make the sling is increased, then the speed at which it can be whipped is increased, and the distance that it can expel its projectile is thereby amplified. Given these facts the two scientists had concurred that a body dropped from a height would be ahead of the rotation of the Earth's surface, and thus land slightly to the east. Hooke found that the weights 'in every one of the said experiments fall to the south east of the perpendicular point, found by the same ball hanging perpendicular. But the distance of it from the perpendicular point being not always the same.'[71] The tests had not been very accurate, as they were performed outdoors, and the effect of wind and suchlike meant there was little certainty about the whole thing, the quality of results being relatively poor.

In his letter to Hooke of 3 January 1680, Newton graciously expressed his grattitude 'for the trials you made of an experiment suggested by me about falling bodies, I am indebted to your thanks, which I would have returned by word of mouth, but not having yet the opportunity must be content to do so by letter.'[72]

Later that month, once the discourse had been fully set into motion, Hooke began to push things in the direction of the calculation of gravitational force. He proposed that, as he suspected, 'the Attraction always is a duplicate proportion to the distance from the Center Reciprocal and consequently that the Velocity is a subduplicate proportion to the attraction.'[73] Not getting any answer from Newton, he pressed the matter, writing again

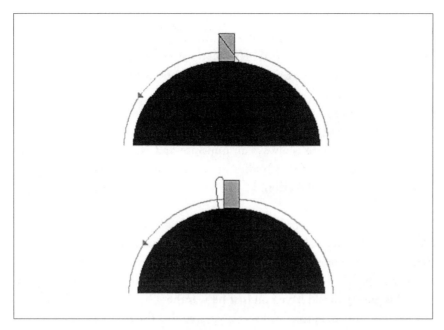

Effect of the rotation of the Earth on a falling body.

on 17 January that 'it now remains to know the propertys of a curve line (not circular or concentrical) made by a centrall attractive power, which makes the velocitys of decent from the tangent line or equall straight line motion at all Distances in a Duplicate Proportion to the Distance reciprocally taken. I doubt not but that by your excellent method you will easily find out what the curve might be.'[74] Hooke disclosed that he knew, or at least suspected, the dependency of gravitational force on the square of the distance between the bodies involved. But the real question is did Newton know it beforehand?

In the book Principia Mathematicae, which was published some five years later, Newton laid down the laws of motion, and described that gravitational force was dependent on the inverse square of the distance between the bodies involved (i.e. if two objects were a distance 'L' apart then the force between them would be proportional to $1/L^2$ or $1/\{L \times L\}$). Upon the discovery of what Newton had put into print, shockwaves passed through the structure of the Royal Society. Hooke alleged that it was he who had brought the inverse square relationship to Newton's attention, and demanded that he should receive acknowledgement for his participation in the physical laws that had been derived from this.

Newton's laws of motion were as follows:

1. Everybody preserves in its state of being at rest or of moving uniformly straight forward, except in so far as it is compelled to change its state by forces impressed.
2. A change in motion is proportional to the motive force impressed and taken place along the straight line in which that force is impressed.
3. To any action there is always an opposite and equal reaction; in other words, the actions of two bodies upon each other are always equal and always opposite in direction.[75]

Newton's first law is certainly reminiscent of what Hooke had stated in his 'Attempt to Prove the Motion of the Earth', but this was derived from basic principals of science that had been founded many years before. If Hooke tried to imply that Newton had pinched from him, then Galileo (or more likely his ghost) would have said that he was there before either of them. As already mentioned, Hooke had also suggested that the Earth would not have the shape of a perfect shape, but it would bulge in the centre, with a greater diameter along its equator than between it poles. This was another pearl of wisdom that was inherited by Newton to place in his meisterwerk, and here Hooke certainly had a valid claim of originality. Newton was clearly not generous when it came to giving out any thanks.

His confidante, Edmund Halley, wrote to Newton on 22 May 1686 to inform him that 'Hook has some pretensions on ye rule of the decrease of gravity, being reciprocally as the squares of the distances from the centre. He says you had the notion from him'[76] and that he 'seems to expect you should mention of him in the preface.'[77]

In Newton's response, five days later, he made clear that there was no need of this, and that in his papers 'there is not one proposition to which he can pretend and so I had no proper occasion of mentioning him there.'[78] In a further letter to Halley, dated 20 June, Newton said he had heard from other sources at the Society that, during its meetings, 'Hook should there make a great stir, pretending I had all from him & desiring that they would see that he had justice done.'[79]

Newton countered his claims saying Hooke had simply guessed at the nature of the attractive force, while he had done all the laborious toil needed to prove this law and establish the truth behind it. He put in all the work, devoted several years of his life to the problem, performing long and tedious observations, and spent countless hours trying to figure out a means to calculate it. He also vindicated his position by making Hooke out to be someone who did not have the necessary savvy to see his ideas through to their conclusion. Newton was adamant that the rights to the discovery lay with the one who could produce mathematical proof of the nature of planetary motion (although we will see later how Newton managed to verify this law), not someone who had just speculated possible solutions. He shrugged off Hooke's continued claims of priority, stating he knew that he 'hath not the Geometry enough to do it.'

He scoffed at the idea that Hooke thought he could claim an equal share of the praise in this defining moment in scientific history. In his letter to Halley, Newton further commented, 'should a man who thinks himself knowing, and loves to shew it in correcting and instructing others, come to you when you are busy, & not withstanding your excuse, press discourses upon you & through his own mistakes correct you & multiply discourses & then make use of it, boast that he taught you all he spake and oblige you to acknowledge it & cry out injury & injustice if you did not, I believe you would think him a man of strange unsociable temper.'[80] This, it must be argued, is very true. Newton had not initiated or encouraged correspondence with Hooke, he would have been quite happy to continue his work undisturbed. He told Halley that philosophy (meaning science) was such 'an impertinently litigious Lady, that a man had as a good be engaged in lawsuits, as have to do with her.'[81] Halley's letter back suggests that Newton had overreacted to Hooke's claims, and that things had been blown out of proportion. He stated, 'as to Mr Hook claiming this discovery, I fear it has been represented in worse colours than it ought: for he neither made public application to the Society for justice, nor pretended you had got all from him.'[82] This may have been Newton's strategy from the start; by taking the stance of the injured party, it allowed him to paint Hooke as the aggressor trying to muscle in on his discovery.

Hooke tried to show that in addition to the letters on the subject he had sent to Newton, he had also alluded to the inverse square relationship between gravity and distance in several of his earlier works. He said that he published a clear statement of the fact that attraction between bodies was dependent on the gap between them squared as far back as his 1674 lecture 'An Attempt to Prove the Motion of the Earth'. The scientific community adjudged that this gave no conclusive proof, and it was just supposition on Hooke's part. Of course, Hooke felt this was yet another occasion when the Society had let him down, but in all honesty he never really published enough material to prove his points.

Newton claimed that he had the basis of his mechanism for the movement of celestial bodies in place by the late 1660s, and though Hooke's lecture in 1666 on 'Gravity and Celestial Motion' predates this (as some writers have chosen to point out), it does not actually mention, or even hint toward the law anywhere within it. Hooke said nothing about the inverse square law in his April 1678 Cutler Lecture on the subject of comets, entitled 'Cometa', though he later claimed that he had known it by this stage. He would go on to imply that his work on springs, namely his Cutler Lecture 'Ut Tensio' 1675, and his discourse to the Royal Society 'Potentia Restituva' in 1678, could be judged as examples of an action having an equal and opposite reaction. Thus he felt he had again given Newton some useful indications. However, it was probably pushing it a bit far to say he had been the precursor of Newton's discovery. Although these works could be applied to Newton's laws, Hooke had not really written them with the intention of being used in such a way, it was purely in hindsight, so they could not warrant serious consideration. The undeniable truth was Hooke had skirted around the periphery of this puzzle, but had not been able to breach its heavily defended ramparts.

Halley heard that in the summer of 1686, Wren supposedly challenged Hooke to produce evidence to back up his claim that he had uncovered the basis of the laws of planetary motion. In response Hooke said that 'he had it, but would conceale it for sometime, that others triing and failing, might know how to value it, when he should make it public.'[83] However, if he honestly meant it, this seems to be completely out of character for Hooke. Had he known, then he would have surely not wasted any time in telling the world about it, he had always thrived on attention, and would not have wanted to miss such a golden opportunity. Besides if he had really worked it out, he was taking a very big risk that someone else would not succeed in solving the problem. He knew that Newton, Wren and Halley, among others, were all keen to disentangle this mystery, and suspected that every one of them was wrestling with the problem at that point. Sooner or later someone was going to unravel the enigma, it would only be a matter of time. Therefore his stalling tactic does not seem rational at all.

So how had Newton managed to derive this law, which had remained hidden from all his fellow scientists for generations? As mentioned earlier, he followed the assumption that the force holding the Moon in orbit around the Earth was the same as that causing an apple (or any other body) to fall to the ground. What kept the Moon from crashing into us was the speed at which it travelled around the Earth. So he began to contemplate what would be necessary for the apple to display the same sort of movement as the Moon possessed. He thought to himself, if he were to throw the apple horizontally it would follow a curved line (a parabola) as the force of gravity began to overtake that of the initial propulsion. Now the harder the apple was thrown the flatter the parabola would be (the closer it would come to being a straight line). Given enough force the distance the projectile travelled would eventually become large enough that the curvature of the Earth would have to be taken into account, and at some stage the projectile would travel with sufficient velocity that the amount it dropped in height along its path would be the same as the amount that the surface of the Earth falls away from the perpendicular, due to its curvature. Okay this means that whoever was throwing this apple would need pretty big biceps, so he took a more practical example. Say you had a cannon placed on a very high mountain. If you fired a cannon ball at such a speed that its trajectory could match the curvature of the Earth then it would never land, because the distance it dropped vertically in proportion to how far it travelled horizontally would be the same as the rate at which the Earth surface fell way – effectively the cannon ball would be in orbit. Newton calculated that the surface of the Earth drops away from the horizontal at a rate of 5m in every 8,000m.

Now a body drops at a rate of 5m per second due to the pull of gravity, so the projectile would have to initially be travelling at 8,000m per second to avoid falling to the surface. This ignores the effect of air resistance of course, but he was trying to compare it with the motion of the Moon, which he rightly assumed was not afflicted by such forces.

So next he looked at the motion of the Moon. It takes one month (27.3 days to be exact) to travel around the Earth, and is 384,000km (or 38 million metres) from us. Newton looked at the orbit the Earth's satellite traversed and applied the same principle that he had to the cannon ball. He took a straight line, and then looked at how much the curve would drop away from this line over a distance. The curve was much bigger, so the amount it fell would be a lot less. The distance that it effectively moved away from the perpendicular line each second would be just 1.37mm (0.00137m). Comparing this with the 5m per second that the cannon ball would drop, showed that the gravitational force did (as Hooke had proposed) reduce greatly with distance, but now Newton was in a position to create a formula to describe by exactly how much.

The ratio of the Moon's vertical movement and that of the cannon ball skimming the Earth's surface would be 0.00137 to 5, in other words 1:3,600.

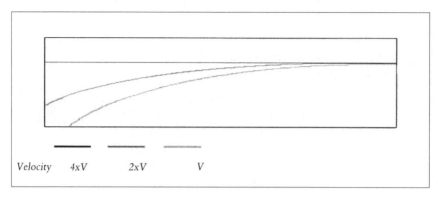

The parabolic paths of an object travelling at different velocities.

So the comparative gravitational force that the Moon would be subject to would therefore be 3,600 times less than that experienced by the cannon ball. The surface of the Earth is 6,350km, and as already mentioned the Moon is 384,000km away, this meant that the cannon ball would be sixty times closer to the Earth's centre than the Moon. Now when he compared the difference in the gravitational forces (i.e. 3,600 times) with the difference in the distances involved (i.e. sixty times), it became crystal clear that $3,600 = 60 \times 60 = 60^2$, so the attractive force was dependent on the distance squared. It was no longer just guesswork it was scientific fact. Newton had achieved his goal; a basic structure with which to describe the motion of the planets, he had effectively discovered the key to the fundamental workings of the universe. What is more he could assume with some certitude that a calculation of such proportions was simply beyond Hooke's capabilities.

Newton would use the dispute as an excuse to dispense with any future references to his Society colleague's work. In the third book of his Principia, he had previously planned to state acknowledgement to Hooke, with regard to the fact that the motion of each planet would consist of an attractive component toward the centre and an orbital component perpendicular to it. Following the conflict between them, Newton decided to scrub any mention of his name.

Newton's fame did not stay rooted within the British Isles, like most of his contemporaries. His work would be read across the whole of Europe. Arouet de Voltaire was among his more illustrious advocates on the Continent, and did a lot to promote the teaching of the Englishman's gospels. The physician, later to become revolutionary leader and martyr, Jean-Paul Marat, translated Newton's Opticks into French. Pierre Signorne, a Gallic philosophy teacher,

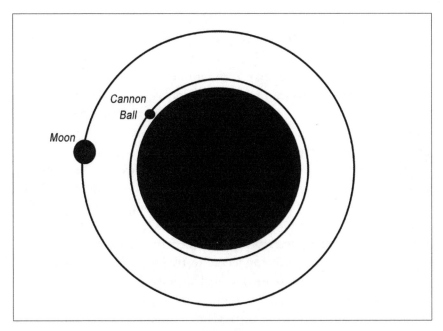

Newton's comparison between the movement of a cannon ball and the Moon.

did the same for some of Newton's early work, while Hermann Boerhaave, professor of medicine at Leiden, was amongst the first to publicise his ideas in Germany. Mathematics professor of the Collegio Romano, Roger Joseph Boscovik, was one of Newton's early supporters in Italy. With the help of his foreign disciples, Newton had placed himself at the zenith of Europe's scientific hierarchy.

Newton made little time for any type of social activity, separating himself from such things. He seems to have formed few strong relationships. He had a close, but rather turbulent friendship with the Swiss academic Nicholas Fatio de Duillier, and it has been debated that they may have had a homosexual affair, but there is no concrete evidence to prove this one way or the other. Another of the few long-lasting associations that Newton was to posses was with the philosopher Joseph Locke, with whom he corresponded on the subject of alchemy.

He finally produced a book on this theme in the late 1690s. Praxis, as it was known, described in detail his experiments in that particular field, though it was never published (well, there would be no use telling everybody such secrets, after all, you would want to keep them for yourself wouldn't you). Newton noted that nature could change the form of things through actions like putrefaction, and thought that if he could unlock how such processes took place, he could change the form of different elements. He conjectured that 'an oak may stand for one hundred years without rotting, but if it be scraped thin and kept twixt moyst and dry it may soon by art by brought to dirt.'[84] By the same principle metals, he conjectured, 'though they may long persist in the earth, without corruption yet duly ordered and mixt due to mineral humidity, by art soon rot and putrefy.'[85]

Despite being snubbed by the scientific community, Hooke continued to be a bountiful source of new ideas. The virility of his mind had not been dampened by his advancing years. His 'Universal Joint' was utilised to allow easy unobstructed movement of the Equatorial Quadrant constructed for the Royal Greenwich Observatory, which Charles II founded in 1675. It basically consisted of a disk connected to two sets of pivots, the axes of which were perpendicular to each other. This 'yoke and spider' arrangement is still used in the transmission systems of cars today, yet another example of how important Hooke's discoveries and innovations were, and how little people realise it.

Grace died in early 1687, at only twenty-six years of age. She was buried on 28 February at Bishopsgate cemetery. It had probably been influenza that took her, although Aubrey mentions a dark bruise found on her chest, which could suggest breast cancer. Hooke was not a man to share his emotions easily. Even during the most traumatic of periods, such as the death of his brother, he showed little sign of what he was experiencing. As Hooke did not keep a journal at the time of Grace's death, his feelings are even harder to establish. However, Waller testified in his recollections of the man, that he became increasingly cynical and melancholy following her passing. It is hard to believe that even as stern a character as his would not have been heartbroken; the two of them had been lovers for over twelve years by this stage. His last two maids, Martha and Mary, both managed to escape the carnal desires of their master, possibly his age meant he had lost interest in such activities, or that he never truly recovered from the loss of his soulmate.

Hooke soldiered on. In late July that year, he demonstrated a mechanically adjustable aperture for telescopes, which was basically a forerunner of the iris diaphragm used in optical equipment for centuries to come, and is still found in modern cameras and microscopes. Although it looked as if his spirit for invention would go on forever, as the 1690s began he finally started to lose his enthusiasm for scientific endeavour, and slowly went into a state of decline.

Though Newton was far younger than Hooke, the years of incessant exploration were hanging heavy upon him too. It was in mid-1693 that he was to go through a full-blown nervous breakdown. It is uncertain as to the exact cause of this, but it is possible that his exposure to certain heavy metals during his forays into the world of alchemy could have had some detrimental effect on his mental condition. He had suffered some sort of psychological ordeal back in 1677, but this second attack was far more severe. Newton would rest for many months, and is said to have never been the same person again. His days of inspired genius were effectively over and he would not return to serious research.

By March 1699, he effectively turned his back on science as a full-time occupation, and became Warden of the Royal Mint. Newton took to his new occupation with considerable zeal. He set in place new methods for thwarting the rising trend of counterfeiting, as well as trying to rid the organisation of the corruption and inefficiency that was crippling it. He had no qualms in punishing offenders either, personally seeing that many of them received the death penalty for their crimes.

He took the role of Master of the Mint at the end of December 1699, which brought with it a substantial salary of £1,500 per annum. In 1701, he finally relinquished his position as Lucasian Professor. In reality he had done nothing to warrant the retention of his post for several years. He had long since stopped giving lectures, and rarely even visited the university. In 1705 he was knighted.

Hooke was not to be so fortunate when it came to gaining high regard from such quarters. He was certainly not a political animal, and was overlooked for any notable honour of this kind. Neither did he mellow with age. Many young scientists would be on the receiving end of one of his tirades, where he would state that he had presented work to the Society in the past that showed his prior claim to the discoveries that they were proposing. More often than not he had made some brief study, or speculated on some theory, but had gone no further. His habit of trying to accuse others of ripping off his work, and telling everybody he had been there first, became increasingly annoying to his colleagues, only serving to tarnish his reputation.

In his last few years he was 'much over-run with scurvy'[86] according to his friend Waller. His eyesight failed almost completely, and left him virtually blind. He had apparently for 'several years been often taken with giddiness in his head, and sometimes great pain, little appetite, and great faintness, that he was soon very much tir'd with walking, or any exercise.'[87]

After a long and rather ungraceful decline, Hooke finally shuffled off this mortal coil on 3 March 1703, a few months shy of his sixty-eighth birthday. Although he had written a will, in which he had separated his possessions between four different parties, they were not named. Clearly he had still been pondering who were to be the recipients, and his indecision meant he died in testate.

Hooke had become increasingly worried that he might exhaust his fiscal reserves before he died. Perhaps the thought of his brother's financial ruin had gone some way towards making him so overcautious. His miserly ways became almost legendary, and everyone who did not know him well assumed he must be penniless. Waller referred to him as a 'hermit', stating that while 'his circumstances, as to his estate were very considerable' he was 'scarcely affording himself necessaries.'[88] But it was not until his belongings were examined after his death that the true extent of the wealth Hooke had amassed during his varied, eventful and bittersweet life became apparent. A chest found in his Gresham College rooms contained in excess of £8,000 (equivalent to £2 million today). Although he never gained the acceptance of the scientific community like his arch-rival Newton, or the notoriety he so desperately sought, he had managed to carve out a very lucrative career in designing and supervising the construction of new buildings after the inferno that had engulfed London back in 1666.

In some ways it is as if history has tried to erase any trace of Robert Hooke. A lot of this is down to Newton's efforts, who despite the fact he had clearly won the day, and gone on to become universally admired, still maintained a high degree of vindictiveness, and looked for further retribution. He attempted to downplay Hooke's contribution to England's scientific renaissance, and using his huge influence, made sure as little record of his achievements as possible endured. In addition to all this, there seem to have been many occurrences that resulted in further eradication of Hooke's memory. For example, it is known that the Royal Society once possessed a portrait of Hooke, but its whereabouts are now uncertain. It is possible that this was also destroyed at the hands of Newton, but it is equally likely that it was simply lost, as the Society's repository and other curiosities were to fall into neglect after Hooke's passing. A stained-glass window found in the Church of St Helen's Bishopsgate, which depicted Hooke and his empirical feats, was completely obliterated by an IRA bomb in the early 1980s. While Newton would be celebrated in sculptures by Dali and Paolozzi, the paintings of Blake, the drawings of Hogarth, the cameos of Wedgwood and the poetry of Wordsworth, Pope and Keats, poor old Robert Hooke would have little in the way of artistic gems to commemorate his extraordinary life.

Although time has managed to make amends for the lack of consideration he was shown, and his talents are now more readily appreciated, Hooke's physical appearance is still a mystery. With no definite visual record of how he looked remaining – not a portrait, sculpture, or even a death mask – the only method that we have of trying to get an idea of his demeanour is from the writings of the men who knew him. Aubrey said that he was of 'mid-dling in stature, something crooked, pale faced'[89] that his head was large and his eyes were grey in tint, as well as being 'full and popping'. He had a 'delicate head of haire, and of an excellent moist curle.'[90] Waller tells us that he 'was always very pale and lean' and in his old age 'nothing but skin and bone, with a meagre aspect.'[91] He also mentions that his eyes were 'grey and full, with a sharp ingenious look'[92], as well as a large forehead and sharp chin.

Hooke was surely one of the most prolific scientific minds with which the world was ever blessed, but he was clearly not the easiest of characters to get along with. He was rather arrogant, highly argumentative, and had a tendency to overstate his own abilities. At first glance of Hooke's life you might get the impression that for all his talents he never stuck to anything long enough to really make an authoritative mark. He postulated a great many things that were actually proved to be correct, and was a precursor for not just one or two, but a whole legion of scientists' work, though because he was involved in so many things, he did not focus on any particular field that would have gained him enduring acclaim. However, when you look a little closer you realise that his many commitments, to the Royal Society, the Corporation of London, the ungrateful Cutler and the more deserving Boyle, made it virtually impossible for him to devote greater time to any one specific goal.

Despite Hooke being almost certainly the most industrious and vision-ary member of the Royal Society during its first few decades, he was for the majority of this time treated as an underling, and did not command anything like the respect that Boyle, Wren, Huygens and others received. This clearly has a great deal to do with class. Wealth and family status were still important factors at that time. Hooke was not part of the social elite, and this humbled his achievements. For example, Boyle was clearly a man of incredible intellect, but in addition to this he had the advantages of noble birth and breeding. So not only did he have the opportunity to concentrate on developing his ideas without having to worry about earning a living for himself, he could also afford to pay others to do much of the donkey work. On top of all that he was in a position where people were more willing to listen to what he had to say. Even when Hooke gained a fellowship, he was still not treated as an equal, it was more that his Society colleagues did it as a token gesture, in order to be justified in placing even greater workloads upon him, and almost certainly to give them more leeway when it came to paying him a salary. His hard work and loyalty to the Society do not seem to have been greatly appreciated by its members, they just started to expect more and more from him. Hooke did prove, however, to be his worst enemy on many occasions. His habit of accusing others of plagiarism became increasingly frequent, and taken with diminishing seriousness. Waller said, 'it must be confess'd that very many of his Inventions were never brought to the perfection they were capable of, nor into practice till some other Persons cultivated the invention.'[93]

Although he clearly never had the time or resources he desired to conclude many of the projects he had begun, it also seems to have been the way he wanted it to be. He could have resigned his post as the Society's experiment workhorse when he started to aid the reconstruction of London. The wages he received as Curator of Experiments were paltry in comparison to what he earned from Wren's employ, or for his private commissions. The fact is it suited him to flit between these many different scientific branches, he enjoyed the varied nature of his work, and it is likely that he would have become bored had he specialised in one particular field. This is not to imply he was lazy, I doubt that anyone could accuse Robert Hooke of lethargy. But when you compare him to someone like Newton, who could focus on one problem for huge time spans, sacrificing everything else in his life to attain his objective, you can see there was a very big distinction between the two of them.

His strategy to do the early running in a particular area of investigation, then move on to something else, kept him motivated, but also meant he could claim his contribution towards a discovery if someone took things further. This is what we have to accept in the case of the Newton-Hooke dispute. Hooke did not have the mathematical ability, or the patient methodical temperament needed, to create such an extensive work as Principia, and so his research on gravitation and the motion of the planets could only serve as a foundation for others. It took a highly focussed, single-minded personality, with a far greater knowledge of calculus than Hooke could ever muster.

Unlike many of the other 'Nearly Men' we will learn about, it was not the case that he was beaten to it. If Newton had not been there it would have been someone else further down the line; sadly it would never have been Robert Hooke who would claim the glittering prize – he simply was not up to the challenge. This is not to try to demean his abilities, or reproach his amazingly fertile and imaginative mind, he just was not the right guy for this particular job.

That he contributed to the development of Newton's ideas is not really possible to dispute, he had postulated that the motion of the planets was made up of two components; the attractive force that was exerted along the line connecting the two bodies centres, and the orbital force perpendicular this. But he probably overvalued the influence he had on Newton's calculation of the all-important inverse square law. If he was guilty of this, well it certainly would not have been the one and only time it happened, he became renowned in his later life for similar occurrences. However, if he did assess the importance of his sway on the formation of Newton's Principia too highly, that was mainly down to Newton himself. He almost certainly knew about the inverse square law before Hooke brought it to his attention, but chose not to mention this in their correspondence. He played dumb, perhaps because he did not want to risk another quarrel.

Newton represented the future. Science had reached a stage where pure observation alone was not going to be enough, theories had to be backed up with mathematical evidence. Newton was the first in a long line of exponents of a new era of science, of which Heisenberg, Fermi, Planck, Feynman and Einstein, among others, would be a part. Hooke was the last of the old school of experimentalists, who based their work on investigative findings. There was no room in an increasingly complex mathematical world for a man like him. Again, this does not belittle him, or make his achievements seem petty. He was clearly a man of most unique virtues, and possessed huge ability across a wide range of scientific disciplines. I think John Aubrey summed up most succinctly what Hooke had missed out on, when he wrote, 'this is the greatest discovery in nature that ever was since the world's creation. It never was so much as hinted by any man before. I wish he had writt plainer, and afforded a little more paper.'[94]

Following Hooke's death the path was clear for Newton to exert his power on the Royal Society. He was elected its president in Summer 1703, just eight months after Hooke's demise. Newton held the position until his own passing, some twenty-four years later. So began a new phase in the organisation's history, the founding fathers were now dead, or as good as. For the next quarter of a century Newton would dominate this fellowship entirely, like no other man before or since has been able to. He had not been a regular visitor to its meetings while Hooke was still alive, but now he attended almost without fail.

As well as waiting to time his entrance with regard to the Society, he had held back the release of his work Opticks until his old adversary was safely six feet under. Newton did not wish to endure another confrontation with Hooke, and tackling a subject perceived as his 'home turf' was only likely to cause further antagonism. The book was published in early 1704. After Principia, this is seen as Newton's most important and influential work. It persisted with his hypothesis that light was made up of particles rather than waves.

Newton, as already mentioned, would continue to pursue his vendetta against Hooke even after the latter was deceased. It is likely that he was instrumental in eradicating all but a few mentions of Hooke's exploits from the annals of scientific history. Hooke's reputation was not the only victim of the power that he wielded. Another who was to fall foul of Newton's hand on several occasions was Astronomer Royal, John Flamstead. Newton used his position as Society president to push Flamstead into supplying him with data from his unfinished astronomical observations to help him with his work on gravitation, and in what can only be described as an immensely arrogant move, told the star gazer not to concern himself with his own theoretical speculations, but just to concentrate on giving him more information to help formulate his more important scientific research.

The dispute between Newton and Gottfried Leibniz gives a further example, and could easily fill an entire book by itself. The German mathematician had developed his own form of calculus, and a huge quarrel arose concerning who had priority. When the Royal Society was asked to launch an investigation into the matter, it was Newton, as its leader, that picked who would conduct it, and it is likely that the report produced was penned by Newton's own hand. Unsurprisingly, the outcome of the enquiry found strongly in favour of (you guessed it) Newton.

Although he may have bought a reign of tyranny upon the Royal Society, Newton's influence was still of huge benefit to the institution's survival. He took it from a state of total disarray, with huge financial problems and a floundering reputation, transforming it into a stronger more respected body. He allowed it to ride on his considerable international acclaim, and re-fortified its status.

In the early 1720s, Newton first began to suffer from bladder stones. Such medical complaints were commonplace during this period of history, but it was an ailment for which there was no effective remedy other than having them surgically removed, and this was deemed more dangerous than trying to live with the complaint itself. With his increasing infirmity Newton's assistant, John Conduitt, began to take over many of his duties at the Mint, and later succeeded him as its master.

Shortly after a Society meeting on 19 February 1727, Newton became seriously ill. After being confined to bed for some weeks, he finally died at his home in Kensington on 20 March. It was a drawn-out and painful death, his enflamed bladder stones finally getting the better of him. The funeral ceremony took place at Westminster Abbey a week later, and was attended by the Lord Chancellor, representatives of the Royal Society, and various members of the aristocracy. The Latin inscription upon his grave states, 'Here lies that which was mortal of Isaac Newton'.

After his death, Newton took on the guise of a demi-god. He would act as the muse to a multitude of artworks; odes would be composed about him and busts rendering his visage would adorn the homes of the wealthy. His public perception was more in keeping with a military hero or a deity of antiquity than a scientist. To this day monuments are erected in his memory.

In 1784, France's Etienne-Louis Boullee designed a mausoleum for Newton. Those of you who have seen Peter Greenaway's film *Belly of an Architect* will know that very few of Boullee's works actually made it off the drawing board, and perhaps in this case it is not a bad thing. His tribute to Newton's 'sublime mind', as he called it, would have consisted of a 120m-high hollow sphere with a huge eternal flame at its centre. One can only imagine the difficulty and expense involved in constructing such a building.

In late 1999, the *Sunday Times* named Newton as the 'Man of the Millennium'. Three years later BBC viewers voting for the greatest Britons placed him sixth, and when a similar poll was conducted for its international audience soon after, he took the top slot, ahead of Churchill, Shakespeare and Darwin. Perhaps Halley expressed most pertinently how the world should perceive this great man. In the preface to the original version of Principia, Halley says of his friend, 'no closer to the gods can any mortal rise'.

Possibly the reason that Hooke and Newton were never able to put up with each other was the fact that they both had such single-minded person-alities, certainly neither were perceived to be particularly congenial people. Although Hooke was probably at times not the easiest fellow to get along with, he was certainly outgoing and amiable on the whole. He had a loyal circle of friends, and took great joy from socialising – dining at the homes of his Society colleagues, or gossiping at the coffee houses he frequented. For Newton it was a different story altogether. Throughout most of his life he maintained the image of a much more solitary figure, unapproachable and aloof. Hooke is often described as an over-suspicious, secretive character, deeply paranoid and over-sensitive. Yet nobody was more endowed with these traits than Isaac Newton.

In recent times, Newton has received a lot of bad press, and although some of this may be warranted, it is vital not to forget the importance of his contribution to the development of our civilisation. The whole basis of physics and mathematics relies on Newton as its cornerstone. To use the words of Pope in a slightly different context, 'Let Newton be'. The Hooke revival that has taken place in recent times more than adequately redresses the balance. The world now knows what a gifted scientist and engineer he really was. To try to snatch the glory of Newton's magnum opus away from him is overstepping the mark. He put the effort in, so therefore he should get the praise. However, what we can say is that Hooke did do a lot more of the groundwork than Newton would have liked to admit.

There is no doubt that Newton and Hooke both deserve their places in Britain's scientific pantheon. They were each blessed with astonishingly productive minds. Yet, in the end, you can only feel sympathy for these troubled geniuses. Newton gained wealth, high standing in society, and worldwide recognition, but it did not seem to make him any happier. For Hooke, though he seems to have had a miserable life, filled with disappointment and what he would feel had been numerous betrayals, perhaps he would be happy now to know that later generations have begun to learn about his many exploits. His fame has been a long time coming, but he finally got what he wanted.

Sir Francis Bacon, whose principles were so greatly idolised by the Royal Society, and which formed the foundation of what its members were striving for when they first established it, once said, 'as births of living creatures at first are ill shapen, so are all innovations, which are the births of time.'[95] It is apt that the man who most embodied what the Society (that both Hooke and Newton loved so dearly) stood for could recognise the fact that any concept or idea is never perfect to begin with, it takes time and effort, as well as the support of others to make it so, to fashion it into something of true worth. Perhaps if his two followers had been a little less self-involved, then they could have remained true to this idiom.

CHAPTER 3

Making Waves

A tall, thin, rather dashing figure appeared upon the stage ready to address the expectant crowd. His jet black hair was perfectly coiffured and his trademark Hoxton moustache neatly trimmed. He was, as always, immaculate in his attire, although it could be argued that wearing a white tie, tails and dressed gloves would be more in keeping with a musical performance, and bordered on being ludicrously overdressed for a scientific lecture.

His name had already attained legendary status; he was acknowledged as the messiah of the electrical age. His abilities had even brought into question if he was truly of this world, and rumours abound that he had supernatural powers that could be barely confined within a human form. Engaging every single member of the audience with his the piercing, blowtorch blue eyes, he began to speak. In a distinctively high, somewhat shrill, voice accompanied by a strong Slavic accent, he began to impart details of his latest research, possibly the most astounding he had ever embarked upon. He explained how he would use unseen forces to transport energy and information over vast distances through the ether; the future he envisioned would not be shackled by wires.

Though the crowd assembled before him must have emanated reluctance to accept such ravings, and doubted his faculty, the murmurs were soon silenced when he gave a demonstration of the wonders he had foreseen. Holding the bulb of a lamp with his palm, clearly not attached to any source from which it could draw power, he brought life into it as though he was some enchanter.

They were born into the same era, in two towns separated by little more than the Adriatic Sea. Each would gain huge admiration for their work, and there is no question that both of them had a profound effect on the modern age, but they would be judged in very different ways. One would leave this world as a hero responsible for saving countless lives, loved by both his country of birth and the one where he spent much of his adult life, a respected political figure, a war hero, and an inspiration to all. The other meanwhile would pass his final days alone, embittered, almost destitute, his mind verging on the borders of insanity, and forgotten by the outside world.

> *'I do not think there is any thrill that can go through the human heart like that felt by the inventor as he sees some creation of the brain unfolding to success.'*[1]
>
> Nikola Tesla

During the witching hour of 10 July 1856, Nikola Tesla was born in the village of Smiljan, which was then part of the Austro-Hungarian Empire, and now lies in Croatia. His father Milutin, who was an Orthodox priest, and his mother Djuka, were both of Serbian origin. Nikola had three sisters; Angeline, Milka and Marika, as well as an older brother, Dane. Sadly Dane was to die young. At the tender age of twelve he fractured his skull after being thrown from a horse. Nikola was just five years old at the time. He later said of Dane that he was 'gifted to an extraordinary degree' and his 'premature death left my parents disconsolate.'[2]

The death of his brother may to some extent explain the many idiosyncrasies that plagued Nikola, and which we will learn more of soon enough. Dane had been his parents favourite, in his brother's eyes at least, overflowing with promise, and their tragic bereavement seems to have made him feel very guilty that he was still around. He was determined not to disappoint his family, and as a result pushed himself as hard as humanly possible to try to make up for the deprivation they had been forced to endure. Unfortunately he felt that anything he did 'that was creditable merely caused my parents to feel their loss more keenly.'[3]

When Tesla was about eight his family moved to the nearby town of Gospic, some 10km from Smiljan, as Milutin was offered a larger parish to look after. The boy's preparatory schooling took place here, though Djuka probably had greater influence in developing his technical prowess. By all accounts his mother was illiterate, but she still had an incredibly agile mind, giving the young man further instruction to help supplement his classes. Nikola lamented her virtues, saying that, 'she was a truly great woman, of rare skill, courage, and fortitude.'[4]

Tesla would claim in his autobiography that as a child he was troubled by flashes of light that would temporarily obscure his normal vision, and as he described it, 'interfered with my thoughts and actions.'[5] It has not been ascertained if there was some medical explanation for these supposed apparitions, and it may be that they were fictitious in their origin. As will become more apparent later, he had a tendency to claim a great many things that went from just slight exaggeration to downright falsehood. It appears that certain parts of his life story are prone to embellishment, and this sometimes makes it hard to tell what is really true. This particular allegory could perhaps be akin to certain Roman emperors, who feigned epilepsy as this was seen as a condition that signified their divinity (making them like Julius Caesar, who did actually suffer from it). Tesla liked to be special, getting as much attention as possible, and we will witness many other examples of his behaviour that expound this view during the following chapter.

Even as a boy Tesla was blessed with huge reserves of ingenuity and a wild imagination. He dreamt up schemes for building gigantic waterwheels to harness powerful rivers and free up huge quantities of energy for the world to make use of. He envisioned an immense orbiting transportation system allowing travellers to reach the other side of the globe in minutes. Everyone has something that they are born to do, and Tesla was clearly born to invent. He would later state that the exhilaration inventing gave him had been his greatest pleasure, and had meant his 'life was little short of continuous rapture.'[6]

His education continued at the Higher Gymnasium in Karlovac, where one of his aunts happened to work. Though he displayed his mental abilities in school, he was not motivated by his teachers in the slightest, and began to show signs of being awkward to relate too. He stayed at the gymnasium for a total of three years. It was there he first gained an interest in electrical engineering, the experiments his teacher performed having a lasting effect on him. While in college, he acquired a taste for gambling. In what we will soon enough discover to be true to form for Tesla's image of himself, he would boast that as he had a stronger will than normal human beings, he could stop at any time. He did give up in the end, just as he had vowed, but almost bankrupted his family beforehand.

By the time he was seventeen he was no longer troubled by the 'visions' that plagued him as a child. He now felt he could actually control them, and they were more a blessing than a curse. Tesla could use them at his discretion to visualise things with incredible accuracy. He would not require drawings or models, he could simple make his inventions, or at least approximations of them, appear out of thin air.

His father expected that his only surviving son would follow his example and go into the clerical profession, but Nikola had other plans. He already knew his talents would never be fully utilised in the service of the Church. He tried to persuade his parents to relent, but they were unyielding. This changed when he contracted cholera and was, by all accounts, on the brink of death. Tesla recalled that he was 'confined to bed for nine months, with scarcely any ability to move. My energy was completely exhausted.'[7] However, he recovered after his father agreed to let him study engineering rather than going into the clergy, coming back to life, as he put it, 'like another Lazarus, to the utter amazement of everybody.'[8]

At his father's advice, he avoided having to serve in the army, spending twelve months camping in the Velebit Mountains. As well as dodging the draft, his parents insisted he should get a sizeable spell of outdoor life and lots of exercise, to strengthen himself after his protracted illness. Tesla's grandfather, also called Nikola, had fought in Napoleon's army, and his uncle Josip also had connections with the armed forces, teaching at the military academy. It is more likely that this is what helped him to get out of enlisting.

In 1875, he gained a place at the Gratz Polytechnic Institute, studying electrical engineering. Here Tesla continued to prove a difficult character to get along with, frequently disputing what he saw as the blinkered views of his lecturers. One of the few academics he had any respect for seems to have been the German physics master, Professor Poeschl. In his opinion he was 'the most brilliant lecturer to whom I ever listened.'[9] His experiments were 'skilfully performed with clock-like precision.'[10] His highly animated lectures seem to have captured the young Serb's imagination, and may have influenced the development of his own rather flamboyant oratory style. On one occasion Poeschl showed his class a Gramme generator. The machine, developed by Belgian engineer Zenobe Gramme only a few years earlier, was the first mass-produced means of transforming mechanical energy into electricity. Although seen as something of a revelation at the time, it was far from perfected, giving off huge streams of sparks throughout its operation. The precocious, if slightly over-confident, Tesla conjectured that if the machine was used to produce Alternating Current (AC) instead of Direct Current (DC), the commutator needed to change the direction of the current flow, which caused these sparks to appear, could be dispensed with and its efficiency greatly improved. The German responded mockingly, saying that, 'Mr Tesla may accomplish many great things, but he certainly never will do this. It would be equivalent to converting a steadily pulling force, like that of gravity into a rotary effort.'[11] Tesla knew that electricity in its natural form would be an alternating wave, so there had to be a way to translate this to and from kinetic energy. In that manner it would be possible to produce electricity from dynamos and turbines, and conversely alternating electrical currents could be used to drive motors to do mechanical work. Tesla was not one to be snubbed, and set out to prove Poeschl wrong.

Tesla's work rate was phenomenal; as well as ploughing through his course literature, he read a varied selection of authors from Goethe to Descartes, from Twain to Newton. He regularly started work 'at three o'clock in the morning and continued until eleven at night, no holidays or Sunday exempted.'[12] He set about reading Voltaire's complete works, only to learn to his 'dismay that there were close on one hundred large volumes in small print.'[13] His rather obsessive disposition made it imperative for him to drudge his way through the writer's entire catalogue. 'It had to be done,' Tesla would write afterwards, 'but when I laid aside the last book I was very glad, and said "never again".'[14] He graduated in 1879, after taking nine exams in one year so he could finish earlier. His father died, following a stroke, just a few months later.

He began his studies at Prague University in early 1880, though it has been noted that there are no records of him enrolling there. It is now apparent that knowledge of Greek was a prerequisite for all its students, and as Tesla had never received an education in the language, he could not officially join. However, Tesla had promised his father, when he was close to death, that he would continue his studies there. So he stayed in the city for a year, and attended various lectures, as well as making use of the university's library.

At the beginning of 1881, he started life in yet another new city, moving to Budapest. There he began working for the American Telephone Company. Tivador Puskas, who hired him, had worked with the already legendary Thomas Edison at his laboratories in Menlo Park, New Jersey. He told Tesla of the wonders Edison and his team were performing in the United States, and the idea of being a part of this activity greatly appealed to him, but this dream would have to wait. The constant exertion he had put himself through over the previous years had induced a minor nervous breakdown. Even during his illness, Tesla kept himself busy, he was not one to rest. While he was meant to be recovering, he pondered once again the issue of creating an efficient system for alternating current.

Englishman Michael Faraday had received no formal scientific education, but with some borrowed books and an incredibly responsive mind he had become the world's leading authority on the subject of electrical engineering. In the 1830s he had discovered that when a magnet was moved across an electrical circuit a current was induced within the wires. He noted that if the magnet was held still the current stopped, it was only while it was moving that the current appeared. This meant it was a change in the magnetic field, and only a change, that caused electricity to flow. He thought to himself, in the nonchalant way that only a true genius can, that there could be some practical upshot of this phenomenon. Turning the whole set-up on its head, he reasoned that by keeping a wire moving within a magnetic field he could produce a continuous electrical current. He put together a simple apparatus, consisting of a wire loop which rotated using a commutator, and the electrical generator was born.

Though advances had been made on DC generators and motors, little progress had been seen in their equivalents using alternating currents. The AC motors proposed until this stage were inefficient as they kept losing their momentum during the course of each cycle, unable to maintain continuous action. The only apparent way to keep an AC motor in smooth, constant motion would be to get the magnetic fields to rotate, and this was impossible. As Poeschl had stated back in Gratz, ridiculing young Nikola's ideas, it was not feasible to make fundamental forces like electromagnetism or gravity follow a curve, they only went in straight lines.

Strolling in the city park one day, with his friend and co-worker at the exchange, Anital Szigety, Tesla finally saw the light. While walking along quoting a few lines from 'Faust' he was, or at least he would have us believe that he was, struck by one of his infamous visions. It was the Holy Grail he had searched for all this time, a workable AC system made flesh. He could now see how a motor using an alternating source could be realised. Grabbing a stick he marked out a design for its construction on the ground. He later described the experience as being akin to 'Pygmalion seeing his statue come to life.'[15] What Tesla proposed was this; if the motor had different circuits placed out of phase (i.e. positioned at ninety degrees to each other), then when the first was at its lowest ebb the second would be at full strength. Continuity could be maintained, with no loss of power witnessed. The induction motor for alternating current needed no commutator, or any of the other paraphernalia previous machines required. The field current was not 'steady' but revolved, pulling the rotor along with it. Tesla had done what was deemed impossible; he had found a way to make a magnetic field effectively rotate, while still abiding by physical laws.

Puskas felt that the opportunities to promote what Tesla referred to as his 'polyphase system' were limited in the backwaters of Eastern Europe. He suggested that the gifted inventor had a better chance of developing his revolutionary ideas further west. Tesla left for Paris in Spring 1882, to work for Edison's European operation. The Continental Edison Company had been set up two years earlier, and was under the management of Englishman Charles Batchelor, one of Edison's most important lieutenants. He had started with the American inventor back in 1870, serving as a draftsman in his workshop in Newark.

Tesla had fond recollections of his time in Paris. In later life he stated that he would 'never forget the deep impression that the magic city produced on my mind.'[16] Unfortunately, he had a habit of spending his income all too quickly, the 'City of Lights' had a great many places where someone could fritter away their hard-earned wages. When Puskas wrote to him from Hungary, to see how he was fairing, Tesla responded that the 'last twenty-nine days of the month are the toughest.'[17]

In a short period of time Tesla became the star performer at Edison's Paris office. The young Serb was undoubtedly keen to quickly start making a contribution to the company's success. The firm supplied electrical distribution and generating equipment across the whole of Europe, and the fact that the new recruit was versed in several languages offered considerable advantages, in addition to his obvious technical skills. By the time he was an adult Tesla spoke five languages fluently, but of particular importance was his grasp of German. As part of the Austro-Hungarian Empire, it was compulsory that all classroom education within his homeland be conducted in the German tongue. This made him the ideal candidate for assignments in that region.

It was one such assignment that gave him the opportunity to develop a working model for his induction motor. While finishing off the rather prolonged spell repairing equipment installed in Strasbourg (the Alsace region of France has flipped between Teutonic and Gallic rule several times, and between Napoleon's defeat and the end of the First World War was within the borders of Germany), he found time to put together such a device, and demonstrated it to the town's most important citizens. Sadly this did not

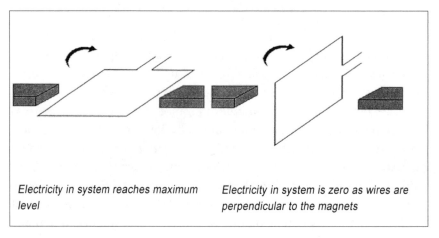

Electricity in system reaches maximum level

Electricity in system is zero as wires are perpendicular to the magnets

Traditional DC generator system.

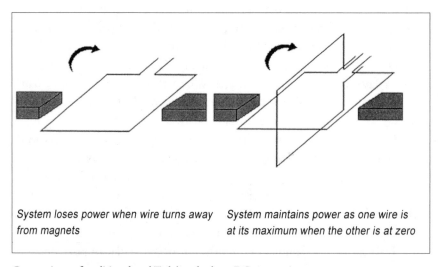

System loses power when wire turns away from magnets

System maintains power as one wire is at its maximum when the other is at zero

Comparison of traditional and Tesla's polyphase DC generation systems.

produce the desired effect. Though impressed with the job he had done of fixing their generator, they had no interest in his invention.

On returning to Paris, he was shocked to discover that the bonus promised to him for successfully appeasing their clients in Alsace was not forthcoming. The management produced a series of lame excuses, and passed the buck between one another. Tesla felt he had been deceived, and decided not to waste his time with people who did not appreciate his abilities. It was not the only occasion that Edison's company got the better of

Tesla, but the next time it would be the proprietor himself who mistreated him, not his employees.

Batchelor recognised the unique gift Tesla possessed, and upon news of his impending departure agreed to give him a letter of introduction which might secure a post in the head office. The Brit's reference was favourable, to put it mildly. It said that he had known two great men in his lifetime, stating that Edison was one of them and he felt Tesla was the other.

With such a glowing testimony to help him, Tesla headed west once more. He reached New York in June 1884. His autobiography would have people believe that he arrived in the 'Land of Opportunity' with just a few cents, the clothes on his back, and a small notebook of his poetry. This is perhaps designed more to pander to American folklore than have any basis in fact. He was not some vagabond, he had made a successful living for himself in Budapest and then in Paris. Despite his god-given knack for blowing money, it seems farfetched to imply he had literally nothing to his name when he left Europe. It is probably another good example of Tesla's tendency to overdramatise. There is little doubt he had a penchant for telling tall tales.

His first impressions of the United States seemed to be filled with disappointment; he found the natives very abrupt, and the environment devoid of either charm or culture. He recalled that what he left in Europe 'was beautiful, artistic and fascinating in every way,' while what he saw in this new domain 'was machined, rough, and unattractive.'[18]

Regardless of this, Tesla set a course for the offices of Thomas Alva Edison, hoping to secure employment there. Producing Batchelor's commendation, he asked the American inventor if he would consider hiring him. Edison was not one to show his enthusiasm without good reason, but with almost stoned-faced indifference agreed to give the young lad a try. Tesla described their first meeting as one of the most memorable events of his life, stating, 'I was amazed at how this wonderful man, who, without early advantages and scientific training had accomplished so much.'[19] He would amend his opinion with time.

Much as he had done at the French works, Tesla was keen to show his worth. 'Our Parisian' as Edison referred to him (almost certainly to Nikola's annoyance), worked long hours, and rarely took any break from his labours. Tesla told Edison of his ideas for an AC motor, but his response was one that contained little verve. Edison was of the opinion that AC was too dangerous and not really efficient enough to make it useful. Such things had little future, as far as he could see. However, he realised that this was a young man with a great deal of promise. As the true extent of Tesla's technical expertise became apparent, Edison supposedly offered the Serb a sum of $50,000 to make improvements to his DC system. Tesla is said to have worked tirelessly on the project, and achieved the required goal, only for Edison to make light of it, claiming he had only meant it as a joke.

Tesla threatened to leave the company if the reward promised to him was not granted, but he knew that his boss was in a powerful position. Alternative employ would be hard to find at this time, and Tesla really could not afford to just up and leave. Edison certainly did not bank on him being bold enough to go through with it. In order to give a balanced commentary on this event, it is only fair to mention Edison's side of the story. He alleged that he offered Tesla the aforementioned $50,000 to buy the rights to his polyphase system. But in truth this does not really add up. Edison had already shown his complete disdain for the use of AC, and continued to do so long after this. So why would he pay for something he had no faith in?

Tesla was far too proud to continue working with a man who, from his perspective, could play such a cruel trick. He resigned immediately. Edison's reneging on his promise would drive a wedge between them, and their differences were never resolved. They would maintain their savage hatred of one another to the grave. These two uncompromising personalities became the bitterest of rivals, to the point that they would even be willing to mar their own success simply to make sure the other came off worse.

This was not a great time to quit, the recession was already starting to bite, and work was in short supply, but Tesla was a man of principle. Edison had made a fool out of him on two occasions, though one was inadvertent; Tesla was not going to hang around for a third. Despite the austere economic climate, he left Menlo Park, never to return. He would be happier to starve than continue serving so ungrateful a master, and that was almost what he did do.

At first everything seemed to be going well. He scraped together enough money to start up a modest laboratory, where he could work on developing his polyphase system. With the help of a few small-time investors, he formed the Tesla Electric Light Company, which produced arc lighting equipment. But his first solo venture did not take off in the manner he had hoped. His backers lacked the cavalier attitude of their Serbian colleague. They did not want to fork out huge quantities of money on his ambitious schemes, they were just looking for a comfortable return on their investment. The use of an AC system, like Tesla was proposing, was too much of a gamble for their liking. They managed to push this foreign trouble-maker out of the company, despite the fact that he had actually founded it.

For the next year Tesla was forced to take whatever work he could get, often resorting to manual labour, such as digging ditches, to put food on the table. The 'American Dream' had not materialised in the form that he had expected. Tesla had left his home and crossed thousands of miles just to end up working as a common navvy.

Just at the point when all seemed lost, a man would enter Tesla's story who would completely transform things. George Westinghouse was an energetic young entrepreneur from Pittsburgh who had made a name for himself in the railroad business. I think that the train service in my native Britain is a bit of a mess today, but that's small potatoes compared to how it was in 1860s America, where derailments and all manner of serious accidents were commonplace. Westinghouse had invented the 'air brake', which used compressed air to slow down trains. It became the industry standard braking mechanism, being far quicker and safer than previous methods. In addition he had developed more efficient and accurate rail signalling systems, as well as a number of other useful gadgets. The Pennsylvanian businessman had heard of Tesla's passion to make the AC system a reality, and offered him the financial backing needed for this to happen. Westinghouse became his most fervent ally, backing Tesla's research and supporting him against Edison in the ferocious discord that would ensue between the direct and alternating systems, a conflict known as the 'War of the Currents'.

Although this particular episode in the young Serb's life is not the focus of this chapter, it offers an interesting detour. First of all it was one of the earliest examples of how corporate propaganda could be utilised. The application of smear campaigns, dirty tricks, possibly sabotage, and basically any means at the protagonists' disposal, no matter how unsavoury, were employed by the two parties involved. Some of the most shameful tactics were made use of in order to secure victory. As a result this may serve as a helpful guide for any budding industrialists out there who want to learn how to go about panning the competition. It also gives us some valuable insights into the ways of Thomas Edison, which will prove useful later in the book when we look in more detail at his life and works. Finally, I personally think it is of prime importance to be aware of one of Tesla's great triumphs before examining some of his blunders.

In principle, a power system based on alternating current had one huge advantage over using direct current. AC electricity could be carried far greater distances, hundreds of kilometres if necessary, while DC could only manage a couple of kilometres at best. But Edison did not want things to go down this route. His company's investment had all been in DC equipment, and he was building up a nice little monopoly here.

Edison set about trying to show the public just how unsafe the utilisation of AC was. To begin with he used his considerable fame (by this stage he had given the world the phonograph, the electric light bulb, and many other important inventions) to get articles published in the popular press, making somewhat unsubstantiated claims of the dangers involved in using an alternating system. But he was not going to stop there, not by a long mark.

Next he had his henchmen distribute flyers warning of the potential risks, and spread rumours of accidents that had taken place. He then hired street kids to bring him stray dogs, at a nickel a head. He put these on public demonstration, having the unfortunate creatures electrocuted using AC (I can only assume that the World Wildlife Fund took him off their mailing list at that point). The opening shots of the 'War of the Currents' had been fired, but Tesla and Westinghouse did not retaliate by using such base methods, they tried to take the high ground, by showing prospective clients the technical benefits of the AC system, rather than having to rely on simply slurring their rival, and though this took longer to carry out, with time it slowly turned the tide in their favour.

Edison decided to up the stakes. What followed was the most telling bombardment in his scare-mongering offensive, and was to lead to the notorious electric chair controversy. The possibility of using electricity as a means of administering capital punishment was starting to be considered at this time, as it was hoped this would offer a more reliable and humane means of termination. Hanging was not an exact science, often the victim's neck did not break, and they would helplessly dangle on the end of the rope until they finally died of asphyxiation. The authorities began to warm to the idea of using electrical currents as a way to carry out executions in a quick, relatively painless and well ordered manner.

Edison saw an opportunity to deride the AC system once more. If he showed alternating currents were capable of being used to kill a human being, then he deduced its public image would be hugely impaired. Simply by association AC would be categorised as lethal. Some of his men, effectively acting as fifth columnists, served on the advisory committee debating the validity of using electricity for execution, and the specific method to be utilised. Edison's infiltrators suggested to the committee that AC was the most obvious way of getting the job done, as it required far smaller voltages than DC. After prolonged deliberation the board came to a decision, and to Edison's barely disguised joy, they favoured the use of alternating current.

On 6 August 1890, William Kemmler's life was taken by use of electricity. However, this initial attempt to mete out retribution did not to prove particularly successful. Rather than being a short and merciful death, the prisoner was effectively 'fried', the whole process taking several minutes to complete. The voltage required had been seriously underestimated. Although it is hard to feel any sympathy for Kemmler – a year earlier he had brutally murdered his wife with a hatchet after all – this was clearly not what the judicial service had intended (I mean who was going to clear up the mess for a start?). As Edison had been called to act as a technical advisor, and proposed the voltage needed, he was to some degree held responsible. The plan to discredit Tesla, Westinghouse, and the AC system, had backfired on him.

Though in principle AC was more dangerous, as the voltages involved would go from the peak in one direction to the peak in the other, effectively doubling the potential difference, what Edison was proposing was in fact a false economy. He was being selective with the truth. In reality the high voltages were only needed to transport the electricity over long distances, through power lines. When it reached the homes or offices of its consumers it was put through transformers and the voltages reduced to safe levels. Regardless of this he would stick to his mantra, AC would kill, and members of the public were in danger of being 'Westinghoused', as he referred to it.

While Edison was doing serious harm to his reputation, Tesla was starting to become a bit of a celebrity. In the space of just a few years he had gone from the humble surroundings of a small Croatian farming community to the splendour of Manhattan, the focal point of a new and vibrant America. He was befriended the cream of New York society: the Morgans, the Astors and the Vanderbilts. The zenith of the country's acting, writing and musical talent would occupy his social circle. The United States had supplanted Victorian England as the industrial powerhouse of the world, the luxury and splendour enjoyed by its wealthy elite easily matching that of the aristocracies of Europe. What was known as the 'Gilded Age' had begun.

Tesla was quite the 'man about town'. Over six feet tall and of very slim build, his eyes were a grey-blue tint, powerful, piercing and vibrant. He was always immaculately attired, in all but the most informal of soirees arriving in a tails and sporting white dressed gloves. He gained the image of being one of New York's most eligible bachelors, but seemed to have little interest in trying to court any of his female admirers.

The World Columbian Exhibition took place in Chicago during Summer 1893, and proved an important coup for Tesla's AC system. The event, which celebrated the 400th anniversary of Columbus' arrival on American shores, was an incredibly ambitious one. It required the construction of a huge building complex, called the 'White City'. Westinghouse managed to secure a contract that meant the whole exhibition was illuminated by alternating current. The signs were becoming clearer; the 'Gilded Age' would be lit by Tesla's electricity.

His alternating system was to take the final and decisive battle in the long-standing war later that same year. The Westinghouse Company was engaged to produce the electrical generators for the first hydroelectric plant to be built in America, at Niagara Falls. The tender for the construction of the plant was both very lucrative in its own right, and also an incredibly high-profile endorsement. Edison, still stubbornly sticking to a continuously less favoured direct system, had been soundly thrashed. But some consolation was afforded him, as his firm was chosen to supply the cabling that distributed the power generated to the towns and cities.

The wily business acumen of Westinghouse and the sublime technical skills of Tesla had proved too great for even the mighty Edison. The AC system had routed the forces of DC. For all his efforts, the use of corruption and adverse publicity, Edison had been vanquished. But for Tesla another conflict was soon to begin. It would be the struggle for the airwaves, and only time would tell if he was destined to be victorious.

The unwillingness Edison had shown to embrace AC, as well as his use of all manner of subterfuge to try to curb its success, was doing a great deal of damage to the image of Edison General Electric. Though he had established the company and it bore his name, the controlling interest belonged to John Pierpoint Morgan, the powerful Wall Street financier. Morgan was not in the slightest concerned about what methods were used for distributing and generating the electricity, as long as they bought it off him, and he made a nice fat profit off it. He could see that Edison was increasingly becoming a liability, his public relations strategy was slightly better than that of Genghis Khan. Mindful of this, he orchestrated a merger with the Thomas Houston Company in 1892. This firm had originally been a DC exponent too, but had seen the way the wind was blowing and shifted allegiances to AC. The amalgamation of the two corporations resulted in the formation of 'General Electric'. It was not just Edison's name that was dropped from the new company, he was squeezed out too. Morgan would be the man in charge, and there was no room for anyone who might oppose his wishes.

Tesla's star was still in the ascendancy. In Summer 1891 he became a naturalised US citizen. The American Institute of Electrical Engineers (AIEE) awarded him the Elliot-Cresson Medal in 1894 for his work on alternating current. His almost theatrical demonstrations of high-voltage electricity wowed crowded auditoriums. Generators were used to produce huge bolts of artificial lightning which he would fire above the audience's heads. Tesla had become a modern-day Merlin, conjuring up fantastic visual spectacles to dazzle the ever excitable public.

A new area of research began to take Nikola's fancy. Hertzian waves had been discovered by Heinrich Hertz (hence the name) in the early 1880s. The German scientist had experimented with the production of electromagnetic radiation using sparks of electricity. His apparatus consisted of a battery attached to a coil which in turn had a copper plate connected at either end. The plates were brought close together so only a small space separated them. The coil was charged up and a pulse of electrical energy passed between the plates, producing a series of waves as it did so. He used a copper ring, with a small break in it, to act as a receiver. When the waves hit the ring, a spark would pass between its ends. The energy from the transmitter had thus been transferred to the receiver, though they were in no way connected.

Hertz was a pupil of Helmholtz at Berlin University, and just as one inspired Bell's work on the telephone, the other would serve a similar role for Tesla when it came to developing radio. It was in 1891 that he first did experiments in this area. Tesla found he could use an alternating current from one wire to induce a current in another, thanks to the electromagnetic waves it produced.

He gave a lecture at the Institute of Electrical Engineers (IEE) in London, during February 1892. In this he looked in depth at the effects of electrical resonance, and propounded the idea that 'transmitting intelligence without wires is the natural outcome of the most recent results of electrical investigations.'[20] He went on to tell the audience, 'I am becoming daily more convinced of the practicality of the scheme; and though I know full well that the great majority of scientific men will believe that such results cannot be practically and immediately realised, yet I think that all consider the developments in the recent years by a number of workers to have been such as to encourage and experiment in this direction.'[21] Though the attendees must have thought that he was deranged, the striking young man had predicted the future with absolute certitude.

Tesla alluded further to the use of Hertzian waves to carry messages in his lecture on 'Light & Other High Frequency Phenomena' delivered to the Franklin Institute in February 1893. In this presentation, he described a wave production system with which he would 'charge condensers, from a direct or alternating source, preferably of high tension, and to discharge them disruptively'.[22] He elucidated, that when 'condensers are charged to a certain potential, the air or insulating space, gives way and a disruptive discharge occurs. There is a sudden rush of current,'[23] this produced the desired electromagnetic waves. At the venue, Tesla made basic demonstration of the transfer of a Hertzian wave. Using one circuit he set off a spark, then with another circuit he picked up the signal, passed from the antenna to the vacuum tube.

The vacuum tube had evolved from work done by, that man again, Thomas Edison. During the 1880s, while making improvements on his light bulb, he noted that hot lighting elements give off electrons. The Edison effect, as it became known, would be utilised to create detectors for electromagnetic waves for many decades to come, and later served as high-speed switching mechanisms for all manner of electronic equipment (as we will discover later in this book).

At a meeting of the British Association in Oxford, an Englishman by the name of Oliver Lodge became the first man to successfully transmit a simple message using these waves, though it was only across a room. By late 1894, he had managed to pass a Hertzian signal a distance of 40m. Nikola could see that these phenomena could be used on a far grander scale. He started to concentrate his thoughts on how to go about this.

Back in 1891, he had created the 'Tesla Coil'. This device, which was based on the inductance coil that Rumkorff had developed some years earlier, could be used for many different purposes. At first it had been applied in high-energy physics experiments (like those he did to impress his adoring fans), but it would also prove itself as a way of producing wireless signals. The basic Tesla Coil consisted of an oscillating transformer, a spark gap, and a condenser (which held electrical energy). A source of energy was made to charge up the condenser, and when the difference in potential at the terminals of the latter reached a certain level, the spark gap would be bridged.

This permitted the accumulated energy to discharge through the rest of the circuit under resonant conditions.

In the coil the potential would gradually increase 'with the number of turns towards the centre, and the difference of potential between the adjacent turns being comparatively small, a very high potential impractical with ordinary coils, may be successfully obtained.'[24] Thanks to this arrangement, 'those portions of the wire or apparatus which are highly charged will be out of reach, while those parts which are liable to be approached, touched or handled will be at nearly the same potential as the adjacent portions of the ground' to ensure 'personal safety'.[25] He pointed out that although 'extreme pressures of many millions of Volts have been for a number of years continuously experimented with no injury has been sustained neither by myself or any of my assistants.'[26]

In 1896, Tesla managed to construct an instrument for receiving radio waves. It had two condensers and a coil which were utilised to produce the electrical energy required. This was passed through a spark gap and a five kilohertz signal thereby emitted. By spring of the following year, he was capable of picking up signals from 40km away.

Tesla was not limiting his enquiries to simply sending messages; he could see that it would be possible to utilise these waves as a method of transferring power. With time he came to the conclusion that in principle the use of wires was superfluous in transporting electrical energy, if the right methods were employed it could be conveyed by simple induction. He remembered how, during his speech at the Franklin Institute four years earlier, he had given a small demonstration of how energy could be passed through the air to illuminate a light bulb held in his hand, and that there was 'almost a stampede in the two upper galleries, the audience thinking that it was some part of the Devil's work.'[27]

His idea was to use the upper layers of the Earth's atmosphere to transmit electricity over huge distances. He argued it would be possible to transmit electricity through the stratosphere, 10,000m above the Earth's surface. At altitude the air became so thin that passing an electrical current through it would be relatively easy. The same principle is used in plasma lamps, or in arc welding; the current taking the path of least resistance. The system Tesla proposed would dispense with the need for the expensive, unsightly, and somewhat inefficient power line network already beginning to build up across the country. It meant that energy could effectively be distributed cheaply and simply over the entire planet.

It is well documented that Tesla had a great many foibles. There was his excessive phobia of germs, and his fear of people touching him. Also he liked everything to be divisible by three, for example at restaurants he always expected to have nine napkins put at his dining place. While he was working in Paris, every day he went to a bathing house on the banks of the Seine, swimming exactly twenty-seven lengths. Likewise the numbers of his hotel rooms always had to be some sort of multiple of three.

The long-running battle with Edison had depleted Westinghouse's reserves, the pricing war he became entangled took the corporation to the verge of bankruptcy. The royalties Tesla was due came to over two bucks per horsepower of electricity produced, and could effectively drag the firm under. Westinghouse approached him and explained the situation; he made it clear that Tesla's fees were going to cripple the company, and his Serbian colleague needed to relinquish them if he was to stay in business. In a selfless gesture Tesla agreed to forego the payments in order to bail out Westinghouse. In return for the sum of $200,000, he waived any further rights to the system.

He effectively said goodbye to tens of millions worth in commissions, but was willing to do this to ensure the polyphase system would be adopted by the world. At the time money did not seem to be an issue, he had no shortage of interested parties eager to back his schemes. However, this affluence would dry up soon enough, and his willingness to help Westinghouse out with such a noble sacrifice proved unwise.

In the early hours of 13 March 1895, Tesla's laboratory on Fifth Street caught fire. The building was completely gutted, much of the research that had taken decades to perfect was lost amongst the ashes. He tried to remain stoic, despite the huge setback, managing to establish a new headquarters on Houston Street that summer, and began to rebuild his body of work.

In 1898, he received a patent for a remote control system that used Hertzian waves for guiding unmanned vehicles (US Patent 613,809). The concept of using automatons, such as robot aircraft, has been used in recent military activity to provide reconnaissance in war zones, but back then it was not given any serious consideration by the armed forces. Tesla was frustrated by the military's lack of interest.

Thinking that his experiments on wireless communication and electrical power might be facilitated in a more remote environment, Tesla opted to move away from the big city. He relocated to Colorado Springs in May 1899, setting up a new laboratory there. The funding for the project had been appropriated from John Jacob Astor, with whom he frequented many of New York's gaming clubs.

In the wilds of Colorado, Tesla continued to work on how oscillators, lamps and electromechanical equipment could be made to operate by means of induction through the air, rather than direct connection. He managed to light sets of hundreds of light bulbs remotely over long distances. He began performing similar experiments to those with which he had stunned the crowds back on the East Coast, but this time with far greater magnitude. He would send enormous blasts of electricity into the air, lighting up the dark night's sky. The excessive energy drain his equipment created eventually blew the generators of the local power station, causing a blackout across the whole area. The power company was far from amused by Tesla's antics. It flatly refused to supply him from then on. In the end he had to agree to pay in full for the repairs before being put back online.

The press were eager to find out what the eccentric scientist was up to at his new laboratory, following him out west. Pictures of him sitting on a chair underneath huge charge-generating apparatus, while bolts of lightning whizzed past his ears, were published in *Century Magazine* in June 1890, but they were just for show. The shots had been doctored, the photographer taking double exposures. Nevertheless, they added to myths that surrounded Tesla; the suspicion he might be some kind of superhuman being with baffling, unnatural powers.

The patent application for his 'System of Transmission of Electrical Energy' (US Patent 645,576) was submitted on 2 September 1897. In this he further developed his concept of using the air at high altitude as a means of conduction. Tesla observed that gases 'are in a large measure deprived of their dielectric properties by being subjected to the influence of electromagnetic impulses' and thus 'the conductivity imparted to the air or gases increases very rapidly both with the augmentation of the applied electrical pressure and with the degree of rarefaction.'[28] He explained that as air became 'rarefied' (i.e. its pressure was reduced) its insulating properties would be lost and it 'could be considered as a true conductor.'[29]

Tesla stated that the air's conductivity became so great 'that the discharge emanating from a single terminal behaves as if the atmosphere were rarefied.'[30] He noted that 'this conductivity increases very rapidly with the rarefaction of the atmosphere and augmentation of the electrical pressures, to such an extent that at barometric pressures which permit no transmit of ordinary currents, those generated' by a high-voltage source, such as one of his Tesla Coils, would 'pass with great freedom through the air.'[31] Nikola was convinced that if he applied potential differences of fifty kilovolts the 'atmosphere must break down'.[32]

Tesla also looked at the possibility of passing low-frequency waves through the Earth's crust, and making use of its natural resonance, as an alternative to transmitting messages through the air. He felt the Earth could be used as a transmission system for electrical energy if the correct frequency was applied. Just like a tuning fork, it had an exact pitch, and if this was found then it would resonate in the same way. The idea of resonance had been a recurring theme that featured throughout most of the varied array of research topics he tackled. Some years earlier, he had developed a gadget called the Relatively Simple Oscillating Device (RSOD). During one experiment in New York, he had ended up going a little too far. The tremors from his oscillating system caused all the nearby water pipes to burst, and resulted in neighbouring buildings being damaged. When the emergency services turned up, Tesla made light of it all, wondering what the fuss was about.

In late 1897, *Scientific American* published an article on Tesla's exploits in the field of wireless transmission. It informed readers that he had:

> completed his wireless telegraph to such an extent as to permit of telegraphy through the Earth for a distance of twenty miles or more, and his experiments satisfy him of the feasibility of wireless telegraphy on a much more extended scale. In fact, he aims at nothing less than the establishment of a system of telegraphy that shall include the whole Earth, and by which items of news may be distributed from one political or commercial centre to every other.[33]

His time in Colorado did not appear to achieve much. He had managed to spend a lot of money, and gained considerable media attention, but it resulted in little of commercial use. His unbridled lust for creating was still apparent, but was becoming progressively more devoid of direction; he wasted valuable time on projects that never went anywhere. His talent for burning up his investors' cash was gaining more renown than his inventions, and this made it increasingly difficult to get backers for his endeavours. He moved back to New York in the closing weeks of 1899.

While Tesla was meandering along, trying to cover a number of bases at once, across the Atlantic a younger and seemingly less gifted individual was taking a more direct route. Before long he would be making waves of his own.

Guglielmo Marconi was born on 25 April 1874 in the Palazzo Mareschali, on the Via Tre Novembre, which lies in the centre of Bologna. His father, Giuseppe, was a native Italian, while his mother Annie was originally Irish. He was their second son, his brother Alphonse had been born in 1865. Annie was one of the four daughters of Andrew Jameson, of Daphne Castle in County Wexford. Her father, along with his brothers, had emigrated from Scotland in the 1820s, and gone on to establish the Dublin Brewery Company. He had then struck out on his own, and set up a distillery in Fairfield, the now famous Jamesons.

In her youth Annie had aspirations of becoming an opera singer, and had gone to Italy to receive tuition. There she became a student of the Bologna Conservatory. Her father had reservations about allowing the young girl, who had lead a somewhat sheltered life, to go abroad on her own, so asked one of his business associates Philip de Renolis, who lived near Bologna, to take care of her. While staying with the de Renolis family, she was introduced to Giuseppe Marconi. He was a reasonably wealthy landowner, originally from Umbria, and had been married to one of their daughters. Sadly she had died in childbirth, leaving Giuseppe to take care of their son, Luigi.

The widower clearly charmed the young Irish heiress, and she became completely smitten. However, her parents did not approve of the idea of their daughter marrying some Latin lothario, especially one nearly twice her age. They flatly refused her request to wed him. But Annie was not going to give up that easily, it was not in her nature. Ignoring the wishes of her mother and father, she did not break contact with Giuseppe. They continued to correspond between Italy and Ireland in a clandestine manner. Then when Annie came of age, she ran off to the Continent to join her Italian sweetheart, and the two of them were married in Northern France in April 1864.

Guglielmo was born into post-unification Italy, prior to this it had been made up of separate states based on the dukedoms of the Middle Ages. The young boy enjoyed a life of privilege. His family had two residences; their Bolognian townhouse, Palazzo Mareschali, and their country estate, Villa Griffone, located in the hills outside the city.

It was later said of Guglielmo, that he had received 'from his father, that independence of spirit, which is the mark of mountain men' and 'from his mother, a will as stubborn as his father's but unmatched with poetry and music and grace. And to her radiant complexion and fair colouring he owed his blue eyes.'[34] Though his mannerisms and speech gave away his

Mediterranean ancestry, to look at him you would assume that he was of Northern European origin. His pale, almost pasty skin, meant that he stood apart from the average Italian boy.

At best, Marconi was a moderately good student. He did his preparatory schooling in Bologna, then at the age of twelve he started his studies at Florence's Cavallero Institute, concentrating on science-based subjects. Whatever constraints there may have been in his swiftness to understand, were easily balanced out by his seemingly limitless determination. Much like his parents, he was far from weak willed. He took full advantage of Villa Griffone's library, and was an avid reader of the electrical experiments of Benjamin Franklin and Michael Faraday. In 1887, he went to the Institute of Technology at Livorno, where he took physics under the guidance of Professor Giotto Bizzarini. He later attended lectures at Niccolini College, conducted by Vincenzo Rosa.

Rosa, a relatively young electrical engineer from Turin, had a big effect on Marconi's methods of research, and helped inspire the teenager to embark upon a scientific career. Another important mentor for young Marconi was Professor Augusto Righi of Bologna University. He had made a name for himself with regard to the study of Hertzian waves. Annie arranged for her son to receive some tuition from Righi, and the concept of utilising these unseen spirits, of which his teacher evangelised, for something of purpose seems to have caught his imagination.

It was in the tranquil setting of Villa Griffone that Marconi began to experiment with wireless. He had read every detail of Hertz' work on the transmission of electromagnetic waves, but knew that the German-Jew had only been able to gain a range of a few metres, which offered little real use. Unfortunately Hertz died in his mid-thirties, before he could take his work any further. Marconi wanted to see how far it was possible to extend their reach. He began to carry out what he later described as 'experiments with the object of determining whether it would be possible, by means of Hertzian waves, to transmit to at distance telegraphic signs and symbols without the aid of connecting wires.'[35] His first forays were simple enough; he sent signals across the attic of the villa in order to ring a small bell. With time he became more ambitious, starting to perform trials in the grounds of the estate, gradually expanding the transmission span until eventually, in mid-1895, he reached distances of 2km.

The Frenchman Edouard Branly had developed the 'coherer' in the early 1890s. It consisted of a glass tube, filled with metal filings, which were kept in motion. The container was connected at each end to an electrical circuit. If the tube was in contact with an electromagnetic wave, the metal particles would collect together and hence bridge the gap between the two ends of the circuit, allowing the current to flow. The system could thus act as a detector for these waves. Lodge had used a similar device in experiments he undertook at Liverpool University during the mid-1890s.

Alexander Popov, a professor at the Electrotechnical Institute of St Petersburg, used a coherer similar to the one Lodge had shown at the Royal Institution in 1894, and attached it to an aerial in order to detect when thunderstorms were approaching – nice trick, but what practical use it really had is still a mystery. For all the work that had been done at this stage, nobody had really found a way to do something of practical benefit with Hertzian waves, they were still just viewed as an interesting quirk.

Marconi borrowed the design that had been created by Branly, managing to make a more optimised version which had far greater sensitivity. This meant he could pick up weaker signals, and increase the range immensely. He also added a piece of equipment that continuously hit the tube, thereby breaking the contact again, and allowing more signals to be detected, rather than just being a 'single shot' affair.

The young lad slaved over his transmitting apparatus for many months, his brother Alphonse serving as his assistant. By early 1897, Marconi could make wireless transmissions over distances of 6km, and before spring was at an end he had doubled that range.

It was in February 1896 that Marconi travelled to England, accompanied by his mother. There they rented a house in the London suburb of Bayswater. Using her family connections would enable him to gain audiences with many important individuals within the upper echelons of English society. Among these was William Preece, the Post Office's Engineer-in-Chief, who had been introduced to them by Annie's nephew, Henry James-Davis. He was greatly interested in Marconi's work, and following their meeting small-scale trials took place at the organisation's site at St Martin's-Le-Grand, in London's East End. Preece had done experimentation with wireless communication some years back, and was impressed with the improvements the young Italian had made in this field. As he had already been booked to talk on the subject of wireless telephony at Toynbee Hall on 12 December, he felt that this would be the perfect opportunity to introduce his young protégé to London's scientific community. Over the next two years Marconi gave speeches and demonstrations at nearly every major scientific venue in the country, including the IEE and the Royal Institution.

The system Marconi was using at this stage consisted of a small induction coil which built up electrical energy, that was then passed between the spark gap. This gap was maintained between two brass spheres roughly a centimetre apart. He noted the greater the diameter of these, the greater the range of communication achieved.

Preece suggested Marconi continue tests back in his native Wales. Guglielmo decided to follow his advice, and on 12 May 1897 the first official transmission took place. A signal was passed from Lavenock on the principality's southern coast, to Flatholm Island in the Bristol Channel, traversing a distance of 5km. In June, Marconi took out his first patent for radio communication (British Patent 12,309) entitled 'Improvements in Transmitting Electrical Impulses'. The following month, he formed the Wireless Telegraph & Signal Company, in Chelmsford, employing some fifty people. It was later re-named Marconi's Wireless Telegraph Company.

He next established a wireless station replete with a 35m antenna outside the Haven Hotel in Poole, Dorset, which was in communication with another station across the water at the Needles Hotel, some 27km away on the Isle of Wight. Here he continued to work on the concept of using resonant signals. His apparatus consisted of a 500-volt battery and an induction coil about 25cm in length. Among the visitors who came to inspect Marconi's work on the South Coast was Lord Kelvin, widely regarded as the most eminent scientist of the age. Plans began to form in his head as to where he would next take this technological wonder, whether the time was right for him to try using it to connect two different countries.

Tests had successfully showed wireless transmission could be achieved at distances up to 60km, so Marconi knew his equipment was technically up to the job. However, he needed to make sure that such a high-profile demonstration went without a hitch, if he was going to raise levels of awareness. In late 1897, preparations began to try to pass wireless signals between England and France. On 27 March 1889, he attempted the first international broadcast, from the port of Dover across the water to the French town of Wimereux. Marconi operated the equipment on the French side, using the call sign 'VVV' to initiate the communication, hoping that it would be V for victory. Suffice to say he did not fail to deliver on his promise.

The first US test was done later that year. The *New York Herald* arranged for Marconi to report on races between the *Shamrock* and the *Columbia* to decide the Americas' Cup. Using his wireless equipment, he transmitted signals to and from the boats, which were some 25km offshore. The challenge here was not so much a technical one, he had already covered the Kingstown Regatta in Ireland a couple of months earlier, this was more of a public relations exercise. It was imperative that Marconi could get acceptance in the United States as well as in Europe, and this gave him the perfect opportunity to, as he put it, acquaint 'the American people with my work.'[36] Following the competition, the *New York Times* published that the 'initial success of Marconi's appeals powerfully to the imagination. It will be the fervent hope of all intelligent men that wireless telegraphy will very soon prove to be more than a "scientific toy".'[37] *Scientific American* was equally generous with its praise, saying that it would be easier to 'sweep back the tide with a broom as to prevent the system of telegraphy which has done such good work off New York Harbour.'[38]

Following his first triumph in America, Marconi was asked to participate in a wireless test for the US Navy. Two battleships, the *Massachusetts* and the *New York*, were used in assessments that took place on New York's East River. The trials were another success. Back in England, he was involved in discussions with Lloyds of London about using radio to communicate with shipping, to help reduce the risk of fatalities, and the Royal Navy was already performing assessments of its own using his equipment. In August 1899, Marconi's system was tested out by the HMS *Juno* and HMS *Europa*, giving good results up to 90km.

The authorities and the public alike all took a shine to the young Italian. In an article published in the *Strand Magazine* in March 1897, he was described as a 'tall, slender young man, who looks at least thirty and has a calm, serious manner and a grave precision of speech which further gives the idea of many more years than are his. He is completely modest, makes no claims whatever as a scientist, and simply says that he has observed certain facts and invented instruments to meet them.'[39] One of his children would later testify that 'he was his own man, an aggregate of opposites; patience and uncontrollable anger, courtesy and harshness, shyness and pleasure in adulation.'[40] Perhaps it was the blend of Southern European hot bloodedness and Irish placidity that made such an appealing mixture.

By this stage Marconi had practically abandoned his original system. Instead of joining the coherer directly to the aerial and earth, he connected it 'between the ends of the secondary of suitable oscillation transformer, containing a condenser, and tuned to the period of the electrical waves received.'[41] He found that this arrangement 'allowed of a certain degree of syntony, as by varying the period of oscillation of the transmitting antennae, it was possible to send messages to a tuned receiver without interfering with others.'[42] Until then he had just been blasting out signals at many different wavelengths, now he began to develop a less crude methodology.

On 26 April 1900, Marconi took out another patent (British Patent 7,777) called 'Improvements in Apparatus for Wireless Telegraphy'. This described the use of a tuned circuit. It no longer emitted a broadband signal that swamped the airwaves, but employed similar techniques to those Tesla utilised. This had the advantage that the power used to generate signals was not spread across all the different wavelengths, but centred on one, making it more efficient. Tesla had always described Marconi's way of doing things as a 'wasteful system'. He also commented that his Italian counterpart's use of very high frequencies, which were more easily absorbed by the air, made it harder to send signals over long distance. He felt that by employing longer wavelengths it was possible to produce signals 'which pass to the opposite point of the globe with the greatest facility.'[43] Marconi paid little attention to the Serb's musings. All he knew was that he was getting somewhere, while as far as he could see Tesla was not.

The next deed he had to attempt was obvious. It would not be possible for people to really see the value of wireless until he could prove it was capable of passing signals over greater distances. If he could successfully communicate between two different continents, then he could finally dispel any queries that still lingered about its limitations. Nonetheless, the sheer magnitude of what he was about to attempt must have filled him with trepidation. All his previous wireless excursions were measured in tens to hundreds of kilometres, this new challenge by contrast would cover thousands of them. It was a bit like thinking that because you could hit the inside of the wastepaper bin in the corner of your office with a scrunched up piece of litter, then you should be able to pull off a Jordan-style shot from one end of a basketball court to the other.

Media attention was at its height, and much of his precious work would be wasted if he allowed it to lull once again. He made ready for what he hoped would be his 'tour de force'. Ambrose Fleming, of University College London, joined the Marconi Company in late 1900. He became Guglielmo's chief technical adviser, and would be a valuable addition to the team, given the gigantic proportions of the job the Italian was undertaking. Fleming helped to design the transmitter being used in this attempt.

First of all Marconi had to decide where to place the two radio stations with which he hoped to achieve this ambitious assignment. The best locations appeared to be the tip of Cornwall, in the western extremity of England, and the coast of Newfoundland. At first he toyed with Cape Cod, a little further down the Eastern Seaboard, but this proved unsatisfactory, the winds were just too furious. Just as its construction was coming to an end, in the final week of November savage storms brought the masts of the American station crashing to the ground. It was judged to be too risky to rebuild at the same location, so a more suitable site on Canadian soil was found, this also reduced the expanse to be crossed. Though this was virtually the shortest distance between Europe and North America, it still represented a vast gap, more than 3,000km had to be bridged if he was to succeed. His best shot to date had been less than 250 kilometres.

Many leading scientists doubted if radio communication beyond the horizon could be done. Even the great Alexander Graham Bell deemed it an 'impossibility'. Based on the principle that electromagnetic waves only travel in straight lines, and allowing for the curvature of the Earth, it effectively meant the two stations had a barrier over 195km in height lying between them. Marconi later admitted, 'all of us, inventors and scientists, were rather ignorant at the time of what an electric wave would or would not do.'[44] The distances he had already managed between England and Ireland some months earlier were long enough that, in principle, they would have been affected by the curvature of the Earth, but he had seen no difference. His instincts therefore told him that somehow the waves were hugging the planet's surface, though he did not know why.

The site of the European transmitter station was to be at Poldhu, on the Lizard Peninsula. The station used a twenty kilovolt circuit to produce the huge thunderclap-like sparks needed to push a signal all the way across the Atlantic Ocean. It ran off a thirty kilowatt power source, sustained by a petrol-driven generator. The transmission system was almost a hundred times more powerful than any constructed before. It cost £50,000 to set up the Cornish wireless station. It consisted of a circle of twenty-four antennas each more than 50m high.

Much of the original structure of the Poldhu station was destroyed in a gale in late September. Marconi hastily reconstructed the site, but simplified the number of masts used to two. Upon these 60m-high wooden structures a mesh of aerial lines was hung. With 'deliverance day' approaching, on 26 November 1901 Marconi set off for North America on the SS *Sardinia* out of Liverpool.

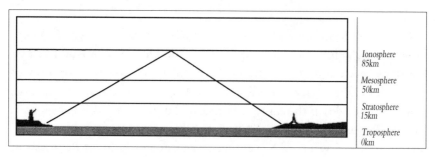

Marconi's 'Atlantic Barrier'.

The receiving station was established at St John's Point in Newfoundland. The aerial at 'Signal Hill', as it became known, was rigged up using kites that were capable of reaching heights of 150m. This meant that time and expense was saved, plus there was no risk of damage from the unsavoury weather conditions that this particular stretch of the North Atlantic coastline was renowned for. By the time Marconi made his appearance at St John's, there was still a lot left to do. All was not going well, his men were wrestling to keep the kites from being carried away. The winter days were getting shorter, and so was his patience. His hopes of getting a favourable result before the year's end were looking unlikely. By the second week of December, some progress was being made. Feeling hopeful that the time had arrived, on Wednesday 11 December, Poldhu started to beam out signals at a frequency of 800 kilohertz, hoping that somehow one or two of them might find their master, expectantly waiting in the icy wastelands across the ocean. All the Italian could do was sit and wait. In the many hours that passed by, he must have begun to think he had finally overreached himself. That this had been battle too far.

The Morse typewriter would not be as sensitive as the human ear, so he decided to switch to the headphones, praying he might be able to distinguish some faint trace of a signal. By the afternoon of the 12th, after losing another kite that morning, things were beginning to turn desperate. Suddenly salvation descended upon the scene; what appeared to be a Morse letter was picked up. In Marconi's own words, 'there sounded the sharp click of the "tapper" as it struck the coherer, showing me that something was coming.'[45] He recalled, 'I listened intently. Unmistakably, the three dots sounded in my ear.'[46] To be completely sure he asked his assistant to corroborate what had happened. 'Can you hear anything Mr Kemp?' he queried. 'Handing the telephone to my assistant, Kemp heard the same thing.'[47]

Following his gargantuan accomplishment, the *New York Times* reported that, 'Marconi, though satisfied of the genuineness of the signals and that he has succeeded in his attempts to establish communication across the Atlantic without the use of wire, emphasises the fact that the system is yet only in an embryonic stage. He says, however, that the possibility of its ultimate development is demonstrated by the success of the present experiment.'[48]

Although he knew there was still a lot of work to be done, the mark had been set. Marconi may have chosen to downplay the scale of his achievement, but its significance was clear. Wireless communication was not just an interesting piece of scientific trickery, it had real potential. The cynics had been smited once and for all.

Fellow scientists were quick to praise Marconi's efforts. Edison, in his inimitable style, said, 'that fellow's work puts him in the same class as me,' and surprisingly even Tesla was among those who publicly congratulated him, stating that he was 'a splendid worker and a deep thinker.'

Even once Marconi had picked up Morse code signal, there was still no shortage of scepticism about how it was possible for radio waves to have made their way over such a distance, and what had enabled them to negotiate the hurdle that the curvature of the Earth created. In the UK, the *Daily Telegraph* was not so easily convinced as other journals, stating that, 'the view generally held is that "electrical strays" were responsible for activating the delicate instruments'[49] Some people chose to brand him a fraud. What we now know is the ionosphere acts like a barrier for radio waves, so they would be bounced back and forth between it and the ocean, effectively acting as a wave guide pushing them to the desired destination. The signal rebounded off these walls like a ball bearing inside a pinball machine. Marconi had managed to prove everybody wrong, even though he was not sure how he did it.

On 18 January 1903, Marconi's system was used to transmit a message from King Edward VII in London to President Theodore Roosevelt in Washington. A permanent transatlantic link was finally put in place, between Cape Breton, in Canada, and Clifden, in Ireland, by December 1907. In time the significance of Marconi's work became more and more apparent. Each wireless antenna erected became another Pharos, protecting those upon the seas from danger, using torches of electromagnetic energy rather than flame. *Punch* magazine's cartoon showed Marconi saving ships' passengers from Neptune's clutches, and stated 'The world's debt to you grows fast.'

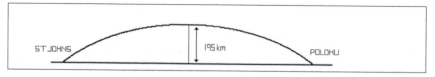

Transatlantic radio waves bouncing off the ionosphere.

Tesla had allowed the impetus to be taken from him. For a long while, he had been the odds-on favourite to bring radio into a phase where it had real commercial potential, but now some badly trained Italian rich kid was basking in all the glory. He wasted no time in trying to gain back some of the lost ground. What he had in mind would make Marconi's achievements look like the amateurish foolings of a schoolboy.

He began to put the wheels in motion for creating his own wireless transmission station, with the intention of showing the world he was still the true voice of authority on such matters. By this point he had gained the support of another powerful backer, namely John Pierpoint Morgan. Would that be the same John Pierpoint Morgan that made mincemeat of Edison I hear you ask? Yes, I am afraid it was.

Morgan had gained a great admiration for the Serb over the years, but was not the only one of his kin to think highly of him. Ann Morgan, the daughter of Tesla's patron, later to become an advocate of women's rights, took quite a shine to him also. She is said to have been bewitched by the Eastern European's charm, but it does not appear these feelings were reciprocated. Tesla kept to his long-standing vow of abstinence.

With Morgan's money he bought a few hundred acres of land on Long Island in Spring 1901, and began constructing a huge wireless transmitter. The tower was over 60m high, and the metal sphere placed on top of it was 25m in diameter. Tesla had managed to prove everyone wrong before, and was confident he could turn the tables on Marconi, as he had on Edison. But the Wardencliffe Tower, as it was known, became another in a long line of very expensive, but seemingly useless, white elephants that filled the later part of his career.

It was while experimenting at Wardencliffe that he started to become aware of what appeared to be radio transmissions that were not from any terrestrial source, but from outer space. Tesla, always eager to cause a stir in the media, proclaimed that it was alien life forms trying to contact us. Suffice to say the papers had a field day, and though Tesla got no shortage of publicity out of it, it was the sort of publicity he could have done without. His comments only fanned the flames of ridicule, and added more credence to increasingly common opinion that he was completely crackers. What scientists have since verified, is that the signals he picked up were electromagnetic waves emitted by distant stars, not intelligent life. Funnily enough, some years later, Marconi managed to get himself in a similar pickle. While experimenting he heard the same signals and as they did not seem to be coming from anywhere on Earth conceded they must be from elsewhere. Journalists decided once again to take this as confirmation there was intelligent life out there attempting to communicate with humanity, which clearly was not the case. Marconi just said he could not explain where the signals came from. In the end he just chose to treat the whole thing as a sign there was precious little intelligent life here on Earth, let alone on other planets, at least when it came to the popular press.

Tesla's ideas became increasingly out of touch with reality. He claimed he could use resonant oscillation to split the world in two, if he so wished. He also bragged of the huge scientific leaps he was making in wireless, but while a decade before he loved nothing better than to demonstrate what he had achieved, now he seemed uncharacteristically shy.

In early 1901, he informed the press that he had already proved 'the practicality of signalling by my system from one to any other point of the globe, no matter how remote, and I shall soon convert the disbelievers.'[50] He boasted that his signals could be 'transmitted to a planet such as Mars with as much exactness and certitude as we now send messages by wire from New York to Philadelphia.'[51] In an article in the *New York Times* later that year, he told the world how his investigation had uncovered the possibility of 'transmission of energy through the air', and how his idea was 'now receiving some attention from scientific men.'[52] They proved to be empty words, Tesla's hopes of transmitting wireless power and information around the globe were nothing more than pipe dreams.

When it became apparent that the competition was doing far better without the need for so much expensive equipment, questions were bound to be asked. Morgan suspected he was being taken for a ride, and wanted to know why so little progress had been made. When Tesla could not provide an adequate explanation, he began to give him the cold shoulder. The Serb carried on regardless, but soon enough his financial reserves were all dried up.

He repeatedly wrote to Morgan, requesting more cash, but was fobbed off. In July 1903, desperate to secure more funding, Tesla, somewhat unwisely, let slip the real objectives of his work at Wardencliffe. He admitted that the tower was not simply for the transmission of wireless signals, but that it was intended to convey electrical power. The system for transporting energy through the air, that Tesla had raved about in the newspapers, was actually underway, and Morgan had been an unwitting accomplice. By this stage it is estimated he had put the best part of $200,000 into the Wardencliffe project.

This plan of Tesla's quite clearly conflicted with the Wall Street financier's commercial interest in General Electric. There was no sense in Morgan cannibalising his existing business. Tesla was foolish to think that this would not be the case. Nobody would be crazy enough to put money into a scheme they knew would only harm them.

Morgan could see that Tesla had been trying to destroy his business, and what was more he was expecting him to pick up the bill for it. His decision not to tell his principal backer of his true intentions was always going to be a risky move, and the gamble had not paid off. Morgan was far from pleased at being duped, and severed the financial umbilical cord attaching Tesla to him completely. From this point on he was on his own.

Without someone to bankroll the endeavour, Tesla's dream of distributing electrical energy to the far-flung corners of the world lay in pieces. He needed to find another backer, but nobody wanted to know. It was a rude awakening for Nikola, back in his heyday he had access to more venture capital than he needed. Before, there was no shortage of people interested in supporting his projects, but now he was yesterday's man. Even his old friend Westinghouse walked away, he was making good money from the AC system, the concept of transmitting power wirelessly did not really add to what he could already offer to his customers, even if Tesla could get it to work. It would just mean further investment and greater risk. As the saying goes, 'if it ain't broke, don't try to fix it'.

Tesla disconnected almost totally from the harsh and unforgiving world he had been left in. He began caring for sick and wounded pigeons that he found in his neighbourhood, taking them to his hotel room and nursing them back to health. It is something of a contradiction that a person so worried about germs should chose to look after creatures that offer such a prevalent method of carrying disease. Tesla was not always a rational guy however.

Morgan died in April 1913, and, as 'the Bard' once wrote, 'the sins of the father are to be laid upon the children.'[53] His heir, J.P. Morgan, had to take over the duty of contending with Tesla's unceasing petitions for money to fund his 'World System', but the requests fell on deaf ears. He was of a similar mind to that of his father. For Tesla, the 'Gilded Age' had lost its shine. The Wall Street high rollers, with whom he had socialised, had all disappeared into the mist. John Jacob Astor had died just before Morgan. He had been aboard the inaugural voyage of the *Titanic*, and through his own arrogance he managed to requisition himself a watery grave. When the incident occurred (for those who do not know about this, the ship hit an iceberg, or something like that) he made light of it all, not fully appreciating the seriousness of the situation, and proceeded to kid around. By the time he finally did understand the peril the ship was in, it was too late, there were no lifeboats left, and he went down with the vessel. Tesla's old partner, Westinghouse, died in March the following year. He was alone, just left with his pigeons for company.

With clear reference to his Italian origins, Tesla said that Marconi had employed strategies reminiscent of the Medici's. He tried to sue Marconi's firm for patent infringement in 1915, but did not have the financial muscle to go through with the long-drawn-out case that would have followed. He was eventually forced to withdraw from the fight before it had really begun.

In November 1915, rumours began to circulate that Tesla and Edison would share the Nobel Prize for Physics, for the contribution to the distribution of electrical power that both parties had made. Finally it looked likely he would get the recognition he had yearned for all those years. But yet again his hopes were to be dashed. It became apparent that Edison would rather go without than share the honour with his old adversary, and turned down the award.

Any hope that he would be able to revitalise his fortunes, and finish the work begun at Wardencliffe, finally disappeared soon after this. His financial situation, which had always been delicate, became increasingly dire. He was eventually forced to hand over the mortgage on the tower to pay his bill at the Waldorf Astoria, which had been piling up for quite some time. For a couple of years the giant edifice remained, almost taunting him, a testament to his failure. Then in July 1917, the tower, which had stood idle for over a decade, was finally dismantled. According to news reports, the authorities suspected 'German spies were using the big wireless tower' as over the previous months 'several strangers have been seen lurking about the place.'[54] Tesla's brainchild had been stillborn. He was heartbroken.

Later that year, the AIEE decided to award him with the Edison Medal, the highest honour the organisation could bestow. True to form, Tesla told them to go to hell. He wrote to them stating, 'you propose to honour me with a medal, which I could pin on my coat, and strut for a vain hour before the members of your institute. You would decorate my body, and continue to let me starve, for failure to supply recognition, my mind and its creative products, which have supplied the foundation upon which the major portion of your Institute exists.'[55] Eventually, with a great deal of persuasion, and perhaps hoping it might result in new interest in his work, he agreed to accept the award.

Tesla wrote his autobiography, entitled *My Inventions*, in 1919. It was originally a series of articles written for the *Electrical Experimenter* magazine, but would later be compiled into a book. In it he told the story of how he had gone from rags to riches, and back to rags again (well actually he left most of that last bit out).

Tesla loved the persona of the 'electrical magician' he had made for himself, and looked for any opportunity to astound people. He craved the attention too greatly, and this overtook his drive to do things for purely scientific reasons. He had been corrupted by lust for fame and adulation. It made it harder for people around him to know what he could actually achieve, and what was purely supposition. Just after his seventy-eightieth birthday, the *New York Times* reported Tesla's claims that he had 'perfected a method and apparatus' which could send 'beams of particles through the free air, of such tremendous energy, that they will bring down a fleet [of] 10,000 enemy airplanes at a distance of 250 miles.'[56] His 'Death Ray' as the papers referred to it, was just another example in a long line of Tesla's sensationalist proclamations. It was perhaps Westinghouse's influence that kept him grounded, and allowed him to achieve so many great things in his youth. Once he was no longer there to keep Tesla's imagination in check, it just went whizzing ever further out of control.

Probably that less said about Tesla's behaviour in his final years the better. In later life he became increasingly eccentric, and was eventually regarded as some kind of mad scientist. Talk of 'Death Rays', 'Earthquake Machines' and suchlike only served to tarnish his scientific reputation. As he got older, his claims became more and more extravagant, but devoid of any concrete research to back them up. Tesla had become a victim of his own narcissism. He had dined out on the name he made for himself following his early triumphs, and the world just believed he could pull off the same stunt again.

Unfortunately, this only served to further pamper his already over-inflated ego. Tesla began to believe he could achieve anything he put his mind to, and did not think of the practicalities involved. Expectations just became too high. He would flit between different projects, looking for one which would restore his fortunes, but never making real progress in any of them.

He had gone from glamorous showman to being regarded quite simply as a crackpot. His early demonstrations had been so well received that he needed to keep on getting bigger fixes of public attention. He was simply addicted to it. As a result his claims became increasingly outrageous. He reverted to the unrealistic notions of his childhood. When a Tesla lookalike began to appear in the Superman comic books of the 1930s, it was unsurprising that he was a mad scientist.

The years brought decay to his physical form, as they had to his ambitions. The lean build he had always been blessed with gave way to almost complete emaciation toward the end of his life. His face became gaunt and pale, and though his eyes still shone brightly, like those of the young Serb who had arrived in New York all those decades ago, his body and his mind were now weathered by years of disappointment and betrayal.

In 1934, he moved into the New Yorker Hotel in downtown Manhattan, a far more humble affair than what he had been used to in the good old days, but by then he was running out of places that would give him credit. It had seemed a little ungrateful of the Westinghouse Company to not have offered him more financial support, considering that his gesture of rendering his royalty claims void had kept it afloat during the years of austerity. However, it could also be argued that whatever the firm did to back him would never be enough, burning up money had always been his forte after all. Nevertheless in the last few years of his life, the company had a slight change of heart, paying the ageing scientist a small pension to help cover his hotel bills.

In July 1937, he was honoured with the Cordon of the White Eagle by the King of Yugoslavia, his nation of birth. Though proud of his American citizenship, Tesla never forgot his Serbo-Croat origins. He once wrote that 'hardly is there a nation which has had a sadder fate than the Serbian. From the height of splendour' it had been 'plunged into abject slavery, after the fatal battle at Kosovo Polje, against the Asiatic hordes. Europe can never repay the Serbians for checking, by the sacrifice of their own liberty, that barbaric influx.'[57] He was of the belief that the only way this troubled region could finally lift the curse of war and famine that had afflicted it was to unite and work together as one country.

Tesla collapsed on the evening of 7 January 1943 in his room at the New Yorker. A maid found him, just before eleven o'clock, but by then it was too late to do anything, he was already dead. He was eighty-six years old at the time. The coroner assessed that he had suffered a heart attack. His funeral took place on 12 January.

The *New York Times'* obituary described him as 'one of the world's greatest electrical inventors and designers'[58] though it went on to mention he 'had been failing in health for two years' and his 'ideas bordered increasingly on fantastical'[59] as he grew older. Nevertheless he had played a part in the development of many news branches of science, including solar power, X-rays, superconductivity and diathermy (use of electricity in medicine). The US Government had Tesla's research notes impounded after his death, all adding to the myth that he knew things far beyond the rest of civilisation.

Fellow telecommunication pioneer Lee de Forest said if he could have been any other man he would have been Nikola Tesla, and Lord Kelvin stated that he contributed more to electrical science than any man that had gone before him. He had close to 700 patents awarded to him during his lifetime, and had received honorary degrees from such esteemed universities as Vienna, Columbia, Paris, Budapest and Yale.

Tesla has been regarded, for the decades since his death, as something of an enigma. The truth about him is often lost in a morass of ill-founded supposition and unsubstantiated rumour. Some believe he faked his death, some even claim he was not of this planet, and returned to his home somewhere amongst the stars. It is of course all complete rubbish.

On 27 June 1956 the International Unit of Magnetic Flux Density was named the 'Tesla', and when the National Inventors Hall of Fame was established in 1973, he was among the first inductees (alongside Bell and Edison). In 1983, the US Postal Service distributed a commemorative stamp in Tesla's honour.

In the years that followed the success of his transatlantic transmission, Marconi was rarely out of the public eye. His notoriety continued to grow all across the globe. To feed this incessant demand, he was compelled to continue a rigorous schedule, splitting his time between the USA, Italy and Britain. In early 1903, he was given Roman citizenship.

He would still visit Poole when possible, often relaxing at Brownsea Island, a popular summer retreat for the younger generation of the English aristocracy. It was at Brownsea, during the summer of 1904, that he met Beatrice O'Brien. Beatrice was the daughter of Irish noble Lord Inchquin. She was only nineteen years old, while he had just started his fourth decade. At first Beatrice seems to have been unsure about him, but before long they became an item. By this point Marconi had gained a bit of a reputation as a lady's man. He had already been engaged twice before. The first time was in Autumn 1899, when he had fallen for Josephine Holman, a native of Indianapolis. They had met while travelling on an ocean liner from New York to England, but this was all over within a year. He had little time for such matters at this stage in his life, and his fiancée soon became disenchanted with him due to his continued absence. His work was clearly the only thing that mattered during this period. He had gained a taste for shipboard romances though. In 1903, he met Inez Mullholland, a young journalist. This time he had been heading in the other direction, from Europe across to North America. This relationship would continue until he met Beatrice, when he quickly broke it off.

Marconi proposed to her on two occasions, the first time Beatrice refused, but when he tried again she accepted. They were married on 16 March 1905 in Mayfair, London. Marconi's brother Alphonse took the duties of best man. Sadly their father was not there to witness the happy event, having passed away twelve months earlier; the wedding was just one week short of the first anniversary of his death. After the ceremony the couple departed for an Atlantic cruise, upon the SS *Campania*. Marconi decided to make use of their time abroad to search for a permanent location for his wireless station in Canada – some honeymoon! It soon became obvious to Beatrice that his research came above all else in his life.

The Marconis' first daughter, Lucia, was born in February the following year, but tragically died just a few days later. It was not until 1908 that they had another child. Their second daughter, Degna, was born in London that September. In May 1910, they had a son, Giulio, who was born in Bologna. Another daughter followed in April 1916; Gioia arrived while they were staying in Rome.

He received the Nobel Prize for Physics in 1909 'in recognition of his contribution to wireless telegraphy'. A string of further awards were bestowed upon him. He received the Kelvin Medal, the Grand Cross of the Order of the Crown of Italy and Chevalier of the Civil Order of Savoy.

George V would appoint him an Honorary Knight Grand Cross of the Royal Victorian Order in July 1913, and he was made a member of the Italian Senate later that year.

In September 1912, while driving in the Italian countryside, Marconi and Beatrice were involved in a car accident. Beatrice escaped virtually unscathed, but her husband was not so fortunate. He had to be rushed to hospital in Turin, suffering with head injuries. His right eye had been damaged irrevocably; it had to be removed and replaced with a glass one.

The outbreak of war came in 1914. In June the following year, upon his country's entry into the conflict, Marconi enlisted in the Italian Army, joining the Engineering Corps. His impaired vision obviously prevented him from being involved in open combat, but also it would be have been an incredible waste of his talents if he had been amongst the 'cannon fodder'. Instead he worked on improving the Italian forces radio apparatus, as well as acting as a powerful talisman with which to fortify the men's morale. He served as a lieutenant at first, but was promoted in 1916 to the rank of captain. Later that year he was transferred into the navy, who had greater need of his technical skill. In the past it had not been so enthusiastic about him, back in his teens Marconi had been turned down by the Italian Naval Academy. Luckily he was not someone to hold a grudge.

Following the Armistice, he served as Italian plenipotentiary delegate at the 1919 Paris Peace Conference. After the war he received still more honours. He was given the title of Commander of the Order of St Anne by the Tsar of Russia, Grand Cordon of the Order of the Rising Sun by the Japanese Emperor, and one of Commander of the Order of St Maurice & St Lazarus was presented to him by the King of Italy.

In 1919, he bought and refitted a 90m steam yacht. The *Elletra*, as he christened it, became his home and doubled as his laboratory. As well as a quiet and distraction-free place for him to continue his research, it also served as a suitable venue for him to indulge in a series of affairs that took place towards the end of his marriage. Marconi had always had an eye (one good eye at least) for the ladies, and put little stall in the sanctity of his marriage vows. His wife seems to have been aware of his rising number of infidelities, but felt that there was little she could do about it.

Though he would continue to do research, for the most part it was now time for others to take the lead when it came to wireless development. In 1906, Canadian Reginald Fessden succeeded in making the first voice transmission by radio. The following year, Lee de Forest made alterations to the electromagnetic valve Fleming had created, to produce the more sensitive 'Audion' tube. It allowed the signal to be amplified as it was picked up, supplying huge boosts to the range that could be covered. In the 1930s, fellow American Edwin Amstrong used de Forest's work to develop the 'regenerative circuit' that allowed signals to be amplified still further, and thus paved the way for FM radio.

In May 1920, the next phase of wireless development began. It had already been used as a means of saving ships in peril, or defending countries from foreign attack, now it would become a source of entertainment. From 'Marconi House' on the Strand, London, the BBC began its transmissions. Dame Nellie Melba, the Antipodean operatic star, being amongst the acts that made up the first day's output.

Marconi's mother, Annie, died in Autumn 1922, and was buried at London's Highgate Cemetery. Even before the death of her husband, she had spent much of her time in England, and once widowed rarely returned to Italy. Many of his family ties were now gone, and what was left of his marriage was soon to disappear too. It finally ended in February 1924, in a relatively good-natured manner.

Surprisingly, no sooner had Marconi got his freedom than he wanted to get hitched again (these impetuous Mediterraneans). His new love interest, just like his last, was a member of the nobility, but this time it was a fellow Italian, Countess Maria Cristina Bezzi-Scali. He had met her in Spring 1925 at the Italian coastal town of Via Reggio. They were introduced by mutual friend Duchess Ravaschieri. The countess, who came from Rome, would later describe his 'mysterious force of attraction which drew people to him.'[60] Even from their first encounter she felt 'there was a kind of electrical current between us, right from the beginning we understood each other perfectly.'[61] As he was regularly in capital to fulfil his senatorial duties, the couple had ample chance to become better acquainted.

Unfortunately, when talk of nuptials began to be initiated, a snag was to appear. As Maria Cristina came from a devout Catholic family, she could not be seen to marry a divorcee. So Marconi had to try to get his estranged wife to agree to have their marriage annulled, rather than just being divorced. She was willing to consent to this, by then she was with the Marchese Liborio so probably did not care what he did. However, it was ironic that Beatrice's adultery was used as grounds for the case, considering she had been faithful to him for years, despite his continued indiscretions, and it was only at the end of their relationship that she looked for affection elsewhere. In early 1927, after some months of waiting, the Sacra Rota (Italy's Ecclesiastical Court) agreed to annul his marriage to Beatrice.

The wedding of Guglielmo and Maria Cristina took place on 15 June 1927, at the Basilica of Santa Maria degli Angeli, in Rome. She was more than twenty years his junior, and had been just a baby when he had done his great transatlantic feat. As we will see, the final years of Marconi's accomplished life would be a mixture of blue blood and black shirts.

Marconi was made a hereditary peer in June 1929, receiving the title Marchese from Italian King Vittorio Emmanuelle III. On 26 March 1930, he was asked to throw the switch that turned on the lights of the World Exhibition in Sydney, Australia. Amazingly, he remained in Italy while doing this; wireless transmission was used to carry the signal across the other side of the world. In July, Guglilemo and Cristina had a daughter, Maria Elettra Elena.

Pope Pius XI was in the Vatican during the 1920s and '30s, and Marconi was to win favour with the Catholic Church by setting up a microwave station there in 1931. He had first started experimenting with these high-frequency waves during the war, using them for ship-to-ship communication. The link was inaugurated on 12 February. It connected the pontiff's official residence with his summer home in Castel Gandolfo, a few miles to the south-west, on the banks of Lake Albano. Papal gratitude was soon forthcoming, Marconi was presented with the Order of Pius a few months after.

Marconi made yet another breakthrough in radio communication later that year. In July, the *Daily Mail* reported how he had used wireless signals to help his crew steer the yacht *Elettra* through a 15km course, marked out by buoys tethered ninety metres apart. It said that 'all of the time the helmsman stood in a curtained wheelhouse and was guided only by signals from a tiny wireless beacon on a hill 300 feet above the sea.'[62] The beacon had been arranged as to send out 'two beams separated by a 'beam of silence'. These beams operate a needle on a dial in a ship.'[63] By staying within the boundaries set by the two beams, and adjusting its course if it strayed out of the path where the airwaves remained empty, it was possible to navigate the ship through the obstacles safely.

As a distinguished member of the scientific world and one of his country's most famous sons, not to mention his wife's high standing in Italian society, Marconi became friendly with Benito Mussolini during the early 1930s. He had originally joined the 'Fasci di Combattimento' back in 1923, following il Duce's rise to power, but had not played an active roll in the organisation. However, for the last decade of his life he would be an ardent supporter of the cause. He gave speeches at fascist rallies, as well as making several radio broadcasts supporting the doctrines of Mussolini, and helping spread right-wing propaganda.

Marconi was unhappy with the way Italy had been mismanaged in the past and perhaps, in the beginning at least, saw Mussolini as a necessary evil if the country of his birth was to gain more economic strength. It may have also been that he simply feared what might happen if he did not do as he was told. The Italian dictator made Marconi the President of the Academia d'Italia in 1930, and later that year he started to serve on the Fascist Grand Council. It has been suggested that Marconi prevented Jewish scientists from entering the Academy, on Mussolini's instruction (according to a report in the *Guardian*, 19 March, 2002).

Following the outbreak of the Italian–Abyssinian War in October 1935, much of Europe chose to politically isolate Italy, placing the country under strict trade embargoes. Marconi wanted to give an explanation of the reasons for his country's invasion of the region, and put forth the Italian perspective, but in November the BBC's board of governors refused to let him speak on the subject. Though, as he pointed out to them, they would not be able to broadcast anything if it had not been for him.

He excused the barbaric acts Italy was engaged in by saying that it would bring civilisation to the backward people of the African nation. In one transmission in May 1936 he proclaimed that 'nobody should forget that our workers and colonists are, and ever will be, soldiers, ready if necessary to defend at any cost the legitimate fruit of their victory.'[64] His pro-war stance and his provoking words gained him widespread condemnation throughout the international community.

The jury is still out as to whether Marconi was really a die-hard fascist, or if he was just forced to go along with it because of his lofty position in Italy's hierarchical structure. His wife was of the opinion that he did not agree with what was going on, but was unable to do anything about it. Though looking back, it would not make any sense for her to say otherwise. Nevertheless, she felt that 'Guglielmo never lost sight of the wellbeing of mankind'[65] and was actually against fascism. It was her belief that Marconi was worried by 'the understanding between Mussolini and Hitler, and thought that it would have serious consequences'[66] warning the Italian leader against allying with the Third Reich.

Marconi died just before four in the morning on 20 July 1937, at his apartment in Rome. He was sixty-three years old. He had been increasingly plagued by heart problems, and had suffered a serious attack only fourteen months before. Maria later said, 'He had a healthy constitution, but his heart was tired.'[67] He was granted the funeral of a national hero in the Italian capital the following day, with memorial services also taking place in Bologna and London. His body was finally interned at a specially built mausoleum underneath Villa Griffone. On 21 July, across the globe, there were two minutes of radio silence, the first time the airwaves had remained quiet since the precocious Italian had brought them to life forty years before.

The *Times'* obituary stated he 'was not the discoverer of electrical waves, and he was not even the first to suggest that they might be used for signalling to long distances, but thanks to his genius and perseverance it is due to him more than any other worker that they are now one of the most important means by which communications are sent by telegraph and telephony to the ends of the earth.' It said that until the end of his life he 'retained the vigour of youth' and 'his zest for experiment had never failed to infect his associated colleagues'.[68] His old friend Fleming, just after Marconi's death, wrote of his 'power for reducing general principles to practice, and devising a form of the apparatus which would give best effect to them.'[69]

In the words of his daughter Degna, 'for me over the years there have always been two Marconis; the scientist, and my father. The first was absorbed in things I could not comprehend. The second, an intricate and fascinating human being I have come to see clearly only by reading backwards, the way the Chinese do, fitting together the pieces of his story with my memories.'[70] She stated he 'was often absent, either in body or in mind, for his dedication to his work was absolute.'[71] Maria acknowledged that 'he never failed to encourage young people to devote themselves to new research and never to give up, but try and try again.'[72] Years later, the Guglielmo Marconi Foundation was founded, to offer scholarships promoting the study of wireless communication.

In April 1943, just nine months after Tesla's passing, the US Supreme Court debated the validity of Marconi Wireless' radio patents. After long deliberation they decided in favour of Tesla, stating that it was he, rather than Marconi, who should be considered the 'Father of Radio', and claiming that his 1904 patent (US Patent 763,772) contained nothing that had not been earlier published by Tesla. The request for a rehearing was denied that October.

The court announced that Marconi's patents were invalidated due to the prior work of Tesla and Lodge, which anticipated it. The Italian's use of tuned circuits, and inductance coils in his design were deemed to have been described in Tesla's patents previously. The Serb finally received the vindication he had begged for, but unfortunately he had not lived to reap its rewards.

His greatest triumph had been the propagation of AC. When you think about your everyday life – the washing machine cleaning your clothes, the television you watch, the computer you work on, the microwave warming up your pizzas – it all runs on alternating current. You could barely do anything today had it not been for Tesla. All this aside, his success only served to make him more deluded, it gave him the impression he was in someway invincible, and could keep beating the odds, in the same fashion he had managed to in the past. This in turn made the pain he suffered, when his 'World System' failed, all the more acute.

Tesla's eagerness to fight for his beliefs and not be deterred was commendable. It had enabled him to persevere when it came to securing acceptance for his AC system, after being told time and time again there was no future in it. However, in later life it proved to be his undoing. He would tend to stubbornly cling to ideas that he simply could not get people interested in, or rather that they were not willing to put money into.

He was a true workaholic, if ever there was one. He put in as much as eighteen to twenty hours a day in order to get his research completed, but it would not come without a penance. It meant he would always be seen as a rather solitary and unapproachable figure. He once said that 'I am credited with being one of the hardest workers and perhaps I am, for if thought is the equivalent of labour I have devoted to it almost all of my waking hours.'[73] He became a junkie, his need to be creative increasingly distancing him from ordinary life. As he once put it, 'such sensations make a man forget food, sleep, friends, love, everything.'[74]

A bit like Newton, he did not really have many close relationships during his lifetime. Both of these men were so focussed on their work that they ended up sacrificing all personal comfort so they could invest more of their time into it. It has been debated that both Tesla and Newton could have suffered from obsessive-compulsive disorder. Certainly Tesla's behaviour tends to back this up. As we have seen, he was highly preoccupied with certain numbers, as well as being fixated with cleanliness, having a huge fear of germs. He also had many little habits and customs, which could hardly be described as completely normal.

In other ways his life more closely mirrors that of fellow 'Nearly Man' Robert Hooke, his early successes gave high hopes of what he could achieve, but his promise soon faded. In much the same manner, his twilight years were filled with unfinished projects with which he hoped to initiate some sort of renaissance, but these ventures became increasingly farfetched and unrealistic.

Tesla blurred the lines between fantasy and reality. Talk of communicating with extra-terrestrial lifeforms, 'Death Rays', and all the other outrageous comments, which too often adorned his aura towards the end of his life, have only helped to endear him to a whole legion of 'New Age' freaks. It sometimes makes it hard, when examining his life, to judge what is true from what is fabrication. The border that demarcates between a prodigious visionary and a foolish dreamer is only very slim, and though he had not realised, at some stage, Nikola Tesla slipped across it.

Marconi just dealt with the facts, he never claimed anything that could not be backed up there and then. His strategy was one of building up to things a bit at a time. He had started by transmitting signals across the Bristol Channel, then the English Channel, before attempting to transmit over the Atlantic. It gave his financial supporters something tangible to put their faith in, not just hearsay. It also meant his achievements continued to receive praise in the press, keeping his company's share price high, and catching investors' imagination. Tesla went for an 'all or nothing' strategy, rather than tackling things step by step. The outcome of this was his backers lost confidence, not knowing if he was making any real progress.

Tesla was in some respects an idealist. He thought of things that would be good for his personal acclaim, and indeed benefit humanity, but did not consider whether they had any real commercial value, and sadly that was what counted then, the same as it does now. Passing electrical energy around the globe without the use of wires would have been an amazing achievement, but in a business sense, who would have actually gained from it? Not his investors, nor the local power companies, nor the providers of the incumbent cabling infrastructure. The concept was an interesting one, but the practicalities of such a system would have been impossible to implement. How would such a system be metered? What size would the electricity generating plants to support the huge single transmitter (or at most a handful of transmitters) have to be? How would the natural resources needed to run such enormous plants be transported? Not to mention the fact that Tesla never really considered the effect it might have on weather conditions, or human health. In the last ten years there have been several reports made by environmental groups on the increased risk of leukaemia in residents of houses close to overhead power cables. If Tesla's ambitious plan had been realised, this sort of danger could have been exposed to the entire population of the planet.

Marconi, by contrast, knew his limitations. He was not looking to impress people for the sake of it. He was not aiming to change the structure of civilisation, and create some sort of utopian society. While Tesla looked at radio waves as a means of transmitting the energy needed to power equipment, or control defence mechanisms, Marconi just concentrated on the simple things, namely using this technology to communicate over distances. He was just trying to make a bit of a name for himself, and generate some money in the process. Tesla talked big, but it was Marconi who had successfully managed to flip a signal across the Atlantic (albeit using bits of other people's equipment). He had given people something substantial, not just theory and small-scale demonstrations. Often he had to fight above his weight, but was able to do so by utilising the skills of other engineers, such as Fleming. Tesla was often too self-reliant for his own good.

In hindsight it is perhaps easy to say that Tesla should have stuck to the job in hand, namely pure uncomplicated wireless communication, and set about trying to beat his Mediterranean rival. Then maybe he would have been able to persuade Morgan or another financier to back his 'World System' in full. But his eagerness to get the project off the ground was too great to consider using an indirect course. Tesla wanted it all. By trying to do everything at once, in the end it meant he achieved less.

The US courts indeed affirmed Tesla's priority, and thus acknowledged him as the true inventor of radio communication, though there are a couple of points to bear in mind when looking into this. Firstly, if this was the case, why on Earth would the Patent Office have given the patent to Marconi in the first place? It certainly does not bode well with the assumption that it had not done an efficient job of checking his application did not contravene any previous patents. Secondly, and perhaps more perplexing is, why had it taken nearly forty years for this amendment to happen?

There seems to be one clear explanation, which more than adequately answers the two questioned posed here. Basically it suited the government to side with Tesla, rather than Marconi, at this stage, just as it had been better for it to support him instead of Tesla back at the turn of the century. Marconi filed for a US patent in November 1900 (do not forget that regardless of what went on in the United States, Marconi's UK patent preceded any radio patent that Tesla filed), but it was turned down. Over the next couple of years he would make a series of alterations to his application, but to no avail, the Tesla patents were upheld. In 1903, the Patent Office had stated that Marconi's applications were 'not patentable over Tesla patents 645,576 and 649,621'. However, they did a complete u-turn, in 1904, letting Marconi's patent go through.

By this stage Marconi's company could offer a good product, and was still effectively the only game in town. Also it was around that point where the US military started to see the value of radio communication, so it could have been in its interests to bring Marconi into the fold, rather than Tesla. It made the whole situation a lot less complicated if the Italian owned the rights to the technology. Conversely, in the 1940s the company held all the aces, and could exert pressure onto its client base. Military contractors are normally expected to kowtow to their patrons, so it is possible that, once again, the government's change of heart was for political reasons.

Lodge certainly deserves a brief mention. He made the first notable transmissions after all. However, there are a couple of things to keep in mind. Though he also tried to claim Marconi had stolen from him, and was to some extent proved right by the US judiciary, he only saw the utilisation of radio being within relatively short distances. Just like Johan Philip Reis had with the telephone, Lodge did not grasp the true potential of this invention. What is more, he was one of the most verbose when it came to casting doubts on its scope. He was a fervent believer that it was impossible to send wireless signals across the Atlantic. So if it had been left to him, we would have progressed no further with this technology.

Quickly mentioning the 'War of the Currents'; this is often pitched as Tesla's victory over Edison, when in reality they both lost out, each of them sacrificed for the greater good, namely that of the businessmen of Wall Street. These inventors had tried to chart their way between the 'Cylla' of Morgan and the 'Caribdis' of Westinghouse, and were both destroyed in the process. Morgan managed to squeeze Edison out of General Electric, while Westinghouse got Tesla's AC system rights at a bargain price, and then stopped backing the Serb as soon as it became apparent that he was a spent force. The truth was that Edison and Tesla were just puppets, controlled by the corporate high rollers. And thus the pattern was fixed; the 'ruling class', regardless of who they supported at the onset, would end up on the winning side. Someone else would always take the fall.

It is now widely accepted that Marconi utilised a variation of Tesla's equipment in the radio transmission system with which he sent the first signals across the Atlantic, rather than the circuit described in his first patent. Perhaps the big difference between them was Marconi did not just speculate about things like Tesla did, he went out and did them. Much as Bell had done, Marconi brought several different elements together and, by adding a little of his own ingenuity to the mix, managed to create something far greater than its component parts. He was effectively a catalyst, openly admitting that he 'made use of known ideas. My instruments are improvements on my predecessors' with the introduction of a few developments which from my observations seemed necessary.'[75] Although it cannot be denied that Marconi utilised devices first developed by other inventors, namely Lodge, Branly and Tesla, he had gone on to do something with them that had real intent. While the others aimlessly messed around with their creations, Marconi added the direction needed to achieve something of far greater importance. He was also a world-class self-publicist, with style and charisma, plus his family ties gave him greater opportunities to network with influential people that could help him achieve his goals. But at the same time Marconi was a bit of a 'one-trick pony', he could not really take wireless any further than the painstaking transmission of Morse dots and dashes. It would be left to others like de Forest and Fessden to make the transportation of voice possible.

Today over two hundred million radios are in use across the United States, effectively eighty-four percent of the population owning one.[76] Throughout modern history it has been the tool that has held entire nations together and inspired people to strive for a better world. Countless millions huddling around their sets to hear Neville Chamberlain's announcement that began the Second World War, Orson Welles' broadcast of the *War of the Worlds*, Clay's triumph over Liston, or the Beatles' first single. It has touched all our lives; whether it was entertaining, informing, or protecting us. Yet without the foresight of a few gifted individuals we would be without this marvel. Though for some, the objectives could not be reached within the time they lived, their intention to give each of us a greater empathy for fellow inhabitants of this planet, and to make the whole matter of our existence more rewarding and meaningful, cannot really be questioned.

CHAPTER 4

Rogue Elements

How many inventors does it take to make a light bulb?

Few know of Joseph Swan; mention his name, and most people will just stare at you blankly. Nonetheless his legacy is to be seen all around us; used in every home, at every office, on every street, across each continent. The light bulb has become almost synonymous with the concept of invention. All over the world it is used as a cultural symbol to signify a bright idea, and the person famous for creating it, Thomas Edison, regarded as the very embodiment of what it is to be an inventor. However, this whole image is ill-founded, and is in fact based on a hideous lie.

For once again history would have us believe that this most vital part of modern-day life was the responsibility of one man's vision, when the undeniable truth is this was not the case. More than twenty-five years before the American even contemplated the matter, a gentleman in Northern England had already made a workable device. But for him the path to enlightenment would prove to be littered with treachery.

Joseph Wilson Swan was born in Sunderland on 31 October 1828. His father John had married Isabella Cameron in 1820. John was part of a seafaring family, but he had been forced to make a living with his feet firmly placed on dry land. His father had served as the captain of a merchant vessel, which was lost at sea. Worried John might follow him to a watery grave, his mother made him vow not to take up a nautical profession. So obeying her wishes he instead worked onshore making ships' anchors and chains.

The couple had a daughter; whom they called Elizabeth, and two sons, John and Joseph. The Swan clan originally lived at Pallion Hall, a spacious house outside the town. Unfortunately, when John's business venture failed, they had to find a more modest dwelling on Olive Street, Sunderland. Joseph was not over-enthusiastic about his schooling, stating it was 'according to the common rule, rather neglected' though believing he owed much of what he described as his 'true education to that neglect. I was very much let alone, and allowed to roam about and see what I could.'[1] He felt that is was not what he learnt in his classes that had been beneficial to him, but what he 'spontaneously absorbed out of school hours, by keeping my eyes and ears wide open, and cogitating over what I saw and heard.'[2]

After completing his preparatory tuition, he went to Hendon Lodge School. His brother had started there a year earlier, but Joseph quickly advanced to a higher grade, joining his sibling's class. He recalled that both of them were 'glad of the reunion, for we were always inseparable companions, wherever there was an opportunity. We called each other Castor and Pollux.'[3] The brothers left school together, John being fourteen and Joseph still only twelve. He later admitted his 'scholarly equipment was therefore not extensive.'[4] Joseph became an apprentice to local pharmaceutical firm Hudson & Osbaldiston. He was due to serve a six-year stint with the company, but after a little less than half that period he left to join his friend John Mawson's chemists, based on Newcastle's Mosely Street. Mawson was, in spite of his youth, considered a pillar of the community. He was a fervent antislavery campaigner, and believed greatly in the virtues of abstinence, neither drinking alcohol nor eating meat.

Though his education had not been very comprehensive, Swan was not found wanting when it came to possessing an inquisitive mind. He had a keen appetite for knowledge, constantly looking to better himself. The young man became a member of Sunderland's Athenaeum, a gentlemen's club frequented by the intellectually progressive. Here he would befriend John Pattinson, an analytical chemist at Felling Chemical Works, and Barnard Proctor, a pharmaceutical chemist at the Newcastle College of Science. The three young men debated the scientific issues of the day, attended lectures that their club held across a whole gamut of subjects, and conducted amateur experiments of their own.

When Mawson married his sister Elizabeth, Joseph was elevated from his assistant to his business partner. The company would be known as Mawson & Swan Manufacturing Chemists from then on, and proved moderately successful. Swan saw his first photographic portrait in a shop window when he was twenty years old, and became fascinated in this new media. He started to devote his time to creating new developing processes. The problem with photographic reproduction until this time was pictures did not give a completely true representation of the tones that were actually seen. This was because the side of the plate that was exposed to light was also the side to which the harsh processing chemicals were administered.

Joseph reasoned that by using the surface of the plate not in contact with light, there would be less damage done by the chemicals, and a more accurate depiction would result. Throughout the late 1850s, he worked on finding a way to achieve this, taking much from research already done by the likes of Poitevin and Fox Talbot in the progression of photographic science.

Aided by his assistant, Thomas Berkley, Swan worked long into the night at the studio he had created in the attic of the Mosely Street shop. After five years of endless searching, he began to realise he was heading down a dead-end street. Finally he came up with the idea of separating the plate from the picture altogether. By creating a thin film of gelatine on which the photograph could be rendered, then placing it under a negative in a printing frame, and exposing it to light, the picture could be produced on paper. It was the first dry photographic plate. In 1864, he patented what became known as the 'Carbon Process'.

During the course of his experiments Swan had been exposed to corrosive chemical fumes, and developed breathing problems. In order to recover, he temporarily retired to the Isle of Bute. During this short break the young man became friendly with Mary White, the daughter of a Liverpool merchant. Through Mary he was introduced to her younger sister Fanny, a teacher in Streatham, south-west London.

They became engaged in September 1861 and married in July of the following year at Camberwell Chapel. Within no time at all Swan and his new wife had managed to create a small brood of offspring; Cameron arrived in 1863, the next year Mary was born, and Henry followed in 1866. To make room for their rapidly growing tribe, they moved to a larger residence at Leazes Terrace in Newcastle.

Swan's business associate John Mawson was tragically killed in an explosion in late 1867, leaving him as sole proprietor of the company. He made Mawson's widow, his sister Elizabeth, a partner in the firm. Still further calamities beset him though. His wife Fanny was taken ill, just subsequent to giving birth to twins, and did not recover. The newborns joined their mother soon after, both dying within just a few months.

Two of Fanny's sisters, Mary and Hannah, stayed with Joseph to help him with the burden of caring for the children. Joseph would become increasing enamoured of Hannah, and by mid-1870, the two discussed the possibility of marriage. (Yep! I know. Betrothal to your dead wife's sister seems pretty distasteful, but look at the lives of Hooke, Darwin and Wallace, also contained in this book, they could hardly claim to have picked the most suitable spouse either.) At this time the 'law of the land' forbade such practices, so they both departed for Switzerland, and on 3 October, were joined in matrimony, in the town of Neuchatelet. With his new bride he had another five children; Hilda, Isabel, Kenneth, Percival and Dorothy.

Swan would dedicate much of his adult life to the pursuit of one specific objective. This was to create the means to offer cheap, convenient lighting to the entire populace. Until this time Britain, and the rest of the industrialised world, relied on gas lamps to light its houses and streets, but this method was riddled with difficulties. They operated as individual devices, meaning each one had to be turned on and off by hand. Indoors they were far from healthy, as they had a habit of giving off noxious fumes. What was really required was a way to produce more convenient, and less hazardous, illumination.

He had first gained an interest in the concept of using electricity for producing light in the mid-1840s, when he read an article in the *Electrical Magazine* on the American engineer John Wellington Starr and the incandescent lamp he had constructed. This curiosity was fortified further after attending a lecture on the subject, conducted by William Staite, at the Athenaeum later that year. Swan, who was still only in his teens at the time, became completely captivated by the idea. He could clearly see this presented a way of circumventing the inadequacies of gas-based lighting.

In itself using electricity as a means for illumination was nothing new, Sir Humphry Davy had invented the arc lamp back in 1811. By connecting two wires to a battery and placing a strip of charcoal between them, the electric current produced caused the charcoal to emit light, but it was a costly affair. Arc lamps would be suitable for outdoor lighting and so forth, but they would never be economical enough to be deployed elsewhere. They required too much energy for use in mass market applications, such as lighting homes or work places.

Starr's efforts were an evolution of what Davy had done. He left his home in Cincinnati, Ohio, and came to England in the early 1840s. There he created his own system for producing electrical light. It consisted of a short carbon pencil operating in a vacuum above a column of mercury. The problem with this apparatus was that it was still quite power hungry. In addition the inner surface of the glass bulb quickly became blackened by stray carbon particles, blocking out the light (which is pretty debilitating for a lamp really). Though he never succeeded in perfecting his apparatus, he took out a patent for it in 1845 (British Patent 10,919).

Swan drew on the experiences of both Davy and Starr to help him reach his desired goal. Each of them had succeeded in producing illumination by passing electrical currents through carbon-based conductors. This offered a seemingly accommodating media, able to glow white hot without actually catching fire. Following their examples, he employed a carbon emitter, held within a glass container. This would make it cheaper to manufacture than the platinum emitter lamp that his compatriot, Warren La Rue, had developed back in the 1820s. Once again this attempt had simply not been practical for widespread use, as it incorporated a material which was too expensive to utilise. Carbon looked much more appropriate, however it proved extremely hard to get enough battery power to make large elements produce incandescence, so he had to stick to small rather flimsy pieces.

Swan experimented on how to create very thin conducting elements, which would thereby not burn up as much energy as the arc lamps Davy had designed. He tried using carbonised paper, taking strips and packing them in charcoal, then baking them in a pottery kiln. By 1855, he had managed to produce 1cm long carbon spirals, which had enough strength to carry electrical currents without snapping.

The first hurdle was behind him, but what was to really hamper his work was finding a means to produce a strong enough vacuum. A suitable arrangement for this was vital; if the emitter was kept out of contact with air it could be made to irradiate sufficiently without igniting. He patented a partial vacuum light bulb in 1860, but this could not be considered as a real breakthrough as it simply did not have any staying power. The device was incapable of running long enough to make it viable in a commercial sense. It was not until the early 1870s that improvements in vacuum pumps proved proficient enough to furnish the levels of evacuation needed to proceed further.

Unfortunately this is where Swan shifted the focus of his scientific investigation back on to the field of photography, much of his early momentum would thus be lost. However, when Werner Sprengel's mercury pump showed higher levels of evacuation could be attained, Swan resumed his work once again. With the help of Charles Stearn, a mechanical engineer based in Birkenhead, he was able to put together a vacuum system that might have a chance of delivering cost-effective electric lighting. Even then it still was not plain sailing. Other difficulties still had to be overcome. Firstly the carbon strips had a tendency to quickly wear away and snap. Secondly, as with Starr's lamp, the bulb's interior became blackened after a while, as carbon particles were being given off by the element, and deposits built up on the glass.

At first it occurred to Swan that heat created by the strip was causing the carbon to become volatised. If this was the case, he knew his method of producing incandescent light had no future, and his years of toil had been for nothing. However, he later surmised that with the use of greater levels of evacuation this blackening effect could be abated. If everything could be completely dispelled from the bulb, then there would be no residual air with which the carbon particles could be transported. If the purging was absolute, no diminishing of the lamp's ability to illuminate should occur. Getting the vacuum to a lower level proved difficult though, the methods he was already employing were state of the art for that time, there was not an obvious way to extend the evacuation process further.

Swan contemplated these two problems, and finally hit upon a way to overcome both of them at once. He attempted to get rid of the air from the bulb over two stages, first while the element was cold, then when this was completed, to continue while a current passed through it. The plan worked, and he concluded that the vacuum could be maintained and a high current applied to the strip without any detrimental effect on it. Likewise, the stronger vacuum meant the strips were not breaking (the carbon no longer wasting away). By mid-1878 he had a lamp that did not blacken, and was not prone to snapping of the conducting medium. His creation was capable of running for a prolonged period of several hours, without interruption.

Swan demonstrated the invention at a meeting of the Newcastle Chemical Society on 18 December 1878. The equipment used consisted of a glass chamber pierced by two platinum wires, which held a thin carbonated strip. However, on this occasion it could maintain the light for only a few minutes before burning out. This demonstration was followed by another in Sunderland, on 17 January 1879. Soon after Mosley Street became the first street on the face of the planet to be illuminated by electric lighting.

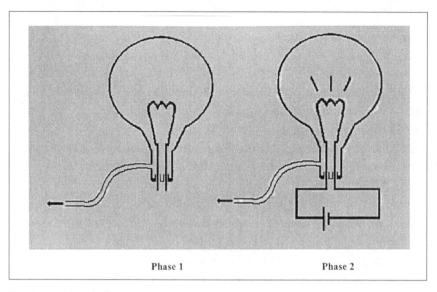

Phase 1   Phase 2

Swan's two-phase bulb evacuation process.

Swan gave a lecture on electric lighting on 3 February 1879, to an audience of 700 people, at the Literary & Philosophical Society of Newcastle. Additional improvements were made to the design. Since the strips had tended to arch when currents passed through them, Swan opted for a hairpin arrangement. By the end of that year, he had replaced paper as the conducting element, and began using cotton, which was dipped in sulphuric acid to harden it, and then coated in carbon. The parchmentised thread process and evacuation method were both patented soon after.

With the help of Charles Gimingham, Swan managed to improve the electrical contacts between the platinum wires and the carbon filament. This was achieved by tubulating the ends of each wire and carbonating the junction points. He patented a method of taking nitro-cellulose dissolved in acetic acid, and discharging this through a small hole into a coagulating fluid. This would allow the formation of a continuous homogenous thread which had greater uniformity, and no fibrous properties whatsoever, as well as being much smaller in diameter. It offered a far more rugged emitter material than those used previously. This squirted filament Swan invented proved to be the winning ticket over the next few years. The Swan Electric Light Company was established in 1881, and soon after Swan set up the world's first electric light bulb factory, in Benwell near Newcastle. Within a few months the first public electric supply was established in the Surrey town of Godalming, using the river Wey to drive turbines that powered Swan's lamps. Before the end of that year, the Savoy Theatre was kitted out with over a thousand of the company's bulbs, and next they were in installed in the Houses of Parliament. In 1882, the British Museum and Royal Academy followed suit. Then in 1884, Wimbledon became the first London Borough to get electric street illumination.

The company's commercial literature informed people that:

> the economical production of incandescent light, which till recently, had completely prevented the general introduction of electricity for lighting purposes has been satisfactorily solved by the invention of the Swan incandescent lamp. By its means separate lights of various powers, applicable to all the uses of ordinary gas burners, and to all the purpose for which artificial light is required, can be produced.[5]

Though Swan had finally managed to get some payback on the years of exploration he had devoted to this subject, he was no longer the only person who saw the potential it held. For many years he had been the solitary researcher in this particular field, but it would soon become apparent that he had not made the most of this privilege. Another individual was about to enter the fray.

Unlike Swan, he was not some part-time dabbler, but a highly focussed professional, with a string of important inventions to his name, and a team of experienced engineers at his disposal. He put some of his best men to work on the task of improving Swan's rudimentary design, and before long had a product that could outperform anything else that was available.

'Everything comes to he who hustles while he waits.'[6]

Thomas Edison

Thomas Alva Edison was on 11 February 1847. His great grandfather, John Edison, had taken the British side in the War of Independence, and fought alongside Lord Howe against George Washington in the defence of New England. Following the American's triumph he and his family were forced to flee across the Canadian border to Nova Scotia. They settled in Bangham, near Lake Erie. John's grandson, Samuel, who would be Thomas' father, married Nancy Elliot in 1828, the daughter of the local Baptist minister. Unfortunately he was equally ill-fortuned when it came to supporting the right faction. When a rebellion broke out in Canada in Spring 1837, he joined the insurgence, fighting against the British. The uprising was rapidly quashed, and Samuel's involvement made it too dangerous to stay. He and his young wife went south into the United States, settling in Ohio in 1839. In the early 1840s Samuel set up a small timber business in the town of Milan.

Thomas was the last of seven children the couple had, three of which had died before he was born. In 1854, the Edisons upped anchor again, moving to Port Huron in Michigan. Samuel made a somewhat ineffectual living there, firstly as a grocer, then turning his hand to real estate. As a young lad Thomas had quite a breathtaking talent for mayhem. On one occasion he accidentally set fire to his father's barn, killing the animals inside. In another incident while swimming with some other children, one of the boys drowned.

Like many of the other great names that are featured in this book, Edison was not seen as a very attentive pupil. Worried that he was not making the necessary progress in school, his mother began to teach him additional lessons at home. Though his methods were far from orthodox, Thomas would try to stretch the bounds of his mental agility. He would perform simple chemistry experiments, and dissect animals in the cellar of his parents' house.

It was in his early teens that Edison first started work with Great Trunk Railroad, whose lines ran through Port Huron heading towards Detroit. He became a newspaper boy, hawking daily editions of the local rag, as well as selling snacks on the side. At seven in the morning, he would catch the train to Detroit, and arrive back home after nine in the evening, taking advantage of the stopover in the city to use the expansive library there.

Edison lost much of his hearing when he was twelve. It is not certain as to the reason for his deafness. He personally claimed it happened when one morning he turned up late for the train. As the locomotive began to pull away from the station, Edison ran after it, before stumbling and nearly getting crushed under the wheels. He later recalled that he 'caught the rear step nearly out of wind, and hardly able to lift myself up. A train man reached and grabbed me by the ears and as he pulled me up I felt something in my ears crack.'[7] Edison described how with time earache gave way to 'a little deafness, and this deafness increased.'[8] He took a philosophical view about the whole thing, saying, 'if it was that man who injured my hearing, he did it while saving my life.'[9] He would also state that he felt his deafness had in some ways been a godsend, enhancing his ability to concentrate on the task in hand, and isolating him from the distractions that plague the common man.

His considerable entrepreneurial talents began to show themselves by early 1861. After a while, peddling newspapers simply wasn't enough of a challenge. Still only fifteen, he started publishing his own newssheet, the *Grand Trunk Herald*. He set up a small printing press in one of the freight cars. The editorial by all accounts had little factual basis, and he did not 'suppose it interested a solitary human being'[10] but that was equally true of the other journals which were purchased on these trains at that time. The sort of items it would run, so he later remembered, were things 'like "John Robinson, baggage master at James Creek station, fell off the platform and hurt his leg. The boys are sorry for John" or it might be "No. 3 Burlinton engine has gone into the shed for repairs"'[11] Joseph Pulitzer and Randolf Hearst were clearly not under threat from Edison's publishing enterprise. However, what he lacked in writing ability, he more than made up for with commercial awareness. When the American Civil War began, Edison used the panic that arose to full effect, making the most of the situation to sell more papers.

In 1862, a small drama took place that opened up a whole new career for the precocious Mr Edison, when he managed to save a boy from being hit by a train. The stationmaster James McKenzie's three-year-old son, Jimmy, had been playing on the platform, and nobody had noticed when he wandered onto the track. A boxcar being shunted onto the end of a train was about to flatten the unsuspecting toddler when Edison jumped off the platform, grabbed the child, and threw them both clear of its path. McKenzie was indebted to the young lad, and to express his gratitude taught him basic telegraph communication. Edison took to this new technology quickly, and within little time became a competent telegrapher, capable of tapping out messages at an impressive rate.

After this hands-on training, Edison worked for a while as a 'tramp' telegraph operator, moving around the whole of the North American continent. He travelled from Memphis, to Montreal, to Louisville, to New Orleans. It was a nomadic existence, and offered little comfort. He would generally have to put up with rather uninviting living conditions, but it was a chance to see the country, and this appealed to his growing sense of adventure.

Edison finally decided to settle down, for a little while at least, in 1867. He took a telegrapher job in Boston, working for Western Union. The following year he patented his first invention, an electrical vote recorder, which he hoped to be able to sell to the government. He got little interest though, and subsequently dropped the idea.

He moved to New York in 1869, with hopes that the creative abilities beginning to ripen within him might find a suitable outlet. At first he stayed with his friend Franklin Pope, who had managed to acquire a well-paid position as chief engineer at the Laws Gold Reporting Company. Edison shared Pope's office for a short period, while he tried to find a job.

The firm's automated reporting system sent up-to-the-minute details of the current gold pricing from the New York Exchange to brokers across the city. By using telegraphic means it allowed all the 'bulls' and 'bears' to keep abreast of what was going on, without needing to employ runners. The system had been developed by the company's proprietor, Dr S. Laws, and had over 300 subscribers distributed across the financial district. As long as the service was maintained, customers would pay handsomely for it. Of course should there be a break in its smooth running, then you would have a lot of wheeler-dealers, not known for their patience, baying for blood. One morning Edison found himself caught in the crossfire of such an occurrence.

He happened to be mooching around the building, when pandemonium broke loose. The indicator machine, the centrepiece of the whole operation, had ground to a halt, cutting all lines of communication with the markets. Employees were running around panicking, and hoards of irate customers had turned up to complain about the inconvenience. While all about him lost their heads, Edison took charge of the situation. He inspected the dormant apparatus, and set about getting it back into action.

Within an hour normal service was resumed, and Laws' clientele quickly forgot about the interruption. Dr Laws was unsurprisingly impressed with the young man who had helped save his business; Edison was quickly hired, and put to work on improving the operation of the system. Then when Pope left, he took over as chief engineer, rewarded with a bounteous salary for one still so young.

In 1870, using the money he made from enhancements to the stock ticker machine, Edison was able to set up his own engineering works, giving him the opportunity to concentrate on inventing full-time. He opened a workshop in Newark, New Jersey. The dream he had harboured since his childhood, of having his own business, was finally coming true, but for Edison this was only the beginning.

On 9 April 1871, Nancy Edison died. She had been a huge influence on him, and for a spell he was inconsolable. Though he went on to accomplish many great things, he always acknowledged that her intervention had been the most important factor in achieving his success. He really felt it was his mother who had made him.

Young Edison showed little interest in the opposite sex, his time being swallowed up by his huge workload. However in late 1871 he became friendly with Mary Stillwell, an employee at his Newark store. There does not seem to have been any real romantic motives behind this association; as far as Edison was concerned, it was more what he thought was expected. On Christmas Day 1871, they married in a simple understated ceremony. The couple had three children in total; Marion or 'Dot' was born in February 1873, Thomas Jnr or 'Dash' arrived in January 1876, and William followed in October 1878.

In 1874, he worked on the challenge of sending multiple telegraph messages at once, developing the quadruplex system for Western Union, which made use of the lines far more efficiently. The system would prove highly useful to the telegraphy company, as it meant the amount of infrastructure that had to be put in place was minimised (less lines, less switches, less repeaters, etc.). It later called upon Edison's services to try to beat Alexander Graham Bell to the prize of domination of the telephone business, but this time he would not be victorious.

Using the money received from the quadruplex telegraph to fund its construction, Edison moved into a new laboratory at Menlo Park, New Jersey, in March 1876. He managed to hire some new engineering staff, and invested in cutting-edge equipment. It was to become a hothouse for creativity; from it would stem many of the innovations that defined the modern age.

Although he was not able to claim the invention of the telephone, he would have an important influence on the field. In early 1877, he worked on a transmitter that could improve the telecommunication network put in place by Bell Telephone. He also created a speaker diaphragm system that allowed better representation of the original sound. This became the standard used by nearly all telephones during that period.

By the end of the year his experiments on the recording and playing back of sounds finally bore fruit. In November he set out the design for the cylinder phonograph, and by early December had a working prototype. He managed to record the words 'Mary had a little lamb' onto a cylinder coated with tin foil. On 7 December, the device received its first public demonstration, at the offices of *Scientific American* magazine in New York. The patent application was granted on 19 February 1878 (US Patent 200,521). Edison Speaking Phonograph was founded soon after. In April he demonstrated his invention to President Rutherford Hayes.

Among Edison's closest friends was another son of Michigan, Henry Ford. Ford had worked for one of Edison's company's for a spell, before turning his talents for mechanical engineering to the nascent automotive industry. He was the first person to see the advantages of using moving production lines, with workers placed on specific tasks in order to raise output levels. By doing so he succeeded in dropping the time needed to complete a car from twelve hours to less than two, making his vehicles affordable to a whole new class of people.

In late Summer 1878, Edison's team began work on creating an electric light bulb. He previously had a brief stab at it in Autumn 1877. He had been looking at carbonised paper to use as emitting elements, as well as for utilisation in battery applications, but had pursued the matter no further due to other commitments. By the time he chose to resume his research, details of Swan's work had been published both in North America and Europe.

His notebooks show early experiments on electrical lighting taking place from August 1878 onward. First of all he tackled the problem of finding an emitting material that would work well without breaking. He looked at boron and silicon as possible emitters, but with little success.

In his draft caveat, produced in September 1878, he states that the 'object of this invention is to produce light for illumination purposes by metals heated to incandescence by the passage of electricity through them,'[12] noting that, 'heat arising from the passage of such current is utilised to regulate the temperature of the incandescent metal which serves to give the light so that it is never allowed to reach its melting point, no matter how strong the current.'[13] He noted that 'wires of platinum iridium, and other metals have been included in the electric circuit.'[14] In later drafts, produced in October, Edison avoided mentioning the material used to produce the 'incandescent foil' employed, suggesting he was still struggling to hit on the right one. By late 1878, he had gone through scores of possible materials; potassium, sodium, lepidolite, lithium, and many others.

Edison had made claims about being in a position to create a workable incandescent lamp, which were publicised in the *New York Sun*, back in mid-1878, but he had not managed to make enough time for the heavy workload that research in this area brought with it. Other projects meant he would not give it serious consideration for another ten months. The Edison Electric Light Company was established in November 1878, but at this point had no product.

As 1879 began, he doubled his efforts on electric lighting. It became his top priority, placing more of his team onto the laborious task of trying out new materials. During that spring he experimented with titanium, zirconium and barium, with some success. The main problem he faced here was not creating a filament strong enough to stay in one piece when large currents passed through it, it was the fact that if it was to be used in common households then it had to be cheap to produce. A lamp incorporating any of the rare metals mentioned above would be out of the price range of most of the general public. He started to realign his research to make use of more practical substances, but much time had already been frivolously wasted.

By late April 1879, several newspapers had lost faith in Edison's ability to live up to the claims he had made the previous year. The *New York Herald* was among those to express its doubts, stating, 'this impulsive man took up the electric light last fall as an entirely new subject of experiment, and allowed himself to believe that he saw a way to make the light useful which others had never thought of. Well informed electricians in New York City do not now believe that Edison is even on the right line of experiments.'[15] Even *Scientific American*, his die-hard supporter of many years, began to question his ability to deliver what he had promised.

In October 1879, he started to work with carbon. Another American, William Sawyer, had produced a carbonised paper filament lamp a year earlier, and it is almost certain that this, and what he learnt from the articles on Swan's work, led him down that path. Until this stage his progress had been sluggish, but with the information gathered from these other sources, and his own impressive gusto, from this point on there was to be no stopping him. By the middle of that month, Edison had a lamp that was capable of running for more than thirteen hours, using a high-resistance carbon filament. He put Menlo Park off limits to all outsiders during this period for fear that his research might be a target for espionage.

On 1 November, he applied for a patent, describing the use of carbonised material to produce electric illumination (US Patent 223,898). He made the first presentation of his lamp on 20 December, but, as we know, his British counterpart had already taken part in several public demonstrations of his invention prior to this.

Swan retorted to the growing publicity that Edison was drawing with a letter to *Nature* magazine in January 1880. Stearn had warned Swan on many occasions of the dangers of not taking out a patent on his invention, and, as will become apparent, this came back to haunt him in a number of different ways. His folly proved to be that he only patented the means he had used to create this light bulb (i.e. the evacuation while a current passed through the filament, and the carbonisation of cotton threads), not the actual design of the bulb itself.

Swan was not simply being negligent or slothful, he just felt that the number of demonstrations and mass of published work he had produced on the subject would prove he had done all the early running, if any dispute arose. Also it is likely that he thought Starr's patent covered this area of research too comprehensively for him, or any other, to legitimately take out further applications. His assessment of the situation proved incorrect.

After two months' wait, Edison received the US patent for his incandescent lamp, based on a carbon filament, in early January, and the British patent (4,576) was granted soon after. By then several hundred bulbs had been put into use both in the Menlo Park laboratory and the streets surrounding it. Before long, the legendary American inventor would manage to create a lamp with a more sustainable vacuum, which would run for over 1,500 hours. Edison had once again shown that if he could not compete with someone when it came to technical ability he would make sure that his business machinations were more than a match for them.

By October 1880 the Edison Lamp Works was turning out thousands of lamps per week. Within the first eighteen months of production, it shipped nearly 150,000 devices. To cover increasing demand, another factory was opened in Newark in late 1881. In September 1882, Edison opened North America's first commercial electric power station for incandescent lighting, on Pearl Street, at the southern tip of Manhattan Island.

Light bulbs from both Edison and Swan were used to illuminate the Paris Exhibition at the Champs Elysees in 1881 and the Great Exhibition at Crystal Palace in 1882. Later that year, another lamp factory, to complement Swan's one at Benwell, and take care of the growing export market, was set up in Lille.

The change from gas to electric lighting was far more languid in Britain than in North America. In cities like London infrastructure was already in place, the Corporation had too much invested in it to implement widespread replacement. Across the Atlantic it was a different story, the country was rapidly expanding, and new development was needed. It was like having a blank canvas, and meant things could be done the right way from scratch. The 'Electric Lighting Act' of 1882 permitted the digging up roads to lay cables, thus allowing Britain to catch up with the Americans. This was also when Siemens began licensing Swan's lamp design, in order to start manufacturing bulbs in Germany. Swan sold his US patents to the Brush Company, of Cleveland, which had established a lucrative business supplying arc lamps for the lighting of public places. The American firm saw the potential of Swan's device, and offered him coast-to-coast distribution that he would not have access to if he tried to tackle such a huge market on his own. The plan had been that the lamps would be sold in conjunction with Brush's proprietary secondary battery. Unfortunately the battery did not live up to expectations, and so the whole venture failed. This meant Swan was unable to put up any sort of fight in North America, and left Edison, for some time at least, with a clear run there.

Edison's British patent had been in place for some time, and with the Swan Company turning out lamps from its Benwell factory, it was not long before it was taken to court for infringement. However, this move was to backfire on Edison. From the start Swan did not seem too worried about what the outcome would be. He was confident that his high-profile demonstrations showed he had been ahead of the game. In a letter to his wife on 22 June 1882, he stated that he had 'made an affidavit denying the contention of the Edison Company and it is expected that today the question will be decided whether they obtain an injunction to restrain our company from making lamps. I do not think there is the slightest doubt the application for an injunction will be refused; our opponents will, therefore, gain no advantage of any kind.'[16]

Edison eventually lost the legal battle, the British courts finding in favour of Swan. The two firms had effectively created a stalemate, blocking each other from exploiting the huge potential market opening up before them. The best course of action now appeared to be amalgamation. Swan's and Edison's companies merged in October 1883, to form Edison & Swan United. The company's lamps used the screw fitting that Edison had introduced, and were capable of producing hundreds of candelas of light. The American, not known for his generosity, was bitterly unhappy at even having to share the company name with Swan, resenting the idea that anybody had equal claim to this innovation. Though they were business partners, Swan and Edison never actually met.

Before the end of that year, the Swan family moved south to Bromley, in Kent, making it easier for him to oversee the day-to-day running of the main Edison & Swan factory in Ponders End, North London. Things were not to remain simple for long though. The problem of the light bulb patents would flair up once more in 1885. While Edison & Swan wanted to enforce its monopoly and stamp out the growing number of manufacturers contravening its patents, Edison's priority began to be questioned. Since Swan had made several public demonstrations of his lamp before Edison's patent had been filed, it could be argued that the document was invalid. Swan would be regarded as what is termed as a 'prior user'. This offered a strong defence to anybody who breached the patent.

This whole mess could have been sorted out from day one if Swan had taken out the patent. Instead he had just part ownership in the company producing the lamps, only getting a cut of the money which really should have been all his. To make matters worse it now became apparent that he had to give up the glory as well. The reason for this is as follows, so pay attention! If the company was to retain Edison's patent, then it had to prove that Swan's work pre-dating it did not interfere. The best way for them to achieve this was to argue that the carbonated conductor Swan used in his lamps was not a filament, like the one which formed the basis of Edison's system. The point was highly contentious, as in fact there was no actual difference in the illuminating element that either party used, but somehow the judiciary bought it. The patent was upheld, and the company's stranglehold on the market was assured. Nevertheless it came with a high price on Swan's part. He had been forced to cheapen his achievements to save the business, and the magnitude of his contribution to one of the most important innovations in modem times had been greatly diminished. Eventually Edison bought him out; the marriage of convenience was quickly annulled.

The market was now in the possession of one man, and one man alone. Edison had almost complete autonomy here. Though Swan had really been the one who made the all-important breakthrough, there was no longer a place for him. Carbon emitters, like those he had used, would eventually be replaced. William Coolidge developed the tungsten filament, which was far more mechanically stable.

In May 1884, Edison was elected the Vice Chairman of the American Institute of Electrical Engineers. His wife Mary died that August, while still only twenty-nine years of age. The reason for her demise is uncertain. It is possible that it was a brain tumour that actually carried her off, however neglect on her husband's part may have been a factor.

Edison wasted little time in remarrying, he was once again in wedlock by February 1886. His second wife was Mina Miller. She was the eldest daughter of the inventor Lewis Miller, who Edison had first met at an industry fair in Louisiana. Lewis had created several machines for use in agriculture, and it was through their friendship that his daughter got to know him. The couple were wed in her hometown of Akron, Ohio. Soon after their nuptials the couple moved in to Glenmont, a large country estate that Edison bought in West Orange, New Jersey. He would have another three children by Mina; Madeleine was born in May 1888, Charles arrived in August 1990, and Theodore came into the world in July 1908.

He soon built a laboratory close to his new home. The West Orange site would supplant Menlo Park as the nerve centre of the 'Edison Empire'. In April 1889, Edison General Electric was formed to build DC power stations and meet the increasing need for electrical energy that was coming from the American people.

Although Edison had made huge advances in the realm of sound and vision, he was not to remain its standard bearer for long. The National Phonograph Company was set up in January 1896, but Edison's preference to stick with the cylindrical format for sound reproduction effectively killed it off in the early 1900s. The cylinders had serious limitations as to how long recordings could be (basically just a couple of minutes), as well as being heavy to carry and inconvenient to store. Other companies looked at methods of recording on to flat discs, which dispensed with such issues. He was to have a similar experience when dealing with the world of film. He demonstrated the 'Kinetoscope', the first motion picture camera, in May 1892. It was something he had been working on for five years. Copyrighted motion pictures went into production within eighteen months. Edison unwisely tried to control the new media himself, and it proved to be a step too far for his abilities. He was an inventor not an artist. What did he know about entertainment? He was an uncultured hick from the Mid-West!

General Electric was incorporated in Spring 1892 following the merger of Edison's firm with rival Thomas Houston, instigated by financier John Pierpoint Morgan. Edison was bought out for the sum of $2 million.

He ploughed the money from the buyout into what proved to be an ill-fated mining enterprise. His idea had been to use electromagnetic technology to improve ore sifting, and create a cheaper and more efficient method of obtaining iron deposits. Huge quantities of rock were dug out of the ground, then passed through rollers and crunched up. The pieces were broken into finer and finer bits, and then taken along conveyor belts through enormous racks of electromagnets. The basic principle being that all the ore would be attracted to the magnets, and hence separated from the other junk. It worked, but the equipment needed was very expensive, and gave a relatively poor yield. In the end it proved much more costly than mining existing seams. The project lost a small fortune.

The depression, mixed with the fallout caused by his badly judged iron extraction venture, forced Edison to make drastic job cuts at the West Orange laboratory. It was a hard time for him, and this was compounded by a death in the family. His father, Samuel, died in February 1896, but it is unlikely that Edison was too sad to see him go. He had not shared his son's industrious nature. After the death of Thomas' mother, Samuel had become steadily more outrageous in his behaviour, gaining a reputation as a drunkard and gambler, and marrying a girl only a third of his age. Edison had been forced to bale him out several times, when his father built up huge debts from over-extravagant living.

In the earlier 1900s, Edison worked on improving the storage battery for electric cars, delivery vehicles and trains. Though the world tends to associate him with more high-profile conceptions, like the light bulb or movie camera, this actually became his most profitable creation.

Meanwhile Swan withdrew from corporate life. In late 1894, he moved to the nation's capital, taking a house in Holland Park. London allowed him greater access to the many scientific organisations he had become involved with. In 1898, he became President of the Institute of Electrical Engineers, and in 1900 took the presidency at the Society of the Chemical Industry. In 1903, he was made President of the Pharmaceutical Society, and the following year took on the same role at the Faraday Society. He also served on the board of the National Physical Society for several years.

Swan relocated once more in late 1908, this time to the quieter setting of Warlingham in Surrey. By this stage he was becoming somewhat frail, so he decided to detach himself from the social whirl, retiring fully from scientific research. In February 1914, the Corporation of Newcastle passed a resolution to bestow upon Swan the freedom of the city.

Swan died in the early hours of 27 May 1914, at his home in Warlingham. The cause of death was judged to be heart failure. He had suffered with coronary problems for most of the previous decade. The *Times'* obituary described Swan as 'a leading authority on electro-chemistry and did a great deal of original work in regard to the practical application of its principles.'[17] His son, Kenneth, said of the man, 'Those who sought his advice never asked in vain, and those who came to consult him never failed to receive the best that he could give,'[18] though, 'beneath his gentleness and courtesy there flowed a tranquil undercurrent of dignity and power.'[19]

Though Swan had been given the praise and gratitude of his nation, it paled into insignificance when compared to how Edison was judged. The respect that he eventually drew reached far beyond his country of birth, and enveloped the globe in its entirety. In 1911, his network of companies was rearranged into one single commercial organism, called Thomas A. Edison Inc. He had learnt his lesson; dealings with the likes of Morgan had left a bitter taste in his mouth. With all undertakings that followed he did not look for help from any of the Wall Street bloodsuckers, who had double-crossed him in the past. To some extent this would stifle the growth of his new ventures, but it suited him better, as it ensured his independence.

On the evening of 9 December 1914, the film processing department at West Orange burst into flame. Rapidly the blaze spread to the other buildings in the complex. It was gone midnight before the fire was finally extinguished, and by then seventy-five percent of the site had been destroyed. A huge proportion of his life's work had perished in the flames, but with characteristic tenacity Edison shrugged the incident off, as if it were a minor trifle, stating, 'although I am over sixty-seven years old, I am starting over again tomorrow'. To everyone's amazement the plant was restored to full operation within six months.

In what must be considered the most truly exceptional act of stubbornness of the entire twentieth century, Edison turned down the Nobel Prize for Physics after learning he would share the honour with Tesla, deciding he would rather go without, than split it with his archrival. He also knew that the Serb was struggling to find investors to back his research work. The opportunity to deprive his old adversary of valuable funds would have been too great a pleasure for him to miss out on.

During the First World War Edison was appointed Head of the Naval Consulting Board, working in particular on systems for detecting enemy submarines. His plant in West Orange became an important supplier of military hardware during hostilities, and its staffing levels would come close to 10,000 during this period. He managed to band together the country's engineering talent to create many new innovations that could aid the navy. His public spiritedness did not go unrewarded; he was given the Distinguished Service Medal for his contribution to the war effort.

Though he had earned universal respect, he did so by his deeds, not through his image. Edison's appearance was shabby to say the least. He was rarely clean-shaven, and did not bathe that often.

At times he did not go home from days on end, and would work for long-drawn-out periods without any respite. Then when he finally had to get some shut-eye, he would just find a quiet part of the workshop, lie down and take a nap. One of his Menlo Park colleagues once stated, 'he could go to sleep anywhere, anytime, on anything'.

In 1922, the *New York Times* conducted a poll of its readers to find the twelve greatest living Americans. Edison was placed at the top of the list, above industrialist J.D. Rockerfeller, automobile millionaire Henry Ford, war hero General Pershing, and former president, Woodrow Wilson.

The thought of relaxing a little, as he moved towards his dotage, does not seem to have crossed his mind. On one occasion Edison exclaimed, 'I am long on ideas and short on time.'[20] He was more driven than perhaps any other man of his era. The need to achieve was as vital to him as the air he inhaled.

He finally abdicated his throne in 1926, handing over the running of Thomas A. Edison Inc. to his beloved son Charles. Even then, he did not retire from his duties completely. He would still make public appearances, holding the position of elder statesman in the realm of science and technology. In May 1928, the US Government awarded Edison with a Congressional Gold Medal.

That October, to celebrate the fiftieth anniversary of his electric lamp, a huge gala was held. The event was attended by such eminent people as France's Marie Curie, American aviator Orville Wright, Ford (of course), and President Herbert Hoover. In conjunction with the celebration, the US Post Office issued a postage stamp commemorating his great accomplishment.

Edison took increasing advantage of his home in Fort Myers, Florida, as he grew older. He tried to distribute his days more evenly between there and New Jersey. He had lost his hearing almost completely by the time he reached eighty, but still continued to work as much as he could. In addition to allowing him a higher degree of concentration, Edison believed his deafness had been an advantage in business too. As he put it, 'the fact that I do not rely on verbal agreements or reports is one reason for this. There would be a chance that I might not hear them perfectly. So I have everything set down in black and white.'[21]

For the last three years of his life he sustained himself almost solely on a diet of milk, to try to combat acute stomach pains. Without warning, he collapsed on 1 August 1931. He slowly regained his strength over the following months, before experiencing a relapse. On 14 October he slipped into a coma.

Edison's light finally went out permanently four days later. He died at his home in Glenmont, his wife by his side. He was eighty-four years old at the time. Unsurprisingly, his death was headline news across the United States. The *New York Times* reported that, 'Thomas Edison, who died at 3.24am yesterday, will be buried on Wednesday, the fifty-second anniversary of his perfecting of the incandescent light, which has been termed by many as his greatest gift to mankind. Today and tomorrow his body will lie in state in the library of his laboratory.'[22] The newspaper estimated that Edison's inventions had been responsible for investment into American industry to the tune of $15 billion. He had almost 1,100 patents to his name at the time of his death.

The funeral was attended by the most important politicians, scientists and industrialists, representing all the areas of society touched by his ingenuity during his long and varied life. Letters of condolence flooded in by the thousand, including ones from heads of state of France and Britain, as well as from Pope Pius XI. Guglielmo Marconi paid homage to his fallen hero, stating, 'he was a constant inspiration to me.'[23] His close friend Henry Ford described him as a 'truly great man. He changed the face of the world in his lifetime, and everything he achieved was beneficial to mankind.'[24] Albert Einstein said that he was 'one of the greatest technical inventors, to whom we owe the possibility of alleviation and embellishment of our outward life.'[25] Still governor of New York at the time, but future president, Franklin D. Roosevelt had known Edison well as they had worked together at the Naval Consulting Board. He proclaimed that Edison 'was not merely a great inventor, but a great citizen, who was constantly thinking of the good of the country.'[26] Even his longstanding foe, Nikola Tesla, managed to find a few nice words to spare for him, telling journalists, 'Edison was by far the most successful and probably the last exponent of the pure empirical method of investigation.'[27] He continued, 'everything he achieved was the result of persistent trials and experiments. His mind was dominated by but one idea, to leave no stone unturned, to exhaust every possibility.'[28]

On the evening of 21 October, all but essential electric lighting was switched off for one minute across the whole of the United States, as a mark of respect for the man who had brought his nation out of the darkness. In 1954, the town of Raritan, the site of the Menlo Park laboratory, was renamed Edison, after its former notorious inhabitant.

For all his faults it is still hard to not have some admiration for Thomas Edison. He may have not been a kind, or particular likeable man; not dissimilar to Newton he had an acidic, vengeful aspect to his personality, but he certainly had many venerable qualities also. He was determined, hard working, and positive thinking, with an almost unmatched will to persevere. He combined incredibly technical and creative talents, with superlative business acumen. Following an interview with the *Pall Mall Gazette*, reporter Robert Sherard said, 'Edison pronounces the words "work" and "working" as some do "prayer" or "religion".'[29]

Edison once commented that he invented to obtain money purely to help him to go on inventing. There may be some truth in this. He built up a sizeable fortune during his career, but would still not go easy on himself, he continued to feed his creative urge right until his death. Perhaps, a little like Tesla, he liked the thrill of innovation more than the actual rewards it brought. During his time at Menlo Park he stated, 'I don't regard myself as a pure scientist, as so many persons have insisted that I am. I do not search from laws of nature,' but his studies were 'conducted entirely with the object of inventing that which will have commercial utility.'[30]

Edison had a very methodical style, which sometimes meant he had to go the long way round to get to the desired objective. Though he could never be accused of not working hard, it is possible to say that he didn't always work that smart. 'In trying to perfect a thing, I sometimes run up against a granite-like wall a hundred feet high. If after trying and trying again, I can't get over it, I turn to something else,' he once wrote, 'then, someday, it may be months or it may be years later, something is discovered either by myself or someone else, or something happens in some part of the world, which I recognise may help me to scale at least part of that wall. I never allow myself to become discouraged.'[31]

His limitation in understanding scientific principles in comparison to some of his contemporaries is often brought up as a possible shortfall in his makeup, but in reality this did not need to be his core competence. As he pointed out, 'I can always hire mathematicians, but they can't hire me.'[32] Tesla had once said of him that if he had to find a needle in a haystack he would go through every single piece of hay one by one, meticulously examining it to make sure it was not the needle, whereas Tesla would have just used a magnet to quickly find it. This was the big difference between him and many of his scientific peers. He was not strong on theory, but got there by slogging it out.

Edison said that of all his inventions the incandescent lamp 'was the hardest one of all; it took many years, not only of concentrated thought, but also of worldwide research.'[33] But what he was not so quick to mention was how much he borrowed from others. It appears that Swan did not patent his work regarding use of a carbon emitter as he just assumed it could not be patented following what Starr had already done. Starr's work did specify a carbon-based conductor and an evacuated tube, as his did, so it is possible he felt that it was not enough of an improvement to the design to warrant the award of another patent. Edison was unfortunately destined to prove him wrong, taking out British and American patents for devices that, just like Swan's, relied on carbonised filaments to produce light. Swan had basically shown too much respect for the work of Starr, and others before him, when he should have just tried to capitalise on what he had personally achieved. Starr was long gone by then, nobody was going to kick up a fuss over him, so Swan should have at least attempted to get a patent for his work.

The fact of the matter is he simply did not have the killer instinct that was so primal in Edison.

Ruthlessness was not coursing through his veins, like it was in his American counterpart. He was an archetypal English gentleman, who believed in fair play, and was perhaps too kind natured as a result to take on such a formidable personality as that of Thomas Edison. He would have rather be a gallant loser than a cheat.

At the Paris Exhibition, Edison received the diploma of honour, the highest accolade awarded to any inventor at the event (Swan being placed in the rank below this). Swan showed his sportsmanship qualities by sending Edison a telegraph to congratulate him. This confirms that Swan was a well-mannered Victorian male sure, but I cannot imagine Edison would have done the same had the roles been reversed. Swan was a little too wholesome to succeed, perhaps if he had been more of a cad he would have done better for himself. He did not know how to play dirty, and this was where Edison was in a league of his own.

Swan once said, 'there are no inventions without a pedigree', meaning that nothing is discovered or developed in complete isolation, and inventors always draw on the work of others to form the foundation of their own findings. Ironically he is someone deprived of much of the credit he was due for his inventiveness.

It must not be forgotten that Swan was doing all his research as a sideline to his full-time job as a chemist. He was just a hobbyist with little time or resources to devote to it. Edison had far more men and equipment at his beckon call. Time and again in this book we will see how Britain lost out to the USA due not to any shortage of ingenuity, but simply through lack of investment. Swan, along with other characters we will look at in the following chapters such as Alan Turing, Charles Babbage and Geoffrey Dummer, all had the right ideas, but could not find the money needed to carry things through to an adequate conclusion.

Incandescent light was not the only time Edison would palm somebody else's invention off as his own. George Bartlett Prescott developed the quad-ruplex telegraph, but it was his boss who took out the patent. It is also likely Edison got the idea of the kinetoscope from English-born photographer Eadweard Muybridge. He made no apologies for his somewhat underhand style, openly admitting 'everybody steals in commerce and industry' and that he had 'stolen a lot' himself. The writer Julian Hawthorne once said of him, had he quit inventing and gone into fiction he would have been a great novelist.

In Edison's defence, it must also be remembered that although Swan had succeeded in creating a lamp, Edison had designed a complete lighting system. This proved very important in the application of these devices. This difference was particularly telling in the US market, where Swan was let down by the company he partnered with, allowing Edison, who was not reliant on anybody else, to effectively clean up.

The *Times* rather magnanimously stated, upon the announcement of Swan's death, that 'various questions of priority both legal and scientific cluster around the invention of the electric glow lamp, but these need not be discussed here. Swan and Edison to mention no others, were working simultaneously and on parallel lines at the problem of producing a satisfac-tory commercial lamp; and if Englishmen like to think that success first came to Swan, it is not unnatural that Americans should do the same for Edison. But it may perhaps be said that Swan was the first to adequately appreciate the conditions necessary for success, whether or not the credit was his of being the first to discover how to realise those conditions in practice and to put the process on record at the patent office.'[34]

By the start of 2003, annual light bulb sales, in the USA alone, exceeded one and a half billion units.[35] These devices have been sold in legion to every corner of the world for close to 120 years. With regard to something so fundamental to human civilisation, it appears rather unreasonable that one person should be accepted as its sole creator. It seems fairer to conclude that Swan and Edison were both highly influential in its development, and are equally deserving of our gratitude. With such a huge prize, there must be enough acclaim to split between them.

CHAPTER 5

A Design for Life

He was bundled out into the courtyard, pushed towards the edifice placed at its centre. Either side of him a jeering crowd hailed abuse, savage-like in its scorn, the feeling of hostility all too apparent. Aggressively now, he was forced to climb the steps onto the gallows, a rope slipped over his head and tightened unceremoniously around his neck. His heart was pounding furiously against his ribs, sweat cascading across his brow. Barely able to keep his mind from racing, he tried to focus his thoughts toward the foolishness that led him here. The time to atone for his sins was long past, all hope was lost. The figure beside him, draped in rough black robes with a heavy wooden crucifix hanging across his chest, took up a scroll, and holding it before him, started to speak. 'Charles Robert Darwin,' he exclaimed, 'you have been found guilty of heresy, and as this is a capital crime you have been sentenced to be hung by the neck until dead.' The crowd were in a frenzy, the noise almost unbearable. Upon the foreman's mark, the trapdoor beneath him gave way, sending him plunging down towards his doom...

He woke on the morning of 22 September 1836 after a troubled night's sleep. Charles Darwin was a man with a lot to think about. Within the recesses of his mind lay an abominable secret that he knew could change the world forever, and rock Christendom to its very foundations. Somehow, he had to find the strength to share it with the populace, but knew that damnation was all he was likely to receive in return.

Okay, so you are thinking, Meucci, Tesla, maybe there is something in those cases, but Darwin and evolution? There is no dispute there surely? What needs to be realised, and is often overlooked, is the fact that Darwin was not the first to present a theory that living things evolved, there had been several others before him. In fact it would have to be said that this particular discovery was more of a team effort than any of those previously discussed in this collection, and as a result this chapter warrants the bestowing of not one, but two 'Nearly Men', namely France's Jean-Baptiste Lamarck and Britain's Alfred Russel Wallace.

> *'Knowledge that is not the product of observation or of the results from observation is altogether without foundation.'*[1]
>
> *Jean-Baptiste Lamarck*

Jean-Baptiste Pierre Antoine De Monet Chevalier de Lamarck was born on 1 August 1744 in the village of Bazentin-Le-Petit, Picardy (which now lies within the borders of the Somme arrondissment), Northern France. Though his family was of noble origin, it was not particularly wealthy. Lamarck left very little in the way of papers, manuscripts or diaries to make the investigation of his story easier, so some finer details of his private life have been lost completely. His father, Jacques Philippe the lord of Bazentin, and his mother, Marie Francoise de Fontaine, had married in 1721. Jean-Baptiste was the last of their eleven children.

He came from a long line of military men. His grandfather, Philippe de Monet, had been a Knight of the Order of St Louis, and his eldest brother was killed in action at the Siege of Bergen-op-Zoom in 1747. However, Jean-Baptiste was not considered physically robust enough to join the armed forces. Instead he was initially coaxed into ecclesiastical training. It was quite customary in the past for the nobility to send their less able children into the employ of the Church. He entered the Jesuit seminary in Amiens during his teens (dates are sketchy, but he is thought to have gone there in 1855 and stayed for around five years), though he was not enthusiastic about his theological pursuits, soon realising this was not what he was destined to spend the rest of his life doing.

After the death of his father in Spring 1760, the young Jean-Baptiste was free from the parental interference that had compelled him to join the priesthood. He decided to rethink his career options. Despite the supposed shortcomings of his slight build and somewhat delicate physique, he was far from lacking in spirit, and keen to live up to the reputation of his courageous lineage. He decided to cast aside a life of worship in favour of a more adventurous one as a soldier.

At the time, the Seven Years War was still raging; Austria, Russia and France all trying to take advantage of the ensuing chaos to seize German territory. Lamarck headed for the French vanguard in July 1761, joining the Grenadiers, serving under Marshal Broglie.

He seemed to find his calling in the ranks. The French anatomist, Georges Cuvier, widely regarded as the 'Father of Palaeontology' and one of Lamarck's most illustrious peers, wrote a little on the man's military exploits. He had this to say about Lamarck's first experience of combat: 'At the break of day Monsieur de Lastic (the colonel of the regiment) rode along the front of his corps, and the first man that met his gaze was the new recruit, who, without saying anything to him, had placed himself in the front rank of the company.'[2]

During the battle at Fissinghausen, which took place only a couple of days after he signed up, Lamarck gained his first opportunity to show his valorous nature. When his company was virtually wiped out by a barrage of enemy fire, and his commanding officer killed, the precocious lad took charge of the surviving soldiers. According to Cuvier, 'there remained only fourteen men, when the oldest grenadier, seeing that there were no more of the French troops in sight, proposed to the young volunteer, become so promptly commander, to withdraw his troops. "But we are assigned to this post," said the boy, "and we should not withdraw from it until we are relieved".'[3]

The incredible composure he had shown under fire and his swiftness to assume command while more experienced men panicked did not go unnoticed. He was recognised for his bravery, being promoted to an officer's rank soon after. He had risen to lieutenant commander by 1765, and once the conflict was over he took charge of a garrison stationed at Toulon. He was later transferred to the fort at Monaco, where he remained for around two years. Sadly, he had to be decommissioned from military service in 1767, following injury.

After being invalided out of the army he was forced to find whatever work he could in order to supplement his meagre military pension, which came to only 400F a year. In 1768, he went to Paris, and for a while worked as a bank clerk. Then in 1769, he enrolled at the 'Ecole de Medecine' and later began studying natural sciences at 'Le Jardin du Roi' (the Royal Botanical Gardens). He took courses conducted by Bernard Jussieu, one of France's leading botanists, gaining a great deal of interest in this field. Lamarck clearly took to scientific disciplines with far greater zeal than his undertakings in the service of the Church, and seemingly better fortune than his time in service of the king. He published his first book *Florae Francais*, a three-volume collection characterising plants found within his native France, in 1778. It was a huge success, and won him much attention in academic circles. It was brought to print with the aid of George-Louis Leclerc, Le Compte de Buffon, who was in charge of Le Jardin du Roi at this time. He managed to get the organisation to fund the production of Lamarck's work, and became one his most staunch allies, something (as we will soon discover) he could have done with a lot more of. The notoriety the book derived earned him admission to the Academie Scientifique in 1779.

Between 1781 and 1782, Lamarck travelled across Europe, under the title of 'Botanist to the King'. The role may have sounded grandiose, but did not furnish him with much kudos. It had been given to him by Buffon, in order that Lamarck could escort his son on a tour of the Continent. Effectively he was just acting as a babysitter. It did however offer him some opportunity to study the biological specimens not found in his homeland, something that was impossible for many of his contemporaries to do. Following his return to France, Lamarck was appointed 'Keeper of the Herbarium' at Le Jardin du Roi. It is believed that he was poorly paid, and continued to endure the conditions of abject poverty he had been subject to ever since first arriving in Paris.

In July 1789, the Bastille was stormed, and the French Revolution initiated. So began years of uncertainty and fear. Following the execution of Louis XVI in 1793, the revolutionary government made sweeping changes to the organisation of 'Le Jardin des Plantes', as it was referred to following the establishment of the Republic. Lamarck was given the post of Professor of Insects & Vermes (or worms) at the Musee Nationale d'Histoire Naturelle that neighboured Le Jardin. It was not the most sought-after professorship in this academic establishment, carrying little regard amongst the scientific elite, nor was it seen as a particularly interesting field of investigation. Perhaps most worrying of all it was something Lamarck knew precious little about. He had only modest experience of creatures that moved about or ate one another. All his prior work had been in botany, and, on the whole, zoology was a complete mystery to him. Nevertheless it was still a considerable improvement on his previous position, so he decided to make the most of the opportunity.

The young novice knuckled down to the task of learning as much as he could about this completely new area of research, quickly managing to absorb a considerable amount. Given his previous ignorance of the subject, it is surprising how rapidly he got to grips with it. It was not long before he started making constructive actions in this sphere, and soon he began to look at ways in which its study could be better performed.

He was the first to recognise that the way this particular part of the animal kingdom had been designated made it effectively a mish-mash. It had been where the scientists that had gone before him had basically just thrown all the phyla (groups of species) that they could not fit anywhere else. It contained types of animals that were not directly related to one another. The classification of species had only really commenced in earnest a couple of decades beforehand, much of the work being done by Swedish botanist Carolus Linnaeus. However, Lamarck could see it had been a botched job, he became 'convinced that worms form an isolated group, including animals very different from those which make up the radiates and polyps, and that the arachnids could no longer be a part of the class of insects.'[4] This new broom began to sweep away the years of neglect endured by this particular branch of zoological studies. He decided to put some sort of order in place, and consequently divided those creatures contained within it into four distinct groups Annelids (worms), Arachnids (spiders), Crustaceans (shellfish), and Insects.

Lamarck was the first to describe these organisms using the word 'invertebrates', something that is now a common scientific term. Despite what would have normally been expected, the chance to study these less fashionable species proved to be a blessing. It exposed him to nature in its simplest and most clear-cut form, thus helping him to piece together a new perspective on the dynamics dictating how organisms lived. He commented on the fact that 'the most important discoveries of the laws, methods, and progress of nature have nearly always sprung from the examination of the smallest objects which she contains.'[5]

He managed to bring new gravitas to this particular field of biological science, greatly improving his rank amongst the scientific community in the process. He was made Secretary of the Board of Professors at the Musee in 1794, and elected its treasurer in 1802. It was later that year that he created the term 'Biology', to describe the organisation of all living bodies. Until then, there had been no singular expression that bought both animal and plant kingdoms together.

Lamarck had already shown he was not afraid to question things, he was clearly willing to voice his opinions and deal with the consequences. Next he started to raise doubts about other aspects of the study of nature, and did not fail to generate controversy in the process.

He could see that the variation between different living things in any particular grouping was effectively a continuum, and any method used to compartmentalise them was purely artificial. There would still remain a myriad of different characteristic variations within the individual groupings. No matter how far down you went from genus, to class, to species, to variety, you could always go further. This suggested to him that the idea of each distinct species being created in the past and staying exactly the same was simply impractical. It made far more sense that species were being formed by a constant divergence from common ancestors. As he put it, 'the assumption almost universally admitted that living things make up externally distinct species on account of their invariable characteristics, and that the existence of these species is as ancient as nature itself, was established when people had not observed nature sufficiently and the natural sciences were not developed. The assumption is contradicted every day in the eyes of those who have followed nature.'[6] He could see that the compilation of different species was a structure that we had created for our own benefit, not because it reflected what existed in nature. Much of what we accepted was based on the preconceptions made in our more primitive past. The concept of the 'Creation' had been formulated several thousands years prior, and back then there had been no comprehensive examination of the finer details of different species. To them a horse was a horse, a wolf was a wolf, a snake was a snake. Lamarck could see that species 'arranged in series and set in order according to their natural affinities, exhibit such slight differences to those next to them. These species merge more or less into one another, so that there is no means of stating the small differences that distinguish them.'[7]

The crux of his argument was this: every individual closely resembled those it came from, but still had differences, and these differences meant, almost by definition, that the species could not stay unchanged. The scientific world had upheld 'the assumption that the individuals who make up a species never vary in their specific characteristics and that therefore the species has an absolute constancy in nature,' but Lamarck said that, 'it is precisely this assumption that I propose to contest.'[8] As he saw it, the principles on which we based our understanding of the natural world had been written by scribes and holy men, not biologists, a number of millennia ago. They were in complete contradiction with what he and other scientists actually observed.

He pointed out that naturalists found it increasingly difficult to decide what should be regarded as a particular species, and what should not be. They came to arbitrary decisions about what level of variation constituted a species, and what should be thought of as a variety. In his opinion, 'the determination of species therefore becomes increasingly defective.'[9] The concept of each species being created at the dawn of time, staying constantly the same in its form and behaviour, seemed just too simple to Lamarck, and not in anyway corroborated by what was seen everyday. Assigning a name to a specific class, order, or genus was quite simply a fudge. The whole process was a purely discretionary one; who was to say if one particular difference in characteristics warranted a new species or not? As he saw it, 'what we call species have been created in this way imperceptibly and successively among them; they have a constant which is only relative to their condition and cannot be as old as nature.'[10]

He believed that in the cases where we only knew of an isolated species, it was purely due to the fact that we had still to gain knowledge of their relatives, and that these 'phenomena occur only because we are missing other closely related species which have not yet been collected.'[11] Lamarck thought living things made up a 'series with irregular gradations, something which has no discontinuity.'[12] If, as he postulated, there was a continuum of different species, then it implied that groups of these species had shared ancestors.

By this stage in history, remains of creatures were being discovered which were clearly of species no longer found alive anywhere in the world. Lamarck asked what had happened to them. If all species remained constant, why was it some had died out and not others? Cuvier, Lamarck's colleague at the Musee, an expert on the then budding scientific study of fossils, believed that species were completely fixed, and plants and animals had maintained the physical form they had been given on the day God created them. As a result he preferred to explain the fossilised records of extinct species as the consequence of catastrophic events (floods, earthquakes, etc.). His markedly more conservative views were of greater appeal to the vast majority of naturalists, and gained wider acceptance than those of Lamarck.

In his book *Philosophie Zoologique*, published in 1809, Lamarck described his theories on how living things had evolved over vast periods of time, rather than simply being created. He postulated two laws:

1. A frequently used organ gradually strengthens and enlarges with time, and by contrast lack of use of an organ will result in its deterioration, and eventual disappearance.

2. The attributes acquired by one generation are then passed onto the following generation, and likewise this generation adds further to the transmutation process.

So let's take an easy to visualise example such as a giraffe. Under Lamarck's laws, over the course of its adult life (i.e. after it had reached maturity. It was completely separate to its growth, while still in its youth), an individual giraffe would gradually develop a longer neck, making it more suited to its place of habitation, as this would give it a greater bounty of food, since it would be able to reach leaves on higher trees. It would effectively become better optimised for living in its environment. Then, in compliance with the second law, the attributes that one generation of individuals gained would be given to their offspring, and this new generation would carry on the development of these attributes even further (i.e. they would develop even longer necks during their lifetimes, and this would be likewise inherited by their progeny).

Individuals who lived long enough to reproduce would have experienced changes within their physical form, which helped them to survive, and they would hand these changes on. This would continue through each generation and with time distinct species would form. He saw it as an involuntary transmutation taking place in all living things, the changes happening to the individuals themselves.

He pictured the evolutionary process as some sort of ladder, with generations of species evolving toward a perfect being. He assumed that humans represented the top rung of this ladder (or at least, that we were further up than any other species). His theory was based on the principle that each generation slowly moved closer and closer to this ideal state. But if all species were over time moving up the evolutionary ladder, then what about those at the bottom? If the simple species placed at the foot of this ladder were climbing up, then why were there still lower life forms like these around today? How come they had not disappeared? Lamarck postulated there had to be some sort of continuous creation of the lowest forms of life, to back-fill the gaps left by those who had moved on.

Unsurprisingly, the book made a considerable stir, it stood against all established theory. Those who took it seriously made him out to be a revolutionary lunatic, and everyone else just implied he was incompetent. This seemed a little bit harsh, to say the least. He was just attempting to discover the truth, nothing more. In this work he did not attempt to cast dispersions on God's existence, and stated that, in his view, nothing had been brought into being except 'by the will of the Sublime Author of all things, but can we set rules for Him in the execution of His will and fix the routine for him to observe? Could not his infinite power create an order of things which gave existence successively to all that we see as well as to all that exists but that we do not see.'[13] His religious past does not appear to have been affected by his observations. He did not try to use his theories on evolution to disprove religious ideology. He simply suggested that this 'Sublime Author' was responsible for the creation process, and had formed the course of action all living things were expected to follow. Regardless of this, some thought what he proposed bordered on heresy.

Cuvier's work was far better received by the learned men of the time, as he had told them what they wanted to hear. They eventually gave him the chair of the College of France. In the meantime Lamarck's theory came under constant bombardment, and Cuvier would be one of his strongest critics. The ridicule continued for many years, his radical ideas considered preposterous.

Lamarck, formerly the 'golden boy' of research in natural science, started to gain a reputation for being an old quack, and other areas of scientific investigation in which he dabbled were only to help reaffirm that judgment. At the start of the nineteenth century he began publishing a series of meteorological annuals, with which he attempted to predict weather cycles for the following year. Lamarck's forecasts did not prove remotely accurate, and his theories on the subject were mercilessly slammed. On one occasion, during a reception at the Tuilleries, Napoleon himself fiercely rebuked him, stating his work on meteorology was 'discrediting your old age'.[14] Lamarck was said to have been reduced to tears by his Emperor's remarks. He finally abandoned this area of research completely. In 1815, he published the first volume of his *Histoire Naturelle des Animaux Sans Vertebrate*, in which he described details of the entire invertebrate phyla. Further volumes followed.

He had a long-term relationship with Marie Delaporte, which spanned some fifteen years, and was blessed with six children. The couple first met in 1777, but did not marry until 1792, when she was on her deathbed. He married again in 1795, but this was short-lived. His second wife, Charlotte, died after they had been together for just two years. He married for a third and final time in 1798, being joined in wedlock to Julie Mallet. She died in 1819.

In his *Philosophie Zoologique*, Lamarck had written that, 'our faculties, regarded as dispositions are innate or contemporaneous with our existence; they are the product of our organisation, and cannot exist apart from it. Without the organ of sight or touch we should have no knowledge of ideas resulting from these organs, and consequently we should be deprived of the faculty of seeing and understanding.'[15] It is of course ironic that Lamarck would suffer the handicap of blindness in later life, and this had a huge detrimental affect on his ability to complete his work. It is likely that overuse of magnifying lenses in the study specimens caused his vision to deteriorate. For the last ten years of his life he was without sight. His disability meant he was obliged to rely on his colleague Pierre Latrielle to take his lectures for him. Lamarck is said to have treated Latrielle like a son. He also came from the clergy, but following the revolution decided the study of the natural world was a safer profession to be in.

For the remaining years of his life Lamarck became almost completely dependent on his children to look after him. He led an increasingly secluded existence, his daughters reading to him served as one of his few pleasures. The final instalments of his *Histoire Naturelle des Animaux Sans Vertebrate* had to be dictated to his daughter Cornelie. She, along with her sister Rosalie, looked after him as his infirmity deepened. Though he had many children, his son Auguste, who went on to be a successful engineer, was the only one of Lamarck's offspring to carry on the family line.

Lamarck died on 18 December 1829, the funeral ceremony taking place at the church of St Medard two days later. He was eighty-five years old, outliving three of his children and all his wives. Due to his family's lack of fiscal clout, his body was consigned to a pauper's grave. The corpse was wrapped in a sheet and indignantly hurled into a lime pit at the cemetery of Montparnasse, in the southern quarter of Paris. The pit was exhumed after five years, and his bones eventually moved to their final home in the dark, dank and unpleasant catacombs that lie beneath the city. Lamarck's colleagues Cuvier and Buffon were both entombed at the Pantheon, the French capital's resting place for its most eminent writers, artists and politicians.

Latrielle succeeded Lamarck as Professor of Invertebrate Biology, but by then he was also an old and feeble man. He is said to have commented on the matter of his belated promotion, 'They give me bread, when I no longer have teeth.'[16] He died just three years later. It is perhaps important to note that Lamarck published *Philosphie Zoologique* in the very same year

Charles Darwin was born. It appears that at this stage the world just was not ready for something of this magnitude. It would be for another generation to take on the responsibility of preaching the gospel of evolution to the masses.

Lamarck clearly saw the clues to the nature of the evolutionary process, he observed the infinite variation, postulated that species had not remained constant throughout history, and correctly predicted that it was the environment in which species dwelt that initiated these changes. Unfortunately, although he managed to uncover the evidence, the hypothesis he presented based on it was fundamentally wrong. He believed the process was being undertaken by each individual as they grew older, not that the population dynamic of the species themselves was changing. He also thought this process had a kind of pre-ordained direction, that there was a specific goal to it. In fact this was not the case, as we will discover soon enough, Darwin and his followers would show evolution did not have the presence of forethought Lamarck's system required.

Nonetheless, when you look at Lamarck's theories in context to the time in which he lived, and compare them with work of his contemporaries, you start to get an understanding of just how advanced his way of looking at things really was. In the words of Cuvier, it 'requires a high price to be a man of genius, and the farther ahead of his times he is, the higher price.'[17]

Like so many others, it was not until long after his death that Lamarck's powers of perception and the true value of his work would be appreciated. In 1875, the Municipal Council of Paris named a street after him close to Montmarte. His bust is to be seen on the outer wall of the Nouvelle Gallerie at the Musee d'Histoire Naturelle, constructed at the end of the nineteenth century. In addition, two genera are named with the Frenchman clearly in mind. These are Lamarckea (a type of grass), and Monentia (a type of mollusc). Today, the entrance to Le Jardin des Plantes is dominated by a statue of the great man, deep in thought. The plinth below him states 'Jean-Baptiste Lamarck. Founder of the Doctrine of Evolution'. The monument, which was inaugurated in 1909, shows how highly the motherland that once mocked his ideas now regards him, albeit somewhat tardy in its response.

My current home in Paris (at the time of writing this chapter), actually lies just off the street that bears his name. It seems fitting that a man who was so far ahead of his time, and who was ridiculed for his outlandish thoughts during his own lifetime, now finally commands such respect in his country of origin. But this of course can be more down to partisan sentiment than true scientific justification. I have mixed feelings about the motivation behind the French people's new-found warmth for their long-dead countryman. Whether it is regret that their ancestors did not see the value in his theories, or just simply to try to gain the glory for their nation, I am unsure. I would be tempted to say that a whole army of statues would not make up for the way that one of their greatest men of science was left to die in ignominy and destitution, but he is far from being the only example of how true visionaries can be mistaken for fools or charlatans.

So let us now move on to another precursor in the formation of evolutionary theory. Though most do not realise it, not only was Charles Darwin not the first person to propose a process of evolution, in fact he was not even the first member of his clan to do so. For the Darwins, evolution was very much a family affair. His grandfather Erasmus Darwin had, to some extent, put forth such ideas many years before. Born in Lincolnshire in 1731, he was educated in the universities of both Cambridge and Edinburgh. He made a very good living as a doctor, and in the 1770s is said to have been offered the post of Court Physician to George III. He did not take the job, because he wanted to concentrate on his own research, and felt that the demands of such a role would hinder this.

His book *Zoonomia*, a monstrous two-volume epic that took him nearly two decades to complete, covered a variety of topics under the broadest of scientific briefs. It included discussions on the nature of animal instinct, the mechanics of creatures' motion and their sensory apparatus, as well as disease, mental conditions and even drunkenness. Darwin felt that a greater knowledge of the animal kingdom would help to further our understanding of human physiology and psychology – that as we were part of a series of products, designed by a single 'Creator', we could learn from his other works more about ourselves. Within the publication he hoped to encapsulate some sort of 'unified theory' of nature, which would 'bind together the scattered facts of medical knowledge, and converge into one point of vein the laws of organic life, would thus on many accounts contribute to the interest of society.'[18]

Perhaps the most interesting chapter to be found within this work is the one on 'Generation', in which he examines the creation of living species and how they had developed. He felt the 'Creator of all things has infinitely diversified the work of his hands, but has at the same time stamped a certain similitude on the features of nature, that demonstrates to us, that the whole is one family of one parent.'[19] He noted similarities between animal embryos of different species, the existence of hereditary diseases which suggested certain biological data could be passed from generation to generation, as well as domestication and crossbreeding, where favourable traits would be selected to be maintained in a plant's or animal's progeny. From this he contended that all animals originated 'from a single living filament'.[20]

He stated that philosophers were of the opinion that during life our 'immortal part' or soul acquired certain modes of behaviour, which became 'forever dissoluble, continuing after death.'[21] In a sentiment similar to that of Lamarck, he proposed applying 'this ingenious idea to the generation or production of the embryo, or new animal, which partakes so much of the form and properties of the parent.'[22] Erasmus believed all animals underwent a perpetual transformation, which was 'in part produced by their own exertions in consequence of their pleasures and their pains, or of associations, and many of these acquired forms or propensities are transmitted to their prosperity.'[23] He also commented that this idea of 'gradual formation and improvement of the animal world seems not to have been unknown to ancient philosophers.'[24] At different stages in the book he remarks on the fact that both Plato and Aristotle had appeared to observe such progression, and Christian-based science had foolishly chosen not to pursue this line of enquiry further.

He gave a few examples of how this transformation manifested itself on different creatures, and the reasons why it had done so. His observations led him to the conclusion that 'the colours of many animals seem adapted to their purposes of concealing themselves, either to avoid danger or to spring upon prey.'[25] For instance the 'black diverging area from the eyes of the swan; which, as his eyes are placed less prominent than those of other birds, for the convenience of putting down his head under water, prevents the rays of light from being reflected in the eye, and thus dazzling his sight.'[26] 'Darwin the Elder' noted that similar methods were used for the concealment of birds eggs, their shells resembling 'the colour of the adjacent objects.'[27] For example the eggs of hedge-dwelling birds were 'greeneth with dark spots; those of crows or magpies, which are seen from beneath, through wicker nests, are white with dark spots.'[28]

Though Erasmus's writings did not seem to cause the same intensity of uproar as Lamarck's had done, neither were they to attain any serious level of acceptance. The scientific world took little notice, possibly because although his observations did have some truth in them, he had no explanation of why and how such adaptation would take place. It would not be until well into the following century that two men would come up with a way of describing this.

One of these individuals was Charles Robert Darwin. Born on 12 February 1809 in Shrewsbury, Shropshire, he was the fifth of six children born to Robert Waring Darwin by his wife Susannah. Robert was a successful country doctor, just as his father before him. He was the youngest son by Erasmus' first wife, Mary. Charles was, it seems, more than a little spoilt as a child. His three older sisters, Marianne, Caroline and Susan, as well as his brother Erasmus, all making quite a fuss of the little chap.

His mother was of an even more notable English bloodline. Susannah was the daughter of Josiah Wedgwood, the famed industrialist and pottery baron. Unfortunately for Charles, she was to be pestered by illness through most of his infancy, and not destined to see her child grow up. After years of suffering she finally died in July 1817, when Charles was just eight years old. Marianne and Caroline were given the responsibility of looking after him and his younger sister Catherine.

Darwin was born into a life of comfort. His own family were reasonably prosperous, and their association with the Wedgwoods significantly accentuated Robert's income as a doctor. As a child Charles would spend a great deal of time with his cousins at the Wedgwood home in nearby Staffordshire.

He recalled that even before his schooling had begun, his 'taste for natural history, and more especially for collecting specimens, was well developed.'[29] In September 1818, he started school in Shrewsbury under the tutorage of Samuel Butler, where he remained until he was sixteen. He boarded there, but as it was only a relatively short distance back home, tried to see as much of his family as he could. Most evenings he would sneak back for an hour or two.

His time at Shrewsbury Grammar School seems to have done little good in motivating the young boy. In his autobiography he stated that 'school as a means of education to me was simply a blank.'[30] He felt that nothing could have been 'worse for the development of my mind than Dr Butler's school as it was strictly classical, nothing else being taught, except a little ancient geography and history.'[31] His scholastic achievements reflect this. He does not seem to have made a particularly positive impression on any of his teachers.

He would later write, 'I was considered by all my masters and my father as a very ordinary boy, rather below the common standard in intellect. To my deep mortification my father once said to me, "you care for nothing but shooting, dogs, and rat catching, and you will be a disgrace to yourself and your family".'[32] Many have suggested that Darwin's father was a very domineering character, who bullied him into trying to make something of himself, but there seems to be little to suggest this was the case. The previous remarks, which are often quoted as an example of this, appear to be the only recorded incident that gives any credence to such conjecture, and is more likely to have been a momentary slip from someone who was otherwise a generous and caring person. Charles certainly always spoke well of his father, describing him as 'the kindest man I ever knew'.[33]

Nevertheless, it appears to be true that this young tearaway does not seem to have wanted to do anything but amuse himself. Back then, he showed little sign of the industrious activity that would characterise his adult life. He later said of his 'wild years' that he did not 'believe any one could have shown more zeal for the most holy cause as I did for shooting birds.'[34] There is clearly a great deal of contrast between the adolescent characters of Lamarck and Darwin; while one was facing mortal danger in the ranks of the French army, the other was entertaining himself by following aimless pursuits and trying to get out of doing any real work.

Though his family were reasonably wealthy, there was no question that the boy would not be able to continue such a lifestyle indefinitely. Charles was expected to earn his keep. He recalled that his father declared he 'should make a successful physician – meaning by this, one who would get many patients.'[35] This seems to be the way things progressed during Darwin's teens – he continued to occupy his time with hunting, and relied on his father to determine what he should do with his life, then waited for other suggestions when he failed to get anywhere. It does not look as if he took an active role in finding an occupation for himself until well into his twenties.

In October 1825, he joined his brother Erasmus in Edinburgh. Charles was far from a model student, having little interest in his medical studies, and for the two years he remained in Scotland did little that could be considered in any way productive. He found the lectures 'intolerably dull' on the whole, particularly loathing those of Professor Robert Jameson on geology (a subject he had the deepest contempt for at this stage. This is in some ways very ironic, as in later life he became Secretary of the British

Geological Society). Darwin would later confess that the only effect that Jameson produced in him was 'the determination never as long as I lived to read a book on geology, or in any way to study the science.'[36] Young Darwin also took an especially strong aversion to anatomy. He later admitted, 'the subject disgusted me,' but regretted not preserving with it as 'the practice would have been invaluable for all my future work.'[37]

While in Scotland he befriended local naturalist Dr Robert Grant, and the two would spend many an afternoon birdwatching and collecting animal specimens. Grant was a follower of the theories of Lamarck, and probably provided Darwin with his first in-depth exposure to the French scientist's work, though it seems he was not particularly interested in it.

In April 1827, convinced he was not destined to be a healer and that there was no future in this profession for him, Darwin finally quit medical school. To his subsequent shame, he remembered his father was 'vehement against my turning into an idle sporting man, which seemed my probable destination.'[38] Robert now proposed that as an alternative to being a physician, he should become a cleric. If he could not cure people, at least he could bury them.

It is certain that he saw this as a career move, rather than a calling. Although until this stage he had no doubts about the teachings of the Gospels, religion was far from being something for which he had any real fascination, he simply 'liked the thought of being a country clergyman.'[39]

It was at this point in his life that Darwin first gained an interest in something that did not involve the killing of defenceless animals. That summer while back in Shropshire, his relationship with Fanny Owen, a rather buxom and lively local girl, first blossomed. Fanny was the daughter of William Owen, a close friend of Robert Darwin, who often joined Charles in his hunting exploits. His daughter was something of a tomboy, and shared Charles' love of the outdoor life.

On 15 October 1827, thanks to a certain amount of parental pressure, he began his studies at Christ College Cambridge, just like his father and grandfather before him. It seemed that Charles was going to continue where he had left off. He was just as lacking in his commitment to pedagogical deeds at Cambridge as he had been in Edinburgh.

His plan was simple enough, if not particularly ambitious. He would become a vicar in some little village out in the sticks and there continue to follow his interest in natural history in his spare time. In his autobiography, Darwin admitted that 'although there were some redeeming features in my life at Cambridge, my time was sadly wasted there.'[40] He also conceded there had been no activity in his time at university that 'gave me so much pleasure as collecting beetles.'[41] In his freshman year at Cambridge, he managed to get a drawing of one of the specimens he collected into *Stephen's Illustrations of British Insects*. He recalled 'no poet ever felt more delighted at seeing his first poem published than I did seeing the magic words "captured by C.Darwin, Esq".'[42]

During 1828, Darwin became strongly influenced by the Revd John Henslow, Professor of Botany at Cambridge. Though Henslow was only in his mid-thirties, he had already gained a considerable reputation. Darwin began to attend his lectures and was greatly impressed. Despite the fact that it was nothing to do with his own curricula, he would often tag along on class field trips.

The young undergraduate spent most of his time with the professor, so much so that other dons began to refer to him as 'the man who walks with Henslow'[43] Darwin said of him, 'his moral qualities were in every way admirable. He was free from every tinge of vanity or other petty feeling.'[44]

While at university, his relationship with Fanny, who was still back in Shropshire, began to fall apart. Darwin had not really shown her much attention since enrolling at Cambridge, and even when he had time to go and visit her he often preferred the company of invertebrates, collecting beetles and such like.

Towards the end of his university career Darwin finally found inspiration. Henslow's guidance and the various works on scientific expeditions abroad he had read, gave him the stimulus needed to consider taking part in such a venture himself. With a group of friends, he 'talked about the glories of Tenerife, and some of the party declared they would endeavour to go there.'[45] Darwin got 'an introduction to a merchant in London to enquire about ships.'[46] He and a couple of colleagues put plans in place to visit the island the following spring, and he even began learning Spanish in readiness for the trip.

He passed his degree in January 1831, coming tenth out of 180 students. Given the almost total lack of effort he put into his studies, the result must have come as a shock to everybody. Perhaps this was the first sign of what he was actually capable of, if he was just willing to apply himself, and although he had not tried too hard to show it until this stage, he possessed a remarkable analytical mind.

Henslow had introduced Darwin to Adam Sedgwick, who lectured in geology. He attended Sedgwick's lectures and for the first time began to gain an interest in this scientific field that had previously been abhorrent to him. When Darwin took a holiday in Snowdonia, following his graduation, Sedgwick joined him and taught him the essentials of geological investigation.

Darwin looked to be heading towards a secure, if uneventful life in some sleepy rustic parish. However, fate was about to deal him a different hand. On returning home from his sojourn in North Wales, he received a letter from Henslow informing him of an expedition that if Darwin was serious about his desire to explore was certain to appeal to him. Henslow, who was well aware of his keenness to travel to foreign parts, had learnt of a naval vessel in search of a 'good naturalist', and that the captain would 'give up part of his own cabin to any young man who would volunteer to go with him, without pay.'[47] For the first time in his life Darwin, this college drop-out and world-class loafer, was about to make a real decision. He had to ask himself if his talk of visiting far-off lands had just been whimsy, or was he really willing to go through with it? He wrote to Henslow affirming his interest, and agreed to join the party.

Robert was not to share his son's previously absent zeal. He was against the idea from the start, and put his foot down firmly on the matter. It seems surprising in some respects that, on finally finding something constructive that his son would do, he poured cold water on the whole thing, but this is what Charles was faced with. Young Darwin, who was eager to take advantage of the opportunity, was greatly disappointed.

Feeling somewhat despondent, Charles paid a visit to Maer, the Wedgwood family residence. Their his 'Uncle Jos' (more formally known as Josiah Wedgwood II) learnt of the planned voyage, and seeing how upset the young man was in not being allowed to go, decided to see if he could change his brother-in-law Robert's mind. He wrote a letter to him listing the reasons Charles had been given that he should not join the expedition, and managed to find a retort to answer each one. The letter was successful in dispelling his objections, and when Jos took Charles to Shrewsbury later that week, Robert agreed to give his consent, albeit grudgingly.

On the morning of 5 September 1831, Darwin arrived in Whitehall for a meeting at the Admiralty with Captain Robert Fitzroy, commander of HMS *Beagle*, the vessel on which (if considered worthy) he would serve. Fitzroy was just a couple of years Darwin's senior, and had only been promoted to captain on the ship's previous voyage to South America, the year before. He believed he could judge the nature of someone's character from the shape of their features. Darwin later commented that 'the voyage of the *Beagle* has been by far the most important event in my life, and has determined my whole career' yet it had depended 'on such a trifle as the shape of my nose.'[48] With all parties in agreement, Darwin made preparations for the voyage.

On 24 October, he arrived in Plymouth, ready for the off, but unfavourable weather conditions meant the ship would remain in port for several weeks. A series of heavy gales prevented them from setting sail on schedule. After much delay he finally left England on the morning of 27 December. Darwin recalled, 'we weighed anchor at 11 o'clock, and with difficulty tacked out. We joined the Beagle about 2 o'clock, immediately with every sail filled by a light breeze we scudded away.'[49]

Maritime living was no picnic. Though life onboard ship had improved from the scurvy-filled, weevil-ridden, rum-soaked environment sailors had been stuck with in the eighteenth century, it was still far from comfortable. The ship was only 90ft long, and carried a crew of seventy-three men. The chart room, which was to be Darwin's home for the duration of the voyage, was just 11ft by 9ft, and the centre of it was taken up by the map table, but in comparison to conditions endured by the rest of the crew, this was the height of luxury. It was not so much the cramped environment that would be Darwin's bane during his prolonged stretch upon the *Beagle* however. Within no time at all, he realised that his constitution was not that of a mariner. His diary entry on 28 December tells how upon awakening he 'soon became sick & remained so during the rest of the day.'[50] Darwin's trip would take five years to complete, at least eighteen months being spent onboard ship. He would be afflicted with sea-sickness for nearly the entire time. In his words, 'the misery is excessive & far exceeds what a person would suppose.'[51] His letter to Henslow in May 1832 recounts that between Plymouth and Tenerife, 'I was scarcely out of my hammock.'[52]

In the early stages of the voyage, Darwin greatly enjoyed the company of Fitzroy, and was filled with admiration for him. Though only slightly older than Darwin, he had risen through the ranks at an astounding pace, and seemed bound for glory. Darwin described the captain's character as a most 'singular one, with very many noble features: he was devoted to his duty, generous to a fault, bold, determined, and indomitably energetic.'[53]

Much to Darwin's disappointment, Tenerife, which was planned as the *Beagle*'s first port of call, was off limits. The local authorities, in view of the high occurrence of cholera in England at the time, put a two-week quarantine on all British craft entering its waters. The delay such procedures would cause to the *Beagle's* voyage meant that Fitzroy had little choice but to forego this stopover.

Instead the ship carried on to the Cape Verde Islands. It then entered the South Atlantic heading towards Bahia, followed by Rio de Janeiro. Darwin stayed there for a little over three months collecting specimens. His first experience of the wildlife from this completely new continent was, as you would expect, quite a departure from what he was used to. He noted 'the existence of a division of the genus Planaria, which inhabits the dry land' and that these animals had been grouped 'with the intestinal worm, though never found in the body of other animals.'[54] This appears very minor, but looking back we can see Darwin was beginning to think the classification of species was somewhat dubious, with no real consistency in how scientists categorised different creatures.

The ship headed for Montevideo, and for the next two and a half months Darwin travelled on horseback across the Pampas to Patagonia. There he made another observation that may have helped plant seeds of doubt in his mind about the formulation of animal lineage. He noted:

> the Molothrus pecoris is a North American bird, and is closely allied in general habits, even in such peculiarities as standing on the backs of cattle, and in appearance, with the species from the plains of La Plata; it only differs in being rather smaller and of a different colour, yet the two birds would be considered by a naturalist as distinct species. It is very interesting to see so close an agreement in structure, and in habits, between allied species coming from opposite parts of a great continent.[55]

As Lamarck had done, Darwin started to wonder what really constituted a species. How could men of science decide that one difference was enough to separate a group of animals from another, the process was a purely arbitrary one.

He returned to meet the *Beagle*, and the ship departed for Tierra del Fuego, before making its passage around the perilous Cape Horn, and into the warmer waters of the Pacific. Though Darwin and Fitzroy became close friends over the long course of their odyssey, it is clear they also had more than the odd falling out. Darwin entered in his diary that 'Fitzroy's temper was a most unfortunate one. It was usually worst in the early morning, and with his eagle eye he could generally detect something amiss about the ship, and was then unsparing in his blame.'[56] Though the captain did on occasion fly off the handle, he was always very apologetic afterwards. On the instances that Darwin was subject to his outbursts, Fitzroy took great pains to make amends.

The *Beagle* headed north along the coast of Chile, towards the Galapagos. This group of islands lie just west of Ecuador, some 900km from the South American mainland. It was in this far-flung setting that Darwin really uncovered the evidence that formed the basis of his life's work. His account tells how 'the natural history of this archipelago is very remarkable: it seems to be a little world within itself; the greater number of its inhabitants, both vegetable and animal, being found nowhere else.'[57] The peculiarities of these islands' wildlife exposed Darwin to spectacles few others had been privileged to see.

He observed that the tortoises of each island had slight differences in form, and on 'certain islands they attained a larger average size than in others.'[58] However, he thought it made no sense for particular species to have been created to live just on one little island. Why go to the trouble, when it would be easy enough to have one species for all of them?

Darwin witnessed similar phenomena amongst the lizards inhabiting the region. For example, on one island in the group the creatures had tails that were 'flattened sideways, and all four feet partially webbed.'[59] These lizards seemed to be much more proficient swimmers than ones found on other islands. He subsequently noted that although the lizards were herbivorous like those elsewhere in the Galapagos, the kind of vegetation they consumed was very different. This made him wonder whether the shorter snouts they possessed were due to adaptation to the type of food that formed their diet. 'I will not attempt to come to any definite conclusions, as the species have not been accurately examined; but we may infer that the organic beings found on this archipelago are peculiar to it,'[60] he wrote. For the moment he was not going to jump head first into producing any half-baked hypothesis, but he clearly observed something was not quite right about what he had been taught about the natural world, and what he actually witnessed.

The *Beagle* travelled across the Pacific via Tahiti, then ploughed on to New Zealand. After a short stay there, it set sail for Australia, followed by Mauritius. The ship stopped in Cape Town before re-entering the Atlantic once again. Over the course of the voyage, Fitzroy had been worried that his longitude readings taken in Bahia were incorrect, and this would put all following positions noted out of whack, so the ship was forced to make a diversion on its homeward journey, heading back across to South America in order to re-take them. This must have been a trial for the crew, who had already been away for over four and a half years, and Darwin was most certainly unimpressed. By this time he had quite enough of sea travel, the confined space, and the constant affliction of 'mal-de-mere'.

The unwilling sailor finally arrived back upon dry land on 2 October 1836. He vowed never to travel by ship again. The *Beagle* docked at Falmouth, Devon. However, the homecoming he longed for did not materialise. As the ship approached, he recalled 'the first sight of the shores of England inspired me with no warmer feelings than if it had been a miserable Portuguese settlement.'[61]

Upon his arrival back in familiar territory, Darwin bade farewell to his shipmates and headed for home. The *Beagle* and its crew still had to reach the docks at Greenwich before the voyage would be completed. Their naturalist was under no compunction to stay with them for this leg, electing to return to Shropshire at the first opportunity. Captain Fitzroy continued to make his way up the 'greasy pole', becoming Director of the Meteorological Office, and eventually taking the post of Governor of New Zealand.

After spending Christmas back in Shrewsbury, with his family, Darwin decided to reposition himself in London. Leaving the ignoble pastimes of his youth behind, he set about making use of the experiences he had gained, and began to indulge in something that had never even occurred to him to try before he left England – work. He stayed with his brother Erasmus at first, before renting an apartment of his own. During this period he collated the records he had made, and examined the specimens accumulated through the course of his expedition.

While sifting through his notes, he mulled over what he had seen during the previous years. He noticed, contrary to what would be expected, that the animal and plant species he had got the chance to examine were not set to specific types of environment. They were not distributed in places with similar terrain, but inhabited a whole region, and appeared to adapt to the different environments within it. In Spring 1837, he started his first notebook on the theme of transmutation. It was a period of energetic fervour for Darwin. He would describe later how 'these two years and three months were the most active ones I ever spent.'[62]

That summer he began to put serious effort into research on this subject. It occurred to him that the correlation was not that certain types of landscape all over the world were inhabited by the same particular species, but that certain species would be found across whole landmasses and different manifestations of these would present themselves in specific areas. Just like the lizards in the Galapagos seemed altered by the circumstances they found on different islands, all living things appeared to have modifications to their form which matched the requirements of their particular location.

In March 1838 he made several visits to London Zoo, where an orangutan had just been introduced. Darwin observed the ape, called 'Jenny', and noted many similarities between its mannerisms and those of human infants. He began to believe that it was only humankind's arrogance that made it think it was in some way a unique creation, above all other living things. Darwin began to suspect that we shared a common bond with all other animals. The evidence seemed compelling to him that 'species gradually become modified' though he admitted 'the subject haunted me.'[63]

During that summer Darwin spent some time back in central England, both at his home and that of the Wedgwoods. It was around this time that he first started harbouring feelings for Emma Wedgwood, his cousin and childhood friend. Unlike Miss Owen, the previous object of his affection, Emma was a delicate bloom. She was a quiet and reserved girl, with an almost introspective nature, but certainly not lacking in intellect.

At the end of the summer, Darwin really started to wrestle with the subject of how species were derived. On 21 September 1838 he had a peculiar dream, in which he was executed by hanging. Maybe this was a sudden realisation of where his research was leading him, and the graveness of what he was proposing.

Although his varied observations all pointed to some kind of transformation occurring, he still had no schematic to describe how this took place. There was no sense in him trying to tear out all the wires of the old system without something to replace it.

That autumn Darwin happened 'to read for amusement Malthus on Population, and being well prepared to appreciate the struggle for existence which every where goes on from long-continued observation of the habits of animals and plants, it at once struck me that under the circumstances favourable variations would tend to be preserved, and unfavourable ones to be destroyed.'[64] This work, by Professor Thomas Malthus, was called 'Essay on the Principle of Population'. Malthus was not a biologist, but of all things, an economist. He had written this thesis back in 1798. It looked at the way human society tended to always lean towards overpopulation, and thus many were forced to starve, or die from disease as a result of poor living conditions. The lack of resources available meant that competition arose, and only those capable of competing would live. This was the turning point; Darwin could see the factors that drove the economic prosperity of humans could apply equally well to plants and animals. It gave him a workable structure on which his ideas on transmutation could actually take shape.

On 11 November, he proposed to Emma and she accepted. He took a house on Upper Gower Street at the end 1839, just prior to their marriage. By this point the Darwin family tree was beginning to resemble a circle rather than having a linear form. His father, Robert, had married a Wedgwood, and one of his sisters had done the same only two years prior.

On 24 January 1839, he was elected as a Fellow of the Royal Society, and a few days later married Emma at St Peter's Church, Maer. The marriage meant Darwin would never have to worry about money again, although he had not really had to before either, for that matter. He received a £5,000 dowry from his father-in-law, and this was supplemented by an annual allowance of £400. Robert Darwin gave the young couple an additional £10,000, and agreed to buy them a home in which to raise a family. The exemption from not having to earn a wage meant he could concentrate whole-heartedly on his scientific endeavours, a privilege Lamarck never received.

However, it was not going to be easy for Darwin by any means. There would be other forces at work that were destined to hamper his research. His health problems had first surfaced back in 1838, and from this point on became more alarming, regularly suffering from dizzy spells and heart palpitations. It was perhaps his internment on the *Beagle*, or the years of intense work he endured upon his return to England, that caused his weakened condition, but there is no detailed information that allows us to ascertain the true culprit of his afflictions.

In May 1839, an account of the *Beagle's* journeys was finally published, but sold badly. It was a large, rather stodgy piece; a compendium of former captain Philip King's exploits between 1826 and 1830, followed by Darwin's and Fitzroy's recollections. Charles later decided to publish his observations as a separate book, and by the end of that summer *Voyage of the Beagle* was released, going on to be a popular read. He would later state that the 'success of my first literary child always tickles my vanity more than that of any of my other works.'[65]

The Darwins were blessed with their first child, William, in December 1839. In March 1841, Emma gave birth to their second, Anne. Faced with a rapidly expanding family and a third baby already on the way, in September 1842 they moved to Downe, in Kent. As he put it, city life 'suited me so badly that we resolved to live in the country.'[66] His hopes that this relocation might do something to relieve his illness all went up in smoke, as no such respite from his infirmed condition would appear. He recalled that 'during the first part of our residence we went a little into society, and received a few friends here; but my health almost always suffered from the excitement, violent shivering and vomiting attacks being thus brought on.'[67]

Charles contented himself with simple pleasures, his constitution not really permitting anything else. He described his scientific research as his chief enjoyment: 'the excitement for such work makes me for the time forget, or drives quite away, my daily discomfort.'[68] Nevertheless the continuous breaks he had to take were thwarting his hopes of formulating a complete theory on species transmutation. He wrote that, 'now it is ten years since my return to England. How much time have I lost by illness?'[69] Soon after arriving in their new home, Mary was born, but lived just a couple of weeks. In September the following year they witnessed the birth of another daughter, Henrietta, or 'Etty', offering some solace from their earlier loss.

Evolutionary theory gained another advocate in the mid-1840s, when *Vestiges of Creation* was published. Although the author remained anonymous, it was thought to be penned by Robert Chambers, whose publishing house brought it into print. The book looked at how the stars and planets formed, the way in which the Earth's geological structure developed, and what had caused the creatures living upon it to actually get there. It put forth a concept of 'Animated Nature' in which species would be in a constant state of development, echoing Lamarck's ideas on all lifeforms gradually moving towards perfection. The book had little scientific value, there was minimal data to back up any of its claims, but it did have some interesting points in it. For instance, it noted that the first mammalian fossil records came from the marsupial family. It stated that since their 'mode of gestation is only part uterine – this family is clearly a link between the oviparous vertebrata (birds, reptiles and fishes), and the higher mammifiers.'[70]

The work of Chambers, Lamarck, and even his own grandfather seemed to have minimal impact on the development of Darwin's theories. He said of *Vestiges* that the book 'made very little impression'[71] on him. His own work was already entering a phase where the musing of others were not really required. None of them had managed to find a real explanation, while he now had the answers.

First Darwin had observed, 'No clear line of demarcation has as yet been drawn between species and sub-species' and 'these differences blend in to each other in an insensible series.'[72] This, as we have already seen, concurs strongly with what Lamarck postulated. Next he noticed that more individuals were being produced than could possibly survive, and so there would 'be a struggle for existence, either one individual with another of the same species, or with the individuals of distinct species.'[73] Thus environmental pressures would dictate, as Malthus had supposed, that only a portion of the population could survive, nature choosing individuals best suited for this. From there the extinction of some species 'almost inevitably follows on' and 'old forms will be supplanted by new and improved forms.'[74]

Based on these facts he postulated that species more suited to the surroundings would thus 'exterminate the older, less improved and intermediate varieties; and thus species are rendered to a large extent defined and distinct objects. Dominant species belonging to the larger groups tend to give birth to new and dominant forms; so that each large group tends to become still larger, at and at the same time more divergent in character.'[75]

The key points, which formed the basis of his theory, were thus:

1. A tendency for species to have more offspring than the environment they lived in could support, due to lack of food, space, etc.
2. This lack of resources resulted in a competition, with only a few offspring reaching a stage where they could propagate future generations.
3. Variation between individuals in particular species, and the fact that these differences in their characteristics might offer certain individuals a competitive advantage, and give them a better chance of surviving long enough to mate.
4. As those individuals who possessed beneficial characteristics were more likely to procreate, the characteristics were passed to their offspring, which in turn would be more likely to procreate in the future.
5. Since individual variation would be favoured by particular environments, more and more distinct variation would be seen with each generation.

Basically the variation in the individuals that made up a species, combined with the inherent 'struggle for existence' resulting from pressures placed on the species, meant only some fraction of the populace survived. Nature was thus keeping the population in check by selecting those most suited, and thus a divergence in the species would take place. Darwin proclaimed it 'the doctrine of Malthus applied with manifold force to the whole animal and vegetable kingdoms.'[76]

The turmoil of knowing what he did, but not being able to tell the world, for fear of the reprisals his family would suffer, must have been a great burden. In early 1844, he confided in Joseph Hooker, a slightly younger, but already very well known naturalist, who had travelled to Tierra del Fuego, and had just published a book on his observations. Darwin was unsure about whether he could trust anybody with his revelations, but felt Hooker was more open-minded than many of his fellow scientists, and would at least give his proposal a fair hearing. Though uncertain whether he was doing the right thing, Darwin was relieved to find Hooker willing to take his ideas onboard.

Also during that year, worried by his ailing health, Darwin wrote a letter to be opened by his wife on the event of his death, asking her to ensure his work on transmutation was published. This may have served as 'safety net' for him. This way he could continue to delay publication indefinitely, and avoid the stigma he was certain to face. As it transpired, the course of events that followed prevented him from taking the easy path. Continued stalling would not be an option. If he was going to claim the discovery of the century (if not the millennium) he would have to fight for it.

In late 1846, he embarked on a classification of Cirripedia (barnacles). The reason for this departure, from the more pressing matter of species transmutation, seems to be this: he wanted to show his validity as a naturalist, by creating a work with true depth, which would demonstrate his commitment to the advancement of scientific research. It would give him a higher standing in academic circles, and thus make his work on evolution, when he finally managed to publish it, harder to refute.

His physical condition had been a constant blight on his work for several years, but by the late 1840s it was proving to be a major concern. Darwin went through a series of water cure sessions over the following years to try to alleviate his complaints. He tried a number of different spas, firstly at Great Malvern, then later in Farnham, and finally Richmond.

Towards the end of the decade the Darwins had yet more children. In July 1845, George was born. Then in July 1847 they had another daughter, Elizabeth. Francis arrived in August 1848. His father, Robert, who had been ill for some time, finally died in November that year, but Charles' own health was so delicate at this point that he was unable to attend the funeral.

Another son, Leonard, was born in January 1850. It was towards the close of this year that his beloved daughter Annie first became ill, suffering with some form of fever that was never fully diagnosed. Her state became so bad that her father sent the girl to Malvern to take a course of water cure treatments, as he had done. The therapy went well by all accounts. Letters from the proprietor Dr James Gully suggested she was on the road to recovery. Sadly not long after, she took a turn for the worse. It appears she picked up a cold during the treatment, and its severity increased at a rapid pace. She died on 23 April 1851. Darwin wrote to his friend and fellow naturalist, William Fox, that, 'she suffered hardly at all, and expired as tranquilly as a little angel. Our only consolation is that she passed a short though joyous life.'[77] The experience drained the last traces of faith from Darwin's body. From this point on he considered himself at best agnostic.

The new decade bought more little Darwins to Downe. Horace was born in May 1851, and Charles Waring Darwin followed in December 1856. To the dismay of the child's parents, he was mentally handicapped, and died of scarlet fever before reaching his second birthday. Darwin had little time to pause for sorrow at the death of another of his offspring. His work seems to have been his only concern at this stage, and possibly the loss of his favourite child had rid him of much of his warmth. In Summer 1854, he completed his work on cirripedes, and gained the Royal Society's Royal Medal for his efforts. Though he had spent eight years in total on this particular work, he estimated that he lost at least two of these through sickness.

Darwin first became acquainted with Thomas Huxley in the early 1850s. Huxley shared both Darwin's and Hooker's experiences of travelling to distant lands. As a young man he was the surgeon upon a naval vessel on an expedition to Australia. While in the antipodes he had collected many species of shellfish, and become an expert on the mollusc phyla. Just like Hooker, Huxley was thought of as a challenger to the old guard that held power in the realm of natural science. Darwin could see he would be another useful ally.

When Charles added Huxley to his group of confidants, just like Hooker, he met with the ideas on transmutation favourably. He wrote to Darwin after reading his M.S. Sketch, 'I think you have demonstrated a true cause for the production of species.'[78] He concluded the letter by saying 'you have earned the lasting gratitude of all thoughtful men.'[79]

With time other important men of learning would be told of Darwin's secret, among them was philosopher and social reformer Herbert Spencer and eminent British geologist Sir Charles Lyell, but it was perhaps Huxley that had been his most important recruit to the cause. When the theory was finally publicised to the world, it was his passion to fight for its acceptance that proved so fundamental to its victory, and won him the nickname 'Darwin's Bulldog'.

Darwin may have not been dragging his heels, but his work on evolution was coming no closer to publication. A mixture of illness and other projects, combined with his habit to collect as much factual evidence as he could, strung the undertaking out for close to two decades. His initial strategy was to create a giant cornucopious work containing countless examples, on a scale 'three or four times as extensive as that which was afterwards followed.'[80] But his plans were overthrown, for in mid-1858 he received a package from a man by the name of Mr Wallace.

Darwin and Lyell had been in correspondence with the aforementioned Alfred Russel Wallace for some time. He was a young naturalist located in Malaya, investigating the natural history of the region. Wallace had been independently developing his own work on evolutionary theory, basing his research on his experiences while in South America, and South East Asia. His work did not have the technical depth of Darwin's, and he was not respected by England's scientific community to anything like the same level, but the reasoning he used was equally valid. He had observed exactly the same things as Darwin; the population kept from rising by environment pressures, the variation in each species giving some individuals an advantage over others, and thus being rewarded with a better chance of survival. It now appeared he had made something of a breakthrough.

'*The Truth is born into this world only with pangs and tribulations, and every fresh truth is received unwillingly.*'[81]

Alfred Russel Wallace

So who was this other character Wallace, and what part did he play in formulating the theory of natural selection? He was, as we will see, natural sciences' 'lost prophet'. He would be instrumental in publicising the definitive hypothesis of evolution to the world. The magnitude of his importance in changing the face of science is without question, despite the fact he remains virtually unknown.

Wallace was born in the village of Llanbadoc, in Monmouthshire, on 8 January 1823, the second youngest of nine children sired by Thomas Vere Wallace and his wife Mary (*née* Greenwell), though only six lived past infancy. Alfred's father was a solicitor, but never made a great success of this profession.

The Wallace family were not particularly close-knit, Alfred later recalled having 'few relations, and I myself never saw a grandfather or grandmother, nor a true uncle.'[82] He openly admitted that after leaving home he did little to stay in touch with his parents, and visited them on very rare occasions.

Just like Darwin, he gained an interest for natural history at an early age. One of his first memories was that of catching lampreys in the river Usk close to his home. The family moved to Hertford, Mary's hometown, in 1836. There Alfred began his studies at Hertfordshire Grammar School, though this was short-lived. His father continued to struggle when it came to making an adequate living, and financial pressure eventually meant Alfred's education was brought to a halt. He was forced to leave school and take an apprenticeship with a builder in London.

Life in the city did not appeal to him, so the following year Alfred started working with his brother William as a surveyor in the South Wales town of Neath. It gave him the chance to develop his observational abilities to a much higher level. He was able to pick up the basic rudiments of mapmaking, and it afforded him the opportunity to improve his drawing skills, as well as methods for recording data accurately. All of these attributes would serve him in his later life. His brother encouraged him to consider utilising his talents to better effect, by studying to become an architect, but Alfred found this career unattractive. He did not like the idea of working in an office, enjoying the outdoor life too much. Nevertheless, Alfred agreed that he could do better for himself. He began to attend lectures on chemistry, botany and zoology while in Neath, and read a variety of scientific books and journals. He made use of the local library, as well as taking night classes in geology and astronomy.

His brother tried hard to support them, but 'the difficulty of finding remunerative work was very great, and he was often hard-pressed to earn enough to keep us both in the very humble way in which we lived.'[83] Alfred decided it was time to start making a living for himself and he left Wales, feeling there were likely to be more options for him back in England. In early 1844 he returned to London, lodging with his brother John, and started looking for work.

Later that year he took the position of drawing master at Leicester Collegiate School. Despite his lack of a formal education, the years of self-tuition meant he did not feel inferior to the other teachers. He was able to converse with his colleagues on most topics, and was an intellectual match for those from far more enviable backgrounds. While in Leicester he became a close friend of another young naturalist, Henry Walter Bates. It was Bates who nurtured Wallace's interest in biological science, introducing him to many seminal works, including the highly controversial Chambers' *Vestiges*.

Wallace felt his time there was a turning point in his life. He would later say that he owed his friend a great debt for broadening his horizons and giving him the courage to attempt some of the things he accomplished in the years to follow. 'My year in Leicester,' he felt, 'I must therefore consider as perhaps the most important in my early life.'[84]

When his brother William died in February 1845, Alfred returned to Neath. Accompanied by his brother John, he set to work winding up his affairs. For a while he took over William's surveying duties. Alfred and John spent their 'leisure time wandering about this beautiful district, on my part in search of insects, while my brother always had his eyes open for any uncommon bird or reptile.'[85] He did not want to remain a surveyor for the rest of his life though. In South Wales he saw how the poverty-stricken were forced to live and it deeply disturbed him. He was also obliged to levy the local farmers for the compulsory tythe-commutation survey. He found it immoral that the State should snatch money from the hands of the country's poorest citizens, and his experiences were to influence the aggressive political standpoint he took as he grew older.

Wallace kept in touch with Bates, and in mid-1847 his friend visited Neath for a few weeks. While he was there, the two of them collected insect species and read of other naturalists who had gone to the Amazon. They began to discuss the possibility of travelling to foreign lands to study more exotic animals. Bates proposed they should go to the tropics to continue their investigative pursuits.

On 20 April 1848, Bates and Wallace sailed from Liverpool, on the naval vessel HMS *Mischief*, bound for South America. It took them a little over a month to cross the Atlantic, finally arriving in Para, an outpost at the head of the Amazon. Within two months the pair of them managed to collect over 400 species of Lepidoptera. They marvelled at the rich diversity of mammals, birds and insects they encountered. At one stage Bates was plagued with sickness and close to death, but managed to recover. Wallace was unfortunate enough to disturb a swarm of wasps and received several scores of stings for his trouble. The two explorers split up in March 1850, deciding that by separating they would be able to cover more ground.

Herbert, Alfred's younger brother, who had gained precious little education and was looking for something to occupy himself, decided to join the expedition to South America, leaving home in June 1849. His skills as a naturalist seem to have been limited though, and he found it hard to share his brother's bountiful enthusiasm. After two years he had seen enough, and made plans to return to England. Alfred continued along the Rio Negro River, while Herbert headed back for Para, to wait for a ship bound for Europe. Sadly Alfred never saw him again. Herbert was struck down with yellow fever while on route, and died on 8 June 1851. He was just twenty-two years old at the time.

Wallace carried on, following the tributaries of the Orinoco to the foot of the Andes. Like his brother he was to catch yellow fever, and also suffered from several bouts of dysentery. This was not the end of his misfortune though. Heading up the Rio Negro, he was injured when a gun, that one the members of the expedition was carrying, went off by accident. Wallace was hit, the shot taking a sizeable chunk of flesh out of his hand.

By Spring 1852, Wallace began to contemplate leaving South America. Over the previous years he had lost a sibling, been blighted with countless illnesses, and almost been maimed. In July he started his return journey to England upon the brig *Helen*, under the command of Captain Turner, taking back with him a wealth of samples of the plant and animal life he had encountered. While heading towards home, disaster struck yet again. On the morning of 6 August, a fire started in the ship's store, quickly spreading throughout the vessel. Wallace recalled that 'Turner had evidently had no experience of fire in a ship's cargo and took quite the wrong way in an attempt to deal with it. By opening the hatchway to pour in water he admitted an abundance of air, and this was to change a smouldering heat into an actual fire.'[86]

The ship was completely destroyed, though thankfully everyone aboard managed to escape. Both the passengers and crew were successfully rescued by another craft, but all of the specimens Wallace had collected over the previous years were lost in the blaze. He eventually arrived back at the port of Deal, on the Kent coast, a little over a week later, with nothing to show for his long and arduous expedition.

The year after he returned to England his first book *Narrative of Travels on the Amazon & Rio Negro* was published. Once again Wallace started to get itchy feet. Despite the trials and tribulations of the previous adventure, he decided to set forth once again for foreign climes. Most authors have implied that the loss of all the work he amassed in South America set his research back, and he would have formulated his theory on evolution earlier had it not been for this turn of events, but Wallace saw the incident as a blessing rather than a curse. If anything it reawakened his hunger for adventure, writing: 'I think the most fortunate thing that happened to me was the loss.'[87] He admitted in later life that had this not happened he would have probably ended up going back to the Amazon, and this would have stilted the progression of his research greatly.

Wallace realised the amount of scientific investigation already done in the region meant he would be able to glean little from it that had not previously been discovered. The lectures he attended while back in London, and the time spent in the ornithological and entomological departments of the British Museum suggested he might find richer pickings if he went to the Malayan Archipelago instead, as while naturalists had now explored South America *ad infinitum*, this was still, as he saw it, a 'virgin country which hardly any naturalist had properly explored.'[88]

In early 1854, he set out for Asia, arriving in Singapore that April. This bustling port was the main commercial hub for the whole of the continent, trading in spices, silks and minerals. There he stayed for several weeks at the French Roman Catholic missionary house. During this expedition, he estimated that he covered 20,000km, and collected in excess of 100,000 biological specimens. He had to tolerate the unwanted attention of snakes, spiders and the occasional restless native, but in comparison to the horrors he had put up with while in the Amazon this was all relatively tame. Regardless of any discomfort, as he had hoped, the region was a veritable treasure trove of new wonders. Plentiful reserves of unstudied varieties of fauna and flora lay there for him to behold. Wallace had found his Nirvana.

In May 1855, while in Sarawak he first started to give serious consideration to the derivation of species. Based on the broad catalogue of naturalists' work he had read, Alfred knew that he had a 'mass of facts as to the distribution of animals over the whole world'[89] and it started to occur to him that nobody had ever really utilised the information available to look into the way in which species came into existence. As he saw it, 'the exact process of the change and the causes which led to it were absolutely unknown and appeared almost inconceivable. The great difficulty was to understand how, if one species was gradually changed into another, there continued to be so many quite distinct species.'[90]

He expected if it was a 'law of nature that species were continually changing so as to become in time new and distinct species, the world would be full of inextricable mixture of various slightly different forms, so that the well-defined and constant species we see would not exist.'[91] The question therefore was 'not only how and why do species change, but how and why do they change into new and well-defined species.'[92] What was causing any intermediaries to be killed off? Why would they not be able to live alongside the other species?

He sent an article containing his musings on the subject back to England later that year. At this stage he had no real explanation of how this system actually worked, it was just observations. Nevertheless he managed to get it published in the *Annals & Magazine of Natural History*. This editorial, entitled 'On the Law Which has Regulated the Introduction of New Species' gained some attention from both Darwin and Lyell, and it was at this stage the former began to correspond with Wallace.

While on the archipelago he started to wonder, just like Darwin had done, if it was the environment in which the creatures lived that was discarding the transitional species. But if so, how was it doing this, and for what reason?

Wallace was certainly on the right track by this stage. He could see that new species only appeared when in close proximity both in geographical terms, and physical form, to an existing species, but he still had no way to describe exactly how the distribution of different species was arising. Two months later all that would change.

His travels took him to Ternate, a small atoll roughly equidistant between Borneo and New Guinea, the most northerly island of the Mollucas. While staying here, Wallace was struck down with an acute fever. For many days he was unable to leave his bed, taking large doses of quinine to counteract the spells of illness he suffered with. Almost completely debilitated by a succession of cold and hot fits he tried to use the time to give more consideration to the species dilemma that had perplexed him for so long. He mulled over the evidence and began to consider how species 'become so exactly adapted to a distinct mode of life; and why do all the intermediate grades die out' to leave only 'clearly defined and well-marked species.'[93]

In this malarial daze something brought to his recollection the work of Malthus on how human populations were kept from increasing by the intervention of external influences, which he had read over a decade before. He thought of the positive checks Malthus had put forth; disease, famine etc. which kept 'down the population of savage races to so much lower an average than that of more civilised peoples. It then occurred to me that these causes or their equivalents are continually acting in the case of animals also; and as animals usually breed much more rapidly than does mankind, the destruction every year from these causes must be enormous in order to keep down the numbers of each species, since they evidently do not increase regularly from year to year, as otherwise the world would long ago been densely crowded with those that breed most quickly.'[94]

He felt the conditions in which a species found itself determined the size of its population. He gave the example that 'very few birds produce less than two young ones each year, many have six, eight, or ten; four will certainly be below the average; and if we suppose that each pair produce young only four times in their life, that will also be below average. Yet at this rate how tremendous would be the increase in a few years from a single pair. A simple calculation will show that in fifteen years each pair of birds would have increased to nearly ten million. Whereas we have no reason to believe that the number of the birds of any country increases at all in fifteen, or in 150 years.'[95] If the population was to retain some sort of balance then 'each year an immense number of birds must perish.'[96] Therefore, he thought, it appeared the need for species to produce such 'large broods are superfluous.'[97] So why did nature require so many offspring if only a handful could survive?

He asked himself what made some individuals within a species live and others die? Then it occurred to him that 'the answer was clearly that on the whole the best fitted lived. From the effects of disease the healthiest escaped: from famine, the best hunters, and so on. Then it suddenly flashed upon me that this self-acting process would necessarily improve the race, because in every generation the inferior would be killed off and the superior would remain – that is, the fittest would survive.'[98] (It was thus actually Wallace who coined this phrase, not Darwin as most assume. Wallace pushed the concept of 'survival of the fittest', whereas Darwin maintained it was just a matter of whichever individuals were best suited to the needs of a specific environment, it could be something like body colouring or extent of plumage, and did not have to be enhanced size or strength. Darwin always personally disliked the use of the term, though his theory of evolution would forever be associated with it.) Wallace, like Darwin, saw life as a struggle for existence.

Wallace spotted the holes in Lamarck's argument. He could see that the giraffe did not 'acquire its long neck by desiring to reach the foliage of the more lofty shrubs, and constantly stretching its neck for the purpose.'[99] It would be the individuals with longer necks that 'secured a fresh range of pasture over the same ground as their shorter-necked companions, and so on the first scarcity of food were thereby enabled to outlive them.'[100]

'The more I thought over it,' he later wrote, 'the more I became convinced that I had at length found the long-sought-for law of nature that solved the problem of the origin of the species.'[101] The synapses within his cerebral cortex had bridged the gap between these two apparently unconnected phenomena, and Wallace could plainly see the system Malthus had applied to human economic prosperity and that of natural progression were one and the same. He compared his hypothesis to those of Lamarck and Chamber's, thinking about the deficiencies in each author's work, and 'saw my new theory supplemented these views and obviated every important difficulty. I waited anxiously for the termination of my fit so I might at once make notes for a paper on the subject.'[102] With great haste he prepared an essay, and soon after sent it to Darwin, asking him what he thought of it, and if in his opinion it was sufficiently important to pass on to Lyell. The bombshell had been dropped.

Notwithstanding the proverb about 'great minds', it seems a little strange that for thousands of years people accepted the principle that all types of plants and animals had been created at one specific moment, with barely a single person questioning it, then all of a sudden in the space of a comparatively short period of time, two men had come up with exactly the same idea. They had managed to uncover the true formula by which all species were derived and could show conclusive proof to back it up. But it was not a matter of pure coincidence. The 'Victorian Age' meant the world had changed, other continents became more accessible, the move away from its feudal past meant society was able to provision for a greater number of people to be engaged in the betterment of our scientific knowledge. However, this must have been little consolation to Darwin, who saw his chance for glory slipping through his fingers.

Wallace sent him the essay he had compiled, entitled 'On the Tendency of Varieties to Depart Indefinitely from the Original Type' on 22 September 1857. Until this point he had just been roaming around in the dark, now all of a sudden he managed to put together the last piece of the puzzle. Already he had observed many of the same things Darwin had with regard to the variety and apparent adaptation of species, and finally he had found the connection, how Malthus' work could be applied to the natural world as well as the socio-economic one.

Suddenly Darwin had no choice but to sit up and take notice. Before this stage Wallace had been just another nondescript amateur naturalist, whose deliberations on evolution were on the same track as Darwin's but not really on anything close to an equal level as his own work. Now this young man was in a position to jeopardise his priority. It appeared that the lack of urgency he had displayed in publishing his findings could prove Darwin's undoing.

He wrote to Lyell, still in a state of shock. His masterpiece, the result of twenty years of toil, looked like it would be stolen away from him in one bold stroke. 'I never saw a more striking coincidence,' he told Lyell, 'if Wallace had had my M.S. Sketch written out in 1842 he could not have made a better short abstract.'[103] In the letter, he further commented that it 'seems your words come true, with vengeance,'[104] that, 'all my originality, whatever it amounts to, will be smashed.'[105] Lyell had, on several occasions prior, warned him of the risk in not publishing his work, though I am sure it gave him no sense of satisfaction at being proved right, to hear of the anguish Darwin was going through. He sent another letter to Lyell a week later, in which he said, 'I should be extremely glad now to publish a sketch of my generation views in about a dozen pages or so, but I cannot persuade myself that I can do this honourably, Wallace says nothing about publication.'[106] He was worried that 'Wallace might say "you did not intend publishing an abstract of your views till you received my communication, is it fair to take advantage of my ideas and thus prevent me from forestalling you?".'[107] His dilemma was a perplexing one. Could he honestly publicise his theory, knowing another man had reached the same conclusions, but that he had hidden it until his own work was released? 'The advantage,' he felt, 'I should take being that I am induced to publish from privately knowing that Wallace is in the field. It seems to me hard to lose my priority of many years standing, but I cannot feel that this alters the justice of the case.'[108] In conclusion, he stated he could not tell 'whether publishing now would be base.'[109]

Darwin clearly had difficulties in reconciling his conscience. He was uncertain about the best course of action. Eventually Hooker and Lyell persuaded him that he could not just let another man seize from him something he had spent much of his adult life compiling, it would be like discarding his birthright.

Hooker was instrumental in rescuing Darwin's work from the abyss. In what would prove to be a masterful piece of political manoeuvring, he proposed both papers be read at a meeting of the Linnaean Society. As he put it in his letter to the Society, 'we are not solely considering the relative claims of himself or his friend, but to the interests of science generally.'[110] In principle both would get equal billing, but it could be argued that this gave Darwin's growing number of supporters the freedom to put whatever slant they wanted to on it, plus his name meant there was little doubt that his work would be perceived as having greater authority.

In the meantime Darwin responded to Wallace, informing him that he could see they 'thought much alike and to a certain extent have come to similar conclusions. This summer will make the 20th (!) year since I opened my first notebook on the question of how and in what way do species and varieties differ from each other.'[111] He explained that he was preparing his work for publication 'but I find the subject so very large that I have written many chapters, I do not suppose I will publish for two years.'[112] Wallace later stated that he was 'of course, very much surprised to find that the same idea had occurred to Darwin, and that he had already nearly completed a large work fully developing it.'[113] In his letter to his old friend Bates, Wallace commented 'I have been gratified by a letter from Darwin, in which he says that he agrees with "almost every word" of my paper. He is now preparing his great work on "Species and Varieties" for which he has been collecting materials for twenty years. He may save me the trouble of writing more on my hypothesis.'[114] He certainly did not seem disappointed to hear that someone else was of a similar mind. In fact it appears he was more confident he had hit on something significant, given that another had noticed the same things, and managed to come to an identical conclusion.

Both men's work was presented to the Linnaean Society on 1 July 1858. Darwin did not attend the meeting, as his son Charles had died just three days before. This was to set the pattern that would be followed from then on. Darwin, mainly due to his poor health, would rely on the younger more confrontational Huxley, and others, to take up the fight for him. The most renowned example of Huxley's passion was to be displayed at the meeting of the British Association for the Advancement of Science, at Oxford University, in June the following year. The meeting was chaired by Henslow, and attended by over 800 people. Bishop Samuel Wilberforce fought for the ecclesiastical corner, scathingly attacking Darwin's theory. He and Huxley exchanged searing verbal blows over the course of the debate.

Although there were some bloody battles in this war, it did not prove as brutal as Darwin anticipated. Chambers had gone to the grave without verifying if he was the author of *Vestiges* for fear of the retaliations that might result, while Lamarck had been discredited by all around him. There would always be resistance, of course, whenever such a groundbreaking theory was put forward, but perhaps the others that had gone before him had chipped away at the battlements of conservative thought, and made his assault a little easier to stage. *Origin of the Species* went on sale on 22 November 1859; it had taken him some two decades to complete this mammoth undertaking. The first print, of 1,200 copies, sold out before the end of the first day. Darwin personally referred to it with the words 'it is no doubt the chief work of my life.'[115] He attributed its success 'in large part to my having long before written two condensed sketches, and to my having finally abstracted a much larger manuscript. By this means I was enabled to select the more striking facts and conclusions. I had also followed a golden rule, namely, that whenever a published fact, a new observation or thought came across me, which was opposed to my general results, to make a general memorandum of it without fail and at once; for I found by experience that such facts were far more apt to escape from the memory than favourable ones. Owing to this habit, very few objections were raised against my views which had I not at least noticed and attempted to answer.'[116]

Another reason for the book's success was what Darwin described as its 'moderate size', which he admitted was thanks to the appearance of Wallace's essay. If he had published on the scale originally intended, the book would have been 'four or five times as large, and very few would have had the patience to read it.'[117]

Darwin acknowledged the research of Lamarck in the prologue of his masterwork, stating the 'justly celebrated naturalist' was the first person whose conclusions on evolution gained real notoriety, and thanking him for 'the eminent service of arousing attention to the probability of all change in the organic, as well as inorganic world being the result of law, and not of miraculous interposition.'[118] But he questioned the mechanism his predecessor had proposed for this, stating 'with respect to the means of modification, he attributed something to the direct action of the physical condition of life' and 'to use and disuse, that is to the effects of habit.'[119] He also disagreed with the Frenchman's concept that this process was heading in some sort of preordained direction, with living things moving up a kind of evolutionary ladder towards perfection. If his model for evolution took the form of a ladder, then by contrast, Darwin's was more like a tree branching out in all directions. It had no presence of forethought, no set goal or hierarchical structure, it just continued to spread out further and further.

As Darwin put it, 'natural selection is daily, and hourly scrutinising, throughout the world, every variation, even the slightest; rejecting that which is bad, preserving and adding up all that is good; silently and insensibly working, whenever and wherever the opportunity offers, at the improvement of each organic being in relation to its organic and inorganic conditions of life.'[120] There was no way the same principles could be applied to the stretching of one particular giraffe's neck, or changes in colour of a specific member of a bird or fish species. The mechanism Darwin and Wallace put forward was far subtler than that of Lamarck. It was based on the principle that attributes were assigned to individuals at birth, they did not develop over their lifespan. Whereas Lamarck thought all individuals in a species would go through some sort of unseen process of development, after they were born, Darwin saw that this was not the case. It was nature that would pick those most suited to survive and flourish, while others were killed off. Certain characteristics would offer them a better chance of reaching adulthood, and thus when they mated, these characteristics would be passed on.

Lamarck's system was more akin to the way similar individuals might develop during their lifetimes, due to exposure to different influences. For example, two brothers might be born with similar physiques, but let's say they gained interests in different sports. The one could choose to be a cross-country runner. He would develop stronger leg muscles, but his torso would not be too bulky, so as not to hinder his progress over long distances. If the other brother became a shot putter, his upper body would over time be built up, his legs would also become more muscular, but only suited to short exertions, not having the same level of stamina as his sibling. Now, and here is the fatal flaw in Lamarck's doctrine, if the brothers had offspring, the extremes of physique types they had moved toward would not be translated to their sons or daughters. They were purely transitory attributes, and could not be inherited by the individual's children. This is where Lamarck made his mistake. He applied the same principles that related the development of certain muscles, or the learning of specific thought processes, to changes in form that could be passed down to following generations.

With time the theory of 'natural selection' gained a greater and greater following. In early 1863, the fossil of a prehistoric bird was

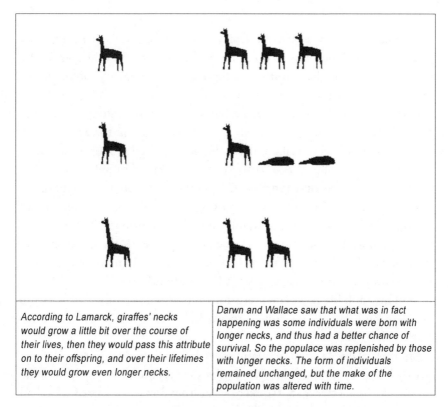

| According to Lamarck, giraffes' necks would grow a little bit over the course of their lives, then they would pass this attribute on to their offspring, and over their lifetimes they would grow even longer necks. | Darwn and Wallace saw that what was in fact happening was some individuals were born with longer necks, and thus had a better chance of survival. So the populace was replenished by those with longer necks. The form of individuals remained unchanged, but the make of the population was altered with time. |

Comparison between the Lamarck and Darwin/Wallace evolutionary systems.

found in Solenhofen, Bavaria. The creature, which was given the name 'Archaeopteryx', displayed characteristics that, even though it looked like a bird, suggested a reptilian origin. The palaeontologist Richard Owen, in his speech to the Royal Society, explained that the beast had feathered wings reminiscent of an ornithological species, but was also armed with reptile-like teeth. In addition, the continued growth of its caudal vertebrae had similar appearance to the tails found in reptiles.[121] It became increasingly apparent that birds and reptiles shared the same ancestry. The timing of this discovery was truly remarkable, it added a lot of weight to Darwin's theory, and greatly increased the already growing throng of his supporters. Darwin described it as 'a great case' for the promotion of his hypotheses. The following year, he was awarded the Copley Medal by the Royal Society, though his evolutionary theory was not mentioned in connection with it, in order that no controversy arose from the bestowing of this honour. While all this was going on Wallace still remained in foreign parts. This in itself was probably damaging to his notoriety. Darwin was in England, surrounded by loyal followers, all pushing his doctrine. Wallace was on the other side of the world, with nobody back home to keep his name in equal prominence to his already more famed counterpart.

In January 1860, Darwin began arranging his notes for a study on how the domestication of plants and animals had played a hand in the evolutionary process. This work would take almost a decade to complete. He put the delay down to 'frequent illness, and partly by being tempted to publish on other subjects which at the time interested me more.'[122] Darwin makes light of the importance of this work here, however it would become the second part of a trilogy in which he described to full extent his evolutionary theory.

In May 1861, Henslow, one of Darwin's oldest friends, the man who had first cultivated his interest in natural history and helped him to obtained the position on the *Beagle*, died of heart failure. Darwin's own health, which for most of his adult life had been precarious at best, was steadily getting worse also. Although he still continued to work, he tried not to put excessive strain on himself.

Another of his longest acquaintances left him in April 1865. Robert Fitzroy, his companion on his five-year voyage around the world, committed suicide. In a letter from Hooker he was told further details of what had taken place. He had apparently 'been very excited for some time, and fancied that many had ill-treated him.'[123] Hooker informed him that Fitzroy had 'taken a house at Norwood for a few weeks – his daughters went to knock at the door on Sunday morning, and receiving no answer the door was forced and he was found dead with his throat cut.'[124]

Darwin's reaction to Fitzroy's sudden death was clearly not one of surprise. He had, despite all his promise and early triumphs, not fulfilled the potential he had originally shown. His career had stagnated, making him increasingly bitter. In his response to Hooker, Darwin wrote, 'I was astounded at the news about Fitzroy, but I ought not have been, for I remembered once thinking it likely; poor fellow, his mind was quite out of balance during our voyage. I never knew in my life so mixed a character.'[125] He continued, 'I once loved him sincerely, but so bad a temper and so given to take offence, that I gradually lost my love and wished only to keep out of contact with him.'[126] In a letter to Henslow some five years earlier, he had said of Fitzroy, 'I think his mind is often on the verge of insanity.'[127]

By the time Wallace returned to Europe, he was approaching forty. He had spent a total of four years in South America, and a further eight years in the Orient, finally he decided it really was time to go home. He left Malaya in mid-1861, and started to make his way back, eventually arriving in England in early Spring 1862. At first he stayed with his sister Eliza and her husband, before finding a place of his own in London. After many years of almost total isolation from the civilised world, and in particular the scientific community, he was now able to contribute to the proceedings at the Zoological, Entomological and Linnaean Societies, attending their meetings almost without fail.

In his second year of living in London he got engaged to the widow of an army officer who had served in India, but this was broken off just a few weeks before the wedding day. In 1864, he got to know the English botanist William Mitten, through mutual friend Richard Spruce. Following several visits to Mitten's house at Hurstpierpoint, in Sussex, Wallace became acquainted with his daughter Annie. She seems to have been particularly impressed with Alfred's knowledge of botany and showed a lot of interest in the subject. Before long they were an item. The couple were married in Spring 1866, he was forty-three at this stage, and she was just eighteen. (Actually it appears Wallace's memory was slightly inaccurate, although in his autobiography he says Annie was eighteen, it seems more likely she was in fact twenty at the time of their wedding. Either way she was still less than half his age.)

They had three children together. Their first son Herbert was born in June 1867, and went on to work in the United States as an engineer. Their daughter Violet, who arrived in January 1869, became a kindergarten teacher in Liverpool. Their second son, William, was born in December 1871, but died in infancy.

During most of 1867 and 1868, Wallace concentrated on preparing his book *Malay Archipelago*. This highly celebrated work, which charted his travels and discoveries in South East Asia, was published the following year. His writing and lecturing activities kept him from impoverishment, but he was never to be considered comfortably off.

While living in London he gained an interest in that most Victorian of past times, spiritualism. He had first learnt of the psychic arts while teaching in Leicester. His friend Bates told him about clairvoyance, and he attended a demonstration on mesmerism conducted at the school. However, life in the metropolis exposed him to a far greater number of these cranks, and Wallace just lapped it up. He attended many séances at Montague Place, under the direction of the medium Miss Cook. Robert Chambers (suspected author of *Vestiges*) introduced him to a Miss Douglas, who was another dabbler in the world of the supernatural. Wallace frequented her house on South Audley Street. He also visited Mr Monk on occasion, who held meetings at an apartment in Bloomsbury.

Huxley, Lyell and Darwin all became close friends of Wallace, and had a great influence on his development as a naturalist. He was finally considered part of the scientific establishment. In 1868, he was awarded the Royal Society's Royal Medal for his 'labour in practical and theoretical zoology', and in the three years that followed he assisted Lyell in the publication of his *Antiquity of Man*. He was elected a Fellow of the Royal Society in April 1872.

By the start of the 1870s he had grown tired of the big city, and relocated to Barking, in Kent. There he found an environment that was a little more tranquil and slower-paced, which was more conducive to his work, and added to the productivity of his writing projects. In 1878, he was unsuccessful in attaining the post of Superintendent of Epping Forest, despite recommendations from residents, scientific men and even members of parliament. He supposed that local merchants had lobbied for someone more open to their exploitation of the commercial opportunities the land offered. Some years earlier he was passed up for a job as Secretary of the Royal Geographical Society, and a directorship of the Natural History Museum had also failed to materialise.

Darwin continued, albeit intermittently, his work on domestication, the second piece of his three-part evolutionary epic. Bouts of illness still limited his progression, but finally, in Summer 1865, he wrote a letter to his publisher, John Murray, confirming that it was almost ready, however he went on to admit his condition was still extremely changeable and it depended 'upon this and upon nothing else whether my book will be ready for the press in the autumn.'[128] The book *Variations of Animals & Plants under Domestication* was finally published in January 1868.

From the beginning he had felt if his theory was correct, and species did go through a process of transmutation, then he could not avoid the belief that human beings must abide by the same natural laws. In March 1871, Darwin published *Descent of Man*. This finally dispensed with any ambiguity about how far the implications of evolutionary theory reached. Until that point he had wisely circumvented this topic, but now felt the time was finally right to tackle it head on. Darwin's softly-softly approach meant he was a lot harder to discredit. He did not go the whole hog from the start, but allowed people to get used to the idea before adding further dimensions.

Within the book he boldly stated, 'Man is constructed on the same general type or models as other mammals. All the bones in his skeleton can be compared with corresponding bones in a monkey, bat, or seal.'[129] He quite clearly picked these three examples for a reason. They showed how wide a number of forms mammalia could take; one living on land, one spending its time in the water, and one being capable of flight. Why would these creatures, which followed such different *modus operandi*, have a similar framework if they had not come from common ancestry? He also noted that 'monkeys are born in almost as helpless a condition as infants'[130] and argued that although it had been 'urged by some writers, as an important distinction, that with man the young arrive at maturity at a much later age than with other animals' in fact 'if we look to the races of mankind which inhabit tropical countries the difference is not that great, for the orang [-utan] is believed not to be adult till the age from ten to fifteen years.'[131] In addition, he pointed out, just as his grandfather had almost a hundred years before, the similarity in the embryonic development of man and other species, stating that in early stages the human foetus 'differs in no respect from the ovules of other members of the vertebrate kingdom.'[132]

In his opinion, there was 'no fundamental difference between man and the higher mammals in their mental faculties.'[133] He conjectured that, just like humans 'animals not only love, but have a desire to be loved'[134] and most of the 'complex emotions are common to the higher animals and ourselves. Everyone has seen how jealous a dog is of his master's affection, if lavished on any other creature; and I have observed the same fact with monkeys.'[135] But he pointed out that what made humankind different was its ability to use its intelligence, and through our mental faculties we were able to 'to keep with an "unchanged body" in harmony "with the changing universe".'[136] We could adapt our habits to new conditions of life, to invent weapons and tools to help us procure food or defend ourselves. We could make clothes and shelters, or even create fire, to protect us from colder climates, while lower animals 'must have their bodily structure modified in order to survive greatly changed conditions.'[137] Natural selection had to allow the species to acquire the attributes necessary to deal with these changes, and if unable to be adequately modified then the species would cease to be.

In autumn the following year, his book *Expression of the Emotions in Man & Animals* was published, and proved to be another great success. It had originally been meant as a chapter of *Descent of Man* but the whole thing had ballooned out of proportion, so he instead decided to release it as a book in its own right. In it he related how 'my first child was born December 27th, 1839, and I at once commenced to make notes on the first dawn of the various expressions which he exhibited, for I felt convinced even at this early period, that the most complex and fine shades of expression must all have had a gradual and natural origin.'[138] He felt similarities in the forms of expression that animals and humans use was yet further proof that they were derived from some mutual ancestor, and 'as long as man and all other animals are viewed as independent creatures, an effectual stop is put to our natural desire to investigate as far as possible the causes of expression.'[139] Certain means of expression that mankind possessed 'such as the bristling of the hair under the influence of extreme terror' he argued 'or the uncovering of the teeth under that of furious rage, can hardly be understood except on the belief that man once existed in a much lower and animal-like condition.'[140]

In September 1876, Darwin's first grandson, Bernard, was born, and in November the following year, he received an honorary doctorate from Cambridge University, where he studied as a young man. It has been noted that despite the enormous effect on the world that Darwin clearly had, he was never knighted, while some less deserving Brits were blessed with this accolade. This is most likely due to the influence of the religious lobby. Even to this day the ruling monarch is charged with the title 'Defender of the Faith' so in some respects it is hardly surprising that he was not chosen as a recipient for fear of the problems it would raise between the Church and the Crown. It seems that he had been considered for the honour, the Whig Prime Minister, Lord Palmerston, apparently proposing to give him a knighthood back in 1859, but the Anglican bishops must have dissuaded the queen from carrying this out. Nevertheless it does appear rather unfair that an Englishman who proved so significant in the history of civilisation, and gained countless gongs from the scientific community, did not receive anything from the State.

Although with age his invalidity became more pronounced, Darwin continued to write. In late 1878, he completed a biography of his grandfather Erasmus Darwin. He still managed to keep a healthy colour in his cheeks, and those who did not know him probably assumed that he was in a state of good health. He once wrote to his friend Hooker that 'everyone tells me that I look quite blooming and beautiful: and must think I am shamming.'[141]

Darwin hoped that, like his father, he would be able to make his way through old age with his faculties still intact. He wrote that he wanted to die 'before my mind fails to a sensible extent.'[142] He did not stay true to the pattern his ancestors followed, in tending to gain weight with the years. His father and grandfather had both become considerably portly in old age, but Charles, perhaps through the effect of his illness, stayed relatively slim.

Despite his ailing health Darwin tried to do as much exercise as his physical condition would permit. He would take strolls along the 'Sand Walk'; a narrow strip of land about one and a half acres in area that lay within the grounds of Downe House, which had a gravel path around it. Everyday he would endeavour to do one walk in the morning, and another in the late afternoon. His son Francis recalled that 'in earlier times he took a certain number of turns every day, and used to count them by means of a heap of flints, one of which he kicked out on the path each time he passed' though in later years 'he did not keep to any fixed number of turns, but took as many as he felt the strength for.'[143]

His brother Erasmus died in mid-1881, and following this Darwin's own will to live began to slip away. Over the following eight months he suffered a number of heart attacks, and remained in a weakened state. While visiting London just before Christmas he had another coronary seizure, and from that point on was confined to his bed. He died on 19 April 1882, at Downe House. He was buried at Westminster Abbey a week later. Huxley, Hooker and Wallace were among his pall bearers. Emma would live on for another fourteen years after him, finally passing way in 1896.

I do not think anyone can really doubt the enormity of the contribution that Charles Darwin made to shaping the world we live in; he along with Copernicus and Galileo, brought us out of the dark ages and questioned the validity of what had been taught through the Bible for many centuries. He showed there was not anything special about humans, we were not some sort of semi-divine creature made in God's image. We were just another variation of living things. But although his part in the development and acceptance of the theory of evolution was huge, it was still clearly an amalgam of the work of others who had gone before him that laid the foundation for it. It was the observations of Lamarck, and even Darwin's own grandfather, that acted as a starting point for his triumphs. It was the courageous efforts of Chambers that gave him a hint of how his work might be judged, and warned him of the necessity to make sure he could back his ideas up with solid, irrefutable proof. Finally it was Wallace, whose work so closely mirrored his, had witnessed many of the same things on his travels, drawn knowledge from similar sources, and arrived at equivalent hypotheses, that hastened Darwin to publish his findings rather than putting it off until a more suitable time which might have never come.

Darwin was certainly not the first to suggest species evolved, he was however the first to come up with a credible mechanism for how this process took place. He could offer compelling evidence that it was no longer possible to rationally deny, as had been done to Lamarck and his contemporaries. I think it is fair to say that Darwin had far greater opportunity to observe nature across the broadest spectrum, travelling around the world and seeing many things that only a bare handful of Europeans had access to at this time, whereas Lamarck had to rely on specimens from his native country, or those he could obtain via merchants. Darwin also married into wealth, which gave him a lot of time to concentrate on the formulation of his theory. Lamarck did not have the level of privilege that was afforded Darwin, he had to worry about the care of his family and putting food on the table first.

Though Darwin was gone, Wallace carried on preaching the doctrine of species transmutation. It was now his time to take up the responsibility of being its primary proponent. In later life he would also become a leading advocate of the need for land reform. He stood up against the low wages, unhealthy working conditions, and overcrowded dwellings of the poorer classes, producing several books on the subject of social change. *Land Nationalisation: Its Necessity & Aims* was published in 1882, and this was followed by *Social Environment & Moral Progress* and *The Revolt of Democracy*, which were both released in 1913. Wallace was one of a growing number of intellectuals who felt that the gap between rich and poor had grown too wide. While a small number of landowners resided in comfort, the masses were trapped in lives of misery. He had stayed true to his working-class origins and tried, as best he could, to bring a more just system into being. When the Land Nationalisation Society was set up he took its presidency, remaining in this post for thirty years.

Despite his advancing age Wallace remained a prolific writer. *The Wonderful Century*, a celebration of the scientific achievements of the Victorian age, was published in June 1898. *Man's Place in the Universe*, which went to print in 1903, looked at the possibility of life on other planets. *My Life*, his expansive, two-volume autobiography, came out in 1905, while *World of Life* was published in 1910.

He continued to be the recipient of a series of prestigious awards. He received honorary doctorates from the universities of both Oxford and Dublin in 1889. The Copley Medal and the Order of Merit were both bestowed upon him in 1908. To mark the fiftieth anniversary of the historic meeting of the Linnaean Society of 1858, where Wallace's and Darwin's works on species transmutation had first been aired to the public, the Darwin-Wallace Medal was inaugurated. Wallace himself was its first recipient. Though he was proud of the honours that others saw fit to lavish upon him, he tended not to take these matters too seriously. On one occasion, after hearing he was to be given the Linnaean Society's Gold Medal, he had exclaimed 'no room for more medals.'[144] In 1901 he took possession of a cottage for him and his wife at Broadstone, on the Dorset coast. 'Old Orchard', as it was known, would be his final home.

Though Darwin's and Wallace's thoughts on the subject of evolution were very much aligned, there was one particular area where they were not in agreement. That was when it came to the origins of one single species, namely *Homo sapiens*. Wallace was not able to believe that our mental and moral qualities could come about through the same route as other creatures had developed. He felt humanity had something extra that the process of natural selection, which he helped define, could simply not explain. The way he saw it was that 'the enormous difference between man and the lower animals must have a cause, and I cannot find that cause in the ordinary processes of evolution.'[145]

While, like Darwin, he believed humankind's physical composition developed through natural selection, Wallace was of the opinion that our intellectual capacity had not been inherited in the same manner. He felt there was 'something in man that is infinite and which differs in nature as well as in degree from anything which is seen in lower animals.'[146] In his opinion, evolution could 'account well enough for the land-grabber, and the company promoter, but it fails to account for Raphael and Wagner. The world has been moved far more by spiritual forces than by material and selfish ones.'[147] If only Wallace had lived to see the beginning of the twenty-first century, I think it likely his views would have needed adjustment.

It seems Wallace's spiritualistic beliefs obscured the truth about the descent of humankind. On the other hand, Darwin did not have any theological baggage to hinder his reasoning. His argument to counter Wallace's views on this matter was that there were a number of higher civilisations across the globe, in the Middle East, India, China, Europe and so on, dispersed amongst regions with less developed peoples. There could be no suggestion that these more advanced cultures had grown out of one common ancestry, so they had to ascend from the lower forms in their specific geographical area. If, as Wallace had suggested, man had been created with all his faculties already present, then how could there be such variation in the levels of civilisation observed? It would mean certain human populations would have retrograded substantially. Darwin pushed his point further, saying 'to believe that man was aboriginally civilised and then suffered utter degradation in so many regions, is to take a pitiably low view of human nature.'[148]

Regardless of all that had transpired, there seems to have been no animosity between Wallace and Darwin. On the contrary, a strong bond developed between them, when bitter enmity is all that could be expected of ordinary men placed in this situation. In one letter Darwin admitted to Wallace, 'very few things in my life have been more satisfactory to me – that we have never felt any jealousy towards each other, though in some senses rivals.'[149]

Wallace once referred to Darwin as his 'kind friend and teacher'[150], and dedicated *Malay Archipelago*, his most important work, to him. Darwin did what he could to help Wallace in return. Before he died, Darwin used his influence to give his old friend a little more financial security, being instrumental in getting Wallace a civil list pension (see *I Will Gladly Do My Best* by Ralph Cop for more details).

Wallace had stayed fit and vibrant long into his eighties, but in his last couple of years his vitality finally ebbed way. Though he was no longer able to stroll in the garden he kept in good spirits, admiring the view from his bedroom window instead. He finally died on the morning of Friday 7 November 1913, only a few weeks short of his ninety-first birthday. The funeral service did not have the pomp and circumstance of Darwin's, but was attended by members of the Land Nationalisation, Linnaean, and Royal Societies. It had been suggested that he should be buried alongside Darwin in London, but his wife did not want this. Instead it was kept as a modest, understated affair. A plaque was unveiled at Westminster Abbey two years after his death. His wife Annie, though considerably younger than him, died within a year of Alfred's passing.

He was 6ft 2in tall and well built, but had a gentle, rather unassuming manner. His daughter Violet recollected that 'he acquired a slight stoop due to long hours spent at his desk, and this became more pronounced with advancing age; but he was always tall, spare and very active, and walked with a long easy swinging stride which he retained to the end of his life.'[151] One journalist described him thus: 'His eyes shine with intelligence, his move-ments are quick and active, there is vigour, force and power in his voice.'[152] Another spoke of him as a man of much modesty, and said that it was 'sel-dom that greatness in this world is allied to humility.'[153]

So did Darwin act un-gentlemanly when it came to taking the vast majority of acclaim instead of Wallace? Like Edison or Bell, does he deserve a place in our 'Hall of Shame'? Not at all – Darwin just sought to protect his life's work, in the same way anybody would have done. He had put many years of toil and personal sacrifice into preparing this theory, and cared passionately about it. Under the circumstances, it is hardly surprising that he wanted to reap the rewards. That said, Wallace showed incredible generosity in not trying to secure a greater share of the glory for himself.

Unlike Tesla, Meucci or Hooke, Wallace did not die embittered. As far as it can be ascertained, he was content with playing a part in the process of convincing the world of the validity of evolution. It may be a testament to his charitable nature or his belief in the magnificence of the human soul that he did not need to be the one with his name stamped across the completed work. He had helped in the development of man's understanding of the world, and that was thanks enough for him. Unfortunately, though in scientific circles his name remains highly venerated, in common historical terms this great man's achievements have been demoted to the level of a virtual non-entity.

In summary, it seems fair to say that Jean-Baptiste Lamarck and Erasmus Darwin did the groundwork, then Charles Darwin and Alfred Russel Wallace were able to build on this platform and piece together a more comprehensive system for the origin of species. Although Lamarck had the basic principles, his final conclusions were wrong. Whereas Darwin and Wallace were involved in expeditions that took them around the world, Lamarck could only make limited excursions within Europe.

Our second 'Nearly Man', Wallace, was completely on the money how-ever. He had seen the same pieces of evidence on his travels in South America and Asia, as Darwin had in the Galapagos, and had taken Malthus' theories on socio-economics and applied them to the transmutation of spe-cies in exactly the same manner. So why is he not given more of the credit?

As already discussed, Wallace was considered an amateur, while Darwin was a member of the establishment, who commanded much greater authority. He did not belong to the same clique, and was effectively an outsider, capable of demanding little respect from his peers. As with Marconi, it was not just what was being said, but who was saying it. Throughout history the well connected have been able to get their ideas across much easier than those without such a distinction. We shall return to this issue later in the book.

Another important factor that should not be overlooked was that Wallace remained on the other side of the world for a couple of years after his work was first publicised. This meant he would not get as much of the limelight. He was basically a lone voice, with nobody else fighting for him, while Darwin had the likes of Hooker and Huxley at his side.

There has been some suggestion that Darwin actually received Wallace's document much earlier, and pilfered ideas from it to enhance his own arguments. Several authors have set forth the notion that Darwin's work, though more capacious, was perhaps not endowed with the same intellectual bite as that of his younger counterpart, and that Lyell, Huxley and Hooker colluded with Darwin in a plot to hide the truth from the outside world. It is of course hard to say whether this was the case, it would certainly explain some of the coincidences involved; how their theories matched one another so perfectly, and how they arrived at such similar conclusions. However, given the fact that Wallace never portrayed himself as an injured party, it is only fair to give Darwin the benefit of the doubt.

Also much has been made of the fact that Darwin submitted Wallace's essay, along with his own, to be read at the Linnaean Society, without first getting the writer's permission. But again this seems to be more a case of looking for conspiracy that was not there in the first place. The transit time of mail sent by boat to Asia would be difficult to estimate. It could have taken months to get the request to him, and as long again to get a response. Also there seems to have been no disappointment on Wallace's part at the idea of both men's work being presented together, in fact from the correspondence we have already seen on the matter, he appeared ecstatic about the whole thing.

Another point that is often bought up is what would have happened if Wallace had not gone to Darwin with his work, but had tried to publish it directly. This book is not going to give lip service to 'what ifs', as they mean nothing, it will only look at events that actually took place. However, if Wallace had chosen this route, there is no question it would have done him little good. His standing in the scientific community was nowhere near high enough to warrant serious consideration, whereas Darwin was an 'Oxbridge' graduate who had moved in the same circles as the most eminent scientific minds since he was in his early twenties, and had a string of published works to his name.

The truth is Wallace and Darwin needed each other. They could not have got the job done in isolation, each had to call on the other to assist in completing this titanic undertaking. Wallace did not have the scientific credibility to get the theory of natural selection accepted on his own. He would not have been able to call on the support of scientific sacred cows like Lyell or Hooker. Neither did he have the methodical disposition, or attention to detail, that Darwin possessed. The shear volume of work and the multitude of examples he could draw from made it a lot harder to disagree with Darwin's ideas, while it would have been easy enough to discredit Wallace as some two-bit hack.

Likewise, there can be little doubt that Wallace acted as a catalyst, forcing Darwin to take action, and stopping him from endless procrastination. Had he not been there it is possible Darwin would have just continued amassing data for the gargantuan epic he had originally planned. It is not out of the question that he would have never finished it, and even if he had (by his own admission) the huge publication produced would have been beyond the intellectual and financial capacity of all but a very select few. Sooner or later the truth would have come out, and evolutionary theory would of course have been accepted, but it could have been several decades further down the line. A whole generation might have been deprived of these facts.

For all the great strides they had made in allowing us to better comprehend the way that nature works, there was one thing neither Darwin nor Wallace could adequately explain. How it was possible favourable characteristics possessed by individuals passed to their offspring. They had shown that natural selection weeded out those less suitable for the environment in which they lived, leaving behind individuals whose attributes were optimised to survive there, but there did not seem to be any obvious vehicle with which these traits could be carried to the next generation. Nor could they explain how some offspring of the same two individuals might inherit a particular trait, while others might not. This would fall to yet another man of the cloth to deal with. Josef Mendel, was a Czech monk with scientific aspirations. He added the final piece to this cosmic puzzle. His work with pea plants in his monastery's garden, during the 1870s, would form the basis of genetics.

As for Lamarck, as we have already established his hypothesis was wrong, but may be not as wrong as once thought. The concept of epigenetic inheritance has started to come to the fore over the last few years. This phenomenon implies that Lamarck's theory had more truth in it than previously believed, with DNA being altered over individuals' lifespans by certain aspects of the external environment, and new attributes getting acquired then passed on to their progeny. Work in this area is still in its infancy, but it proposes that environmental influences on the cytoplasm (the liquid contained in cells of living things) could actually affect the genetic structure. It could mean that resistance to things like chemical pollution might be transferred to new generations of a species, and would supplement the creatures' genetic coding.

Unlike Wallace, Lamarck felt humankind was not to be made an exception from the laws of evolution, they applied to us in just the same way as all living things. Darwin was of course in agreement with Lamarck on this particular matter, but was cautious about saying so. The Frenchman was far more open about his beliefs, and maybe this was his error. He went straight for the jugular, stated that all animals were related to one another. Darwin tried to avoid the whole subject of human descent for as long as possible and it was only when his theory had been accepted with regard to animals and plants that he brought up the evolution of humankind.

It has, on occasion, been suggested that Wallace's affinity to subjects like spiritualism actually degraded his achievements elsewhere. That because he did not limit himself to true scientific research he was branded (in the same way Tesla had been) with the mark of a charlatan. But if every scientist had possessed the mindset of just staying within the realms of what we already had a good understanding of, we would never have learned anything new. Admittedly the existence of supernatural beings has been fairly comprehensively discounted now, but back in Wallace's era it was still seen as a strong likelihood. What had made him open-minded about such matters is what also allowed him to question the established theory of how life was formed. If he had been more conservative about one thing it might have made him so about other things too, and the theory of natural selection might not have developed when it did.

Wallace thought of himself in comparison to Darwin as 'the "young man in a hurry": he the painstaking and patient student seeking the full demonstration of the truth.'[154] Though it is hard not to have anything but the greatest respect for the achievements of Charles Darwin, this can almost be totally eclipsed by the admiration Alfred Russel Wallace deserves. He managed to unlock the secret of how living things are created, and ensured this discovery was presented to the world, but his modesty and sense of fair play prevented him from taking anything but the bare minimum of praise for it. Civilisation owes him so much.

Left: 1 Antonio Meucci. (Courtesy of Meucci–Garibaldi Museum)

Below: 2 Meucci's telephone.

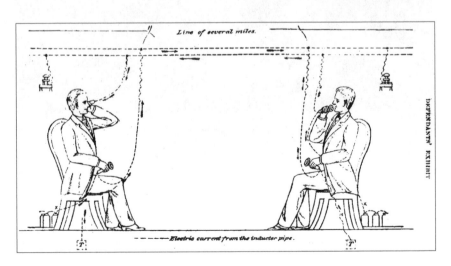

3 Sir Isaac Newton. (Courtesy of the Royal Society)

Above left:
4 Zoe Schieppati Emery's portrait of Robert Hooke. (Courtesy of the Royal Society)

Above centre:
5 Nikola Tesla. (Courtesy of the Tesla Memorial Society)

Above right:
6 Wardencliffe Tower. (Courtesy of the Tesla Memorial Society)

7 Sir Joseph Swan. (Courtesy of the National Gallery)

8 Lamarck's statue in Paris.

9 Charles Darwin. (Courtesy
of the Natural History Museum)

10 Alfred Russel Wallace.
(Courtesy of the Royal Society)

Left: 11 Alan Turing.
(Courtesy of the Royal
Society)

Below: 12 The first
integrated circuit. (Courtesy
of Texas instruments)

13 Jack Kilby. (Courtesy of Texas instruments)

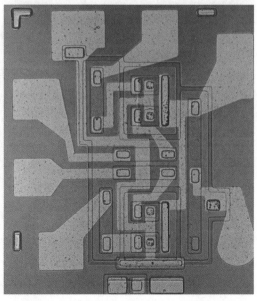

14 Early Fairchild IC, based on the Planar Process. (Courtesy of Fairchild Semiconductor)

15 Robert
Noyce. (Courtesy
of Fairchild
Semiconductor)

16 Kilby with
a collection of
integrated circuits.
(Courtesy of Texas
instruments)

CHAPTER 6

Errors of Judgement

'A torrent of light into our dark world. A new species would bless me as its creator and source; many happy and excellent natures would owe their living to me. No father could claim the gratitude of his child so completely as I should deserve theirs.'[1]

Mary Shelley

In Shelley's book of the same title, Frankenstein contemplated the possibility of 'bestowing animation on lifeless matter'. In this particular case, however, the writer concerned herself with the possibility of restoring vitality to beings from which it had previously been taken, not to form new entities capable of their own thought or action, displaying the very essence of life itself. By the middle of the last century humanity was reaching a point where it would be able to change its categorisation from created to one of creator. As Shelley had foreseen, it would be capable of siring a new species; the work of man not God. The evolution of these children that humanity had spawned had been a long and arduous one; though much of it had gone unnoticed, it took over 250 years to reach its current state. Yet it could be argued that the development of these creatures is still only at its beginning.

These uber-beings would be very simple at first, much in the same way as the single cell organisms that were the first inhabitants of our world had been. But in time the subtlety of their composition would become more apparent; they would be able to fully interact with the environment that surrounded them, and even possess the power of rational contemplation. We are not talking about some Cameron-esque nightmare of the future where machines rule the world, but we have reached a stage now where many of them have attributes which we must regard as analogous to a living thing. The human psyche is just made up of great bundles of neurones which are not dissimilar to the wiring and interconnect buses of a computer, conversely they are now capable of exhibiting clear signs of actual intelligence and certain members of their number even have organic matter in their make up. We are really not that different anymore.

Many people would help to bring about the computer age, and we will look at several of them in detail, though the centre of our attention will be the hapless British mathematician Alan Turing, and the better-fortuned Hungarian polymath John von Neumann. However, we should not get ahead of ourselves; we have some twenty decades of history to wade through before we get to discuss either of these gentlemen. So let's crack on.

Wilhelm Schichard, a professor in mathematics at Tubingen University in Germany, built a mechanical calculator as far back as the 1620s, but in truth he had little success with it. His 'Rechen Machine' was capable of adding, subtracting and multiplying, but could only deal with numbers of relatively small size, and just one device was ever constructed. It consisted of six rolls, each with a set of numbers on them, and a series of nine sliding racks. The user would pick the number that they wanted to utilise and shift the corresponding rack, then by rotating the rolls it was possible to pick a six-digit number to multiply it by. Moving the rolls to that set of positions the calculated number would appear. Unfortunately Schichard's notes were lost for many years, only re-emerging after the Second World War. As a result his work had very little influence on the development of future calculating equipment.

Some twenty years after Schichard, Frenchman Blaise Pascal constructed a mechanical device for adding sums of money. During the 1640s, he managed to turn out about fifty machines. Then in the 1670s, the Saxonian mathematician (co-inventor of calculus, and old buddy of Newton) Gottfried Leibniz created his own mechanical calculating system. This used a rotating cylinder, with a set of different length teeth, each tooth representing a digit from one to nine.

Often referred to as the 'Father of Computing', Charles Babbage was born on 26 December 1791 in London. He was the son of Benjamin Babbage, a reasonably well-off banker, and his wife Betsy Plumleigh. Charles was a delicate child, often troubled with bouts of fever. In order to save him from further risk of illness in the less than favourable surroundings of the big city, his parents decided to send him to a healthier environment. He was taken away from the filth-ridden, polluted streets of the Borough of Southwark and conveyed to Devonshire, where he followed his preliminary schooling.

When his constitution became a little less frail, he enrolled at Forty Hill Academy back in London. Then in Autumn 1810, he entered Trinity College, Cambridge, where he studied mathematics. Babbage later moved college allegiance, transferring to Peterhouse. In 1816, he was elected to the Fellowship of the Royal Society. He was just twenty-four at the time, making him one of its youngest ever fellows. Babbage was not someone fond of taking the diplomatic route, often speaking his mind. He once said that the Society's only function was to simply 'praise itself over wine and give each other medals'.

He is thought to have first contemplated the idea of using machinery to alleviate the problem of performing intricate mathematical tasks while still an undergraduate. On a regular basis he found himself bored to tears, desperately struggling to sift through reams of logarithmic tables, and began to reflect whether there was an easier way to obtain figures from complex calculus. He would later describe how 'one evening I was sitting in the rooms of the Analytical Society, at Cambridge, my head leaning forward on the table in a dreamy mood, with a table of logarithms lying open before me. Another member coming into the room and seeing me half asleep called out "Well Babbage, what are you dreaming about?" I am thinking that all these tables (pointing to the logarithms) might be calculated by machine.'[2] Over the following decade he pondered how to go about achieving this goal, and set to work designing apparatus that could create arithmetic data without the need for human intervention.

At first he 'considered that a machine to calculate the more isolated operations of arithmetic would be comparatively of little value, unless it were very easily set to do work, and unless it executed not only accurately, but with great rapidity.'[3] He could see that whatever preparatory work was needed to make the machine ready for the prescribed task had to be minimal. If the set-up was a long-drawn-out affair that outweighed the time saved by employing the machine in these calculations then all that effort would be for nothing. He had to find a simple way that the task in question and the data required to do it could be transferred to the machine.

Babbage created the first prototype of what he described as the 'Difference Engine' in 1819. It was a mechanical device, an ungainly mass of cogs and gears, but it did offer him the basic functionality needed to perform calculations automatically. This was just a starting point though; the complete apparatus that he foresaw would be made up of more than 20,000 components.

He read his paper 'Application of Machinery to the Computation of Astronomical & Mathematical Tables' to the Royal Astronomical Society in June 1822. By way of thanks he was to be awarded a gold medal from the organisation the following year. At that time he also received government funding to develop a more comprehensive version of the 'Difference Engine'. It became apparent to the inhabitants of the corridors of power that what Babbage was doing could have no end of uses. Britain had swept aside the last remnants of its feudal past and entered into the industrial age. The device could be employed by the navy to create more accurate navigation tables, it might also be utilised for statistical analysis in the fields of economics or insurance, and various engineering activities were also certain to be able to make use of it.

In 1827, Babbage took the Lucasian Chair at Cambridge, the professorship held by Isaac Newton some 150 years earlier. Sadly this triumph was marred by the death of his wife and two of his children a few months later. In 1834, Babbage published a further work on the subject that had occupied so much of his time over the last fifteen years, it was entitled 'On the Economy of Machinery & Manufacturers'. By this stage the State had ploughed nearly £20,000 into the 'Difference Engine' project, and, annoyed at the lack of progress, withdrew its support.

By then Babbage's attention had shifted onto something even more significant, and to some extent explains the tardiness of his government work. His 'Analytical Engine' would have much further-reaching capabilities. Unlike its ancestor, he envisioned this would possess the ability to be given different sets of instructions, and its operation was not rigidly fixed. The design consisted of five key parts:

1. The Store
2. The Mill
3. The Control
4. The Input
5. The Output

The 'Input' and 'Output' are self-explanatory, of course. The 'Store' contained, as he put it, 'all the variables to be operated upon.'[4] It was in the 'Mill' that these operations took place. The 'Control' was where the program (or formula, as he referred to it) by which the process would be defined was situated. The 'Analytical Engine', Babbage wrote, would work thus: 'whenever variables are ordered into the Mill, these figures will be bought in, and the operation indicated by the preceding card will be performed on them. The result of this operation will then be replaced in the Store.'[5]

In the first years of the nineteenth century, a Frenchman named Joseph-Marie Jacquard helped revolutionise the textile industry by creating a loom that used punched cards to store the patterns machines would weave. It meant designs could be implemented quickly across a vast number of weavers. Babbage saw that this method could be used to give instructions to his calculating machine. Just as they had set the pattern that would be used in weaving, here they would define the formula by which the machine would operate. Interestingly enough punched cards were to continue to find use in computers well into the 1960s, when they were finally replaced by magnetic tape.

Babbage had shown he was an astounding visionary, but the problems of how to go about manufacturing such a machine were a barrier he could not negotiate. Not only was this a far more elaborate assignment than the one he had initially embarked upon, but he was now deprived of the government stipend previously at his disposal. Despite these difficulties he carried on.

Unfortunately Babbage died on 18 October 1871, and his hopes of realising such an ambitious machine within his lifetime were not to blossom. In 1991, London's Science Museum built a working model of the 'Difference Engine', like the one he had been trying to build for the British Government for all those years and was never personally able to complete.

Though Babbage had made groundbreaking work, there were plenty of others who helped to take this concept ever closer to some perfected state of grace. By the 1880s things had managed to move on a little further. Herman Hollerith, an engineer at the US Census Bureau came up with the idea of using punched cards, similar to those employed by Jacquard and Babbage, to compile information on the rapidly expanding American population. This proved far more efficient than the rather laborious method of processing data by hand, and was quickly implemented. He went on to form a profitable business selling the systems produced to other government departments and large corporations. Out of his humble enterprise International Business Machines (IBM) was formed, which today is the high-tech industry's biggest corporation, a giant behemoth with annual revenues of over $80 billion, and more than 300,000 employees.

The first mechanical desktop calculating machines were introduced in the 1920s. They consisted of ten by ten grids of keys, and worked in a rather archaic manner when compared to modern calculators. Multiplications were done by successive additions, which meant that the operator had to work out which row and column should be utilised to produce the correct calculation. These machines were energy sapping to use, because of the sluggish system employed, plus they were very prone to human error. Someone had to come up with an item which was a bit more practical if humankind was to beat the numbers game.

The next man to step up to the plate was Harvard University's Howard Aiken. Aiken started looking into the matter of creating a more efficient calculating system back in late 1937. He designed a large-scale machine called the Harvard Mark 1, which used electromechanical relays to perform its operations. The project was undertaken in conjunction with the staff at IBM, and although it was a considerable advance on what had gone before, it still had more in common with Babbage's Differential Engine than modern computers. It was put together at the Harvard campus at first, then transferred to the IBM site at Endicott, in upstate New York. The contraption initially went into service in early 1943. It was a monster, covering over 400sq.ft of floor space, and made up of 800,000 components.[6] Due to its success, new generations of the machine would be constructed in the late forties and early fifties, but because Aiken stuck with his familiar mechanical design they never progressed any further. This methodology soon became outmoded and pedestrian. As a result of this, his impact on the advancement of computing would be negligible. His designs for the Harvard machines represented the limits of how far anyone could go using a mechanically driven machine. A glass ceiling had been reached, the move to electronics was necessary to increase the speed of such devices any further. Other technologies only offered stunted growth at best.

John Atanasoff, a resident of Iowa State University in Aimes, was in fact the first person to develop a truly electronic calculating machine and certainly deserves a mention. He began to work on it just prior to the start of the Second World War, but his success was (to put it mildly) fairly restricted. Though he personally made only minor headway, as we will see later, he did manage to have some sway on future developments.

This brief summary of what went on up to the 1940s suggests that there was some progress being made in this nascent branch of science.

Nevertheless, its pace was not that impressive. A number of people had brought things forward by small measures, but there was still not that much to get really excited about.

Several important advances would come from what can only be described as a highly unlikely origin. The aforementioned Turing was one of the men who would change the way that computers would be perceived. He would help transform these huge, ugly, lumbering pieces of hardware into the sleek, compact, mercurial devices we are now familiar with.

Alan Mathison Turing was born in the Warrington Lodge nursing home in Maida Vale, West London. His father, Julius, worked for the British Government in India. His mother, Ethel Sara Stoney, grew up in Ireland, and had gone to live in the colonies with her father, who was chief engineer of the Madras Railway. She was the granddaughter of George Johnstone Stoney, the nineteenth-century scientist who formulated a theory on the nature of electricity. He proposed that it consisted of indivisible equally charged particles, christening them 'electrons', a name which they are still known by today.

Ethel had met Julius while both of them were travelling back from the sub-continent to Europe. The couple married in the Irish capital in October 1907. Their first son, John, was born in September 1908, and Alan arrived on 23 June 1912. Julius returned to India the following spring, less than a year after his second child's birth, and Ethel followed her husband just six months later. The children were left in the care of family friends Colonel and Mrs Ward, who lived in the sleepy coastal town of St Leonards-on-Sea. They were moved around between several homes over these first few years, a number of different people taking care of them.

His mother described Alan as being 'a very pretty and engaging small boy' attracting a 'good deal of notice from complete strangers.'[7] Back then, if her accounts are taken as accurate, he appears to have been 'quite free from shyness and ready to greet anyone.'[8] Even in his early years, Alan had a virile if slightly misguided mind. At the age of four he broke a wooden soldier that had been given to him. He took the arms and legs of the toy and planted them in the garden, confident he could grow a whole platoon from the pieces.

In a biography of her son, Ethel Turing recollected that, 'when he was about six years old his quite original comments on or description of things led me to suspect unusual gifts and the likelihood of his becoming an inventor.'[9] This may of course be judged as the benefit of hindsight, but it certainly seems that Alan displayed the hallmarks of a brilliant intellect from a very young age.

In 1918, he began his education, attending St Michael's School in St Leonards-on-Sea. He does not seem to have been a particularly studious or enthusiastic pupil. His parents were told of his lack of tidiness and poor level of attention. Despite adverse reports, he made an impression on the headmistress, Miss Taylor. Upon his departure from the school she informed his mother that, 'I have had clever boys and hard working boys, but Alan has genius.'[10]

With this stage of his education completed, he was sent to Hazlehurst Preparatory School. Once again he was not keen to show any signs of excelling in his school work. When his mother returned to England for a brief spell, in mid-1921, she found that her son's persona had altered dramatically. He had transformed from 'having always been extremely vivacious – making friends with everyone' and had become 'unsociable and dreamy'.[11] She was concerned by his apparent detachment from the outside world, but could see no obvious way to rectify the situation, though it probably did not escape her that in all likelihood her continued absence was a contributing factor.

In 1926, disillusioned with his pitiful progress in the organisation, and feeling he was making no serious advance with his career, Julius Turing resigned his post with the Indian Civil Service, and along with his wife headed back to Europe. But any hopes of Alan being granted a more ordinary home life, and perhaps receiving some small measure of parental affection, quickly vanished. His mother and father decided to set up home in Brittany, where they would live as tax exiles, remaining as distant, emotionally if not geographically, as they had before. Alan's isolation from his family continued, he and his brother would visit them for school holidays, nothing more.

He took his Common Entrance Examination in Spring 1926, and was accepted to Sherbourne, a rather unassuming Dorset public school, later that year. In the early stages of his time there he was very cut off and lonely. He was not an average kid, and struggled to fit in with the others lads. His teddy bear 'Porgy' was probably his closest companion during this period. It is almost certain that he was bullied to some extent. In his school report of Spring 1927, Alan's housemaster told his parents that 'his ways sometimes tempt persecution: Though I do not think he is unhappy, undeniably he is not a normal boy.'[12]

He took little interest in class, his work remained untidy, receiving only mediocre grades, though evidently he was capable of far more. One of his schoolmasters at Sherbourne described Alan as possessing a 'periscope mind' seeing beyond certain arguments and hurdles and allowing him to reach a satisfactory though unforeseen conclusion. On the whole, his teachers found him a complete paradox, clearly intelligent, but unable to apply himself. He did not follow the preset model to which the school's pupils were expected to conform, which invited the possibility of being held back a year. On the other hand, his remarkable abilities argued in favour of him being moved up instead. His teachers' hopes of Alan getting a well-balanced education were not fulfilled, he was always going to be something of an exception to any rule.

Though Turing never fell into the category of what Sherbourne regarded as ideal for one of its attendees, he would later manage to discover some semblance of direction. Surprising everybody, he attained very high grades for his School Certificate exams in June 1928. He managed to pass in all subjects including his much-loathed Latin. His newfound zeal for study was probably due to the encouragement of a fellow pupil who had befriended him, Christopher Morcom. Christopher's grandfather was Sir Joseph Swan (Swan's daughter, Isobel, had married army officer Reginald Morcom in 1907). Though their relationship was one of affable brotherhood, for Alan it meant much more than that. This was to be his first love. Alan would, as is well documented, eventually become a practicing homosexual, but at this stage far more innocent feelings were probably being nurtured. Both shared a keen interest in scientific investigation and worked on experiments together. Morcom seems to have been a good influence on the rather wayward boy that Turing was becoming, and helped him to finally knuckle down.

In Summer 1929, Alan passed his higher certificate, and that December took a scholarship examination for England's premier seat of learning, Cambridge University. Unfortunately, he did not get the grades required. He stayed on for another year at Sherbourne, preparing himself to retake the test the following winter, and becoming a prefect in the meantime. His house tutor had been unsure that Turing would be able to cope with the responsibility, or had the temperament needed for this. However, he thought the experience would prove character building for him. He stated that the boy had 'brains: Also a sense of humour. That should carry him through.'[13]

Tragically Christopher died in February 1930. He had contracted tuber-culosis when he was younger, and suffered several periods of illness because of it. Alan was devastated by the loss. In a letter to his mother he wrote, 'I feel that I shall meet Morcom again somewhere and that there will be some work for us to do together as I believed there was for us to do here. Now that I am left to do it alone, I must not let him down, but put as much energy into it, if not as much interest, as if he were still here. If I succeed I shall be more fit to enjoy his company than I am now.'[14]

Alan visited the Morcom family home in Worcestershire several times after Christopher's death, and kept in correspondence with his parents for some years afterward. In memory of their dead son, the Morcoms inaugu-rated a special prize for science, and that summer, motivated by the thought of his lost friend, young Turing proved himself worthy of being awarded it. In December, he succeeded in getting a Cambridge scholarship. He readied himself for starting there the following summer. As March drew to a close, and not long before he finally left Sherbourne, Alan received an Edward VI Gold Medal for Mathematics.

Accompanied by Porgy, he entered Kings College that September, and began studying for a degree in mathematics. He had managed to gain a scholarship of £80 per annum to go there, but had not been able to obtain a place in Trinity College, as he had hoped for. This had been his first choice, as it was where Christopher had just begun his own scholastic career, before it was cut short.

Away from the provincial surroundings of the past, Alan blossomed. He found himself much more at home in Cambridge than he had ever been at Sherbourne. His fellow students offered him the agreeable company that had not been so abundant while he was at school. He was also free of the over-regimented atmosphere he had previously been stuck with.

His university years gave him the opportunity to develop into a more rounded individual. Though his studies would be of prime importance, he cultivated many other interests too. He became part of the college rowing and hockey teams, as well as indulging in less strenuous pastimes like chess and the ancient Japanese board game 'Go'. He read the works of philo-sophical thinkers like Bertrand Russell and Ludwig Wittgenstein. During the summer and winter breaks he and his pals would travel abroad on ski-ing or cycling holidays.

It would be during his stage in his life that Alan was first able to express his sexuality more openly. In June 1933, he took a vacation in Cumbria with one of his college friends, James Atkins. The two of them became lovers, and had an intermittent sexual relationship over the next few years.

Britain's public school tradition of keeping young lads, filled with hormones, in almost complete isolation from female contact made the likelihood of pupils experimenting with homosexuality all the greater. In the universities during this period there was little difference. The environment was one almost devoid of any interaction with the opposite sex. As a result such practices, though not overtly accepted, were in many cases tolerated. However outside the ivory tower of British academia it was still a heinous crime, and considered the depths of depravity. Alan would perhaps become too used to the overprotected surroundings in which he had grown up, and this would eventually lead to serious consequences.

He gained a first class degree in July 1934, receiving a research studentship of £200 per annum soon after. Later that year he acted as best man at his brother's wedding. He was still only twenty-two years old when he was made a Fellow of Kings College, the earliest of all the entrants starting at the same time as he did. He also won the Smiths Prize for his paper on 'Gaussian Error Function' the following year.

For his doctoral thesis, Alan decided to confront the 'Entscheidungsproblem' (Decision Problem). This had been one of the much-famed twenty-three unsolved problems set by Prussian mathematician David Hilbert at the International Congress of Mathematics, held in Paris back in 1900. With these, he had tried to lay down a challenge to the next generation of mathematicians, hoping it would stimulate great advances in the subject. This particular question had been thus; could a definite process or algorithm exist that was capable of allowing all mathematical questions to be adequately decided?

To tackle Hilbert's question, which had perplexed far more experienced minds for over three decades, Turing managed to form within his mind's eye a conceptual machine. It took the form of a small box 'with "tape" running through it, and divided into sections (called "squares") each capable of bearing a "symbol".'[15] The square that was inside the machine at any point would be referred to as the scanned square. He noted that 'in some of the configurations in which the scanned square is blank, the machine writes down a new symbol on the scanned square: in other configurations it erases the scanned symbol. The machine may also change the square, which is being scanned, but only by shifting it one place to the right or left.'[16] In addition, it was possible for the configuration to be changed by the data on the scanned square.

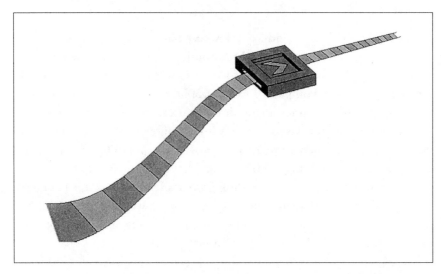

An illustration of the Turing Machine working its way through an infinitely long tape.

The 'Turing Machine' would be capable of doing any task it was specified, given the right set of instructions. It would have a series of configurations, and these would define how it reacted to the information it found in each square (i.e. the input data). A particular configuration might state that if the square had a 'o' in it, then the machine would replace it with a '1', and afterwards move along the tape to the left. If the square contained a '1' then it might leave the square as it was, move to the right, and change to another configuration. The configurations, and what these would entail for a particular input, could be constructed into a table, for example:

	o	1
Configuration 1	Change to 1, move right, keep configuration	No change, move right, take configuration 2
Configuration 2	Change to 1, move left, keep configuration	Change to 0, move right, take configuration 3
Configuration 3	No change, move right, take configuration 1	No change, move right, keep configuration

The question would then be whether all numbers that existed could be calculated using one of these tables? Was it possible for them to form an algorithm by which any number you could think of could be created? On the whole yes they would; a table of finite length could be used to calculate nearly all numbers, even ones with an infinite amount of digits, provided the configuration table was correctly written. However, there would be certain numbers for which this was not possible, and this was the decisive factor. With this Turing had shown there was no general algorithm capable of creating all numbers. He had answered Hilbert's question, the first person to do so adequately.

Turing's paper 'On Computable Numbers' basically confirmed that not all mathematical problems can be solved within a finite number of steps. In this work he managed to bridge the gap between human abstract thought and computing operations. Max Newman, one of Turing's lecturers at Cambridge, later said of the paper, it is 'difficult today to realise how bold an innovation it was to introduce talk about paper tapes and patterns punched in them into the discussion of the foundation of mathematics.'[17] This young man had already started to challenge some of the underlying questions that would characterise the latter stages of the twentieth century – namely those of computer science.

Opportunities now materialised in far-flung lands. Alan became interested in the possibility of trying to get a scholarship to study in the USA. It had been something that he had been thinking about for quite a while. The idea first presented itself in Spring 1935, following a visit to Cambridge by John von Neumann of Princeton University. Turing had read much of his work, and knew that if the circumstances presented themselves he would gain greatly from the chance to work alongside such an intellectual heavyweight.

Princeton provided a fellowship each year to sponsor one Cambridge mathematics student to go to the United States. The Procter Fellowship, as it was known, was worth the handsome sum of $2,000. Alan applied for it, but he was not successful. Not to be deterred, he still managed to scrape together enough funds, and was thus able to go over to America under his own steam.

In September 1936, Turing left Britain, via Southampton, in a third-class berth of a liner set for New York. Upon his arrival in Princeton, he was put under the supervision of Alonzo Church. Church had independently worked on a solution to the Entscheidungsproblem using what he called 'Lambda Calculus'. He came to the same conclusion as Turing had, but actually managed to get his solution into print first. Today, however, it is the Englishman's take on things that is most widely referred to.

In a letter to his parents, once settled in his new home, Alan communicated his excitement at the prospect of working amongst so many remarkable men of science, stating that the faculty came 'fully up to expectations. There is a great number of the most distinguished mathematicians here – von Neumann, Weyl, Courant, Hardy, Einstein'[18], but he did not find their company as appealing as he would have imagined. At a dinner party, held at Church's house, he recalled, 'considering that the guests were all university people I found the conversation rather disappointing, they seem (from what I can remember of it) to have discussed nothing, but the different states they come from.'[19] He found his bearings soon enough, though American brashness seemed difficult to deal with when compared to the painfully polite Englishness with which he had grown up. He managed to stretch his allowance enough to keep him in reasonable comfort, but could not afford to live extravagantly.

Turing's work on 'Group Theory', with which he was primarily engaged at this time, bought him to von Neumann's attention. The highly revered professor came from whatever the Hungarian equivalent of blue-chip stock would be. He had made his way across Europe like an academic Attila the Hun, pillaging a string of degrees and doctorates, before setting his sights on North America.

Alan continued the interest he had gained in skiing, by taking a week-long vacation in New Hampshire, during December 1936. And though his budget was somewhat limited, he still managed to make the most of his time abroad, visiting cities like Washington and Boston. It became apparent that there was a possibility to stay on for another year. Originally he had gone for a nine-month stint, but the chance to extend this seemed somewhat tempting.

In May 1937, he put his prior disappointment aside and applied for the Procter Fellowship again. This time he was successful. It allowed him to continue his sojourn stateside for a further twelve months. In an attempt to secure him the bursary, von Neumann himself had written a recommendation to the Vice Chancellor of Cambridge.

During the second year at Princeton he managed to occupy himself with a number of different projects, most notably his treatise on the calculation of the Riemann zeta-function. This was an infinite series – so you could never actually complete such a task, only continue to produce more accurate approximations. Here once again Turing contemplated the idea of using a machine to do all the tedious arithmetic work. His paper 'Method for Calculating the Riemann Zeta-Function' was presented at the London Mathematical Society in March of the following year. That Christmas he spent in South Carolina with college friends.

By mid-1938, Turing was preparing to return to Europe. The venerable von Neumann had offered him a permanent post as his research assistant, but he decided that even though the United States had been a worthwhile experience, it was not somewhere he wished to remain long term.

When Alan arrived once more in the 'old country' the atmosphere was one of great uncertainty, and undeniable dread at the thought of what lay in store. Conflict looked imminent. The British, Americans and French had not taken action when Germany blatantly ignored the Treaty of Versailles, and began a massive rearmament campaign. They had allowed it to occupy the Sudetenland, but the move had proved unwise; soon after the Nazis decided to annex the whole of Czechoslovakia, and the threat that they would next invade Poland loomed. Europe lay on the brink of war. Alan's parents had both advised him not to leave the USA, but stay at Princeton until things were a little less precarious.

Back in Cambridge life seemed to carry on much the same as it had when he left. Everything was as normal as it was possible to be, under the circumstances. Turing completed a Ph.D. in logic and number theory, entitled 'Systems for Logic Based Ordinals', much of the research for which he had worked on at Princeton.

Turing looked to be set to follow the undemanding life of a Cambridge don, or so he thought. However, unbeknownst to him, the world was going to need his talents for something far more crucial. With fear that war was now inevitable, he was asked by the Government Code & Cipher School (GC & CS) to help with the task of breaking the Nazis' military codes. He began his duties at the GC & CS buildings at Bletchley Park, Buckinghamshire, in September 1938. The Victorian mansion had been built in the early 1880s, and served as the home of stockbroker Sir Herbert Leon, before it was bought by GC & CS in 1937. Alistair Denniston, a veteran of the previous conflict's codebreaking operations and man in charge of Britain's crypto-graphic outfit during the interwar years, oversaw the relocation there. It would be the nerve centre of Allied deciphering, the department moving its entire staff out of London, to safeguard it from the threat enemy attack. The majority of the decryption team were installed at Bletchley by August 1939.

During the early stages Bletchley's full complement numbered just 200, but by the time Alan arrived this was ramping up considerably. The site became completely flooded with people, at its peak it had close to 10,000 personnel working there. Alan was joined by other leading academics from both Oxford and Cambridge. These included John Jeffries, Dillwyn Knox and Gordon Welchman. Due to the number of staff and the limited space available, a series of huts were erected in the mansion's grounds.

Although Germany's land forces and Luftwaffe had been extensively strengthened during the 1930s, and by this time had considerable might, its navy was still weak, and certainly no match for that of the British, who had remained the world's largest naval power for centuries. As a result, the Nazis preferred a strategy concentrating on the use of their Unteseeboots (or U-boats for short). Britain only had meagre resources, and relied heavily on importing supplies. The convoys, which acted as the nation's lifeline, presented a far easier target for German submarines than attacking military ships, and it would prove an effective method of depleting its enemy's strength. Under the command of Admiral Karl Donitz, the U-boat fleet started to prey on Allied shipping. Their hunting ground was the mid-Atlantic, where there was no air cover from their enemies' planes, whose range was limited to not far off the coasts of Britain one side, and Canada on the other. The area became known as the 'Atlantic Gap'. With the fall of France under Nazi rule in early 1940, the Germans secured greater access to the Atlantic coast. British ships would come under serious threat from this point on. The initial losses were enormous, in that year alone over a thousand Allied vessels were sunk, with eight million tons of cargo finding its way to the bottom of the ocean. This increased by a further thirty percent the following year.

Only two years after commencement of the war, the number of U-boats in service had multiplied five-fold. The Germans had only fifty-seven at the start of the conflict, but expanded this to over 300 by the time its third year had begun. From the start of 1942 right through to Spring 1943, Allied shipping lost 600,000 tonnes of cargo each month. The Cabinet Secretary, Lord Maurice Hankey, described the anti U-boat campaign as 'our greatest failure'[20], and even Britain's leader, Winston Churchill, later admitted that the 'Battle of the Atlantic', as it became known, was the one thing that really frightened him.

Britain was going to have to win this most crucial of clashes if it was to avoid Nazi invasion. However, it could not be done with conventional weapons, as no adequate method of combatting submarines had been developed. The only option was for the British to find a way of learning the details of German U-boat movements; this would require the intercepting, and deciphering of, their messages.

Codes have been used since the dawn of civilisation, a symptom perhaps of our untrusting nature. The first historical reference to their use goes back some 2,500 years. The Ancient Greeks used a series of numbers to form a four by six grid, each box on this grid denoting a letter. Thus by using the numbers to reference positions on a grid, the meaning of the message could be unravelled. The Romans used a shift in the alphabet (e.g. all letters were moved four to the left, therefore A would be E, and F would be J). This would be a very simple code to use, but by the same token, it required little investigation to work out its structure. A slightly more complex code was to use substitutions rather than shifts (e.g. A corresponding to L, and Q corresponding to D). Again, something like this would not be hard to decipher. Certain letters appear more frequently than others in nearly all sentences. So with enough coded material, pieces could slowly be put together, and eventually the encoding formula could be retrieved.

The Cambridge professor John Wilkins (who instructed the young Robert Hooke on scientific experimentation) invented a code which utilised five 'bits' of information to signify a letter or number. This information could then by transferred by sound (e.g. the firing of a gun) or visually (e.g. the raising of a lamp). For instance a shot then two spaces then another two shots in succession could correspond to the letter H, while five shots could denote the letter P. In the trenches of the First World War a similar code system, consisting of a series of three-letter words, was devised by the Germans. It was made up of 4,000 code words, which were changed regularly.

The art of codes and ciphers had moved on at a great pace since then, and all of these methods were mere child's play in comparison to what the Allies would have to contend with if they were to defeat the Nazis. The code the enemy was making use of was known as the 'Enigma', and it was more complicated than any previously invented. The Enigma Machine, which was to encode the material, had originally been created in Berlin, during the closing stages of the First World War, the brainchild of Albert Scherbius. This rotary electromechanical enciphering device was used by the German armed forces and secret service throughout the 1920s and '30s, as well as many Teutonic financial institutions. The basic principle was that letters were substituted for others by the electrical connections of the machine, and this would be followed by shifts in the alphabet and further substitutions.

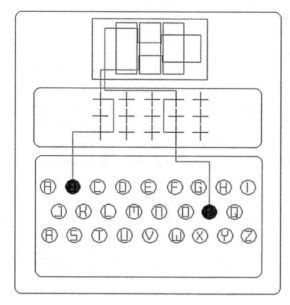

Schematic of the Enigma Machine.

The Poles had managed to crack early versions of the code with their 'Bomba' device. It was called this because of the ticking noise it made during operation (due to the electromechanical relays that occupied its guts). However, this was just a minor setback, the Germans had displayed great prudence in adding further levels of complexity to it by the time war was declared. Enhancements that Willi Korn endowed on the system in the late 1930s meant that a total of seven substitutions could take place. This translated to over 17,000 different possible alphabets being available. The machine itself consisted of twenty-six keys, a plug board (like those used in the old telephone exchanges) on which substitutions could be performed, and three rotors which set the alphabet shifts. The total number of states of the machine was 10^{23} (i.e. 10 x 10 x 10, a total of twenty-three times). This meant there were a huge number of possible settings that the Nazis could 'tune' their machines to, and left the Allies with an almost impossible task. By the outbreak of war, 40,000 of them were in operation. The Germans were confident that above all, the code was unbreakable.

This confidence was to prove misplaced though. The Enigma Machine had two slight, but fatal, flaws. Firstly, it would never encipher a letter as itself (i.e. if you put in A, it always gave a letter other than A out). Secondly, the code was reciprocal in nature (i.e. if it substituted an A for an I, then it would likewise substitute an I for an A). The reason for this was that it used the same configuration for both sending and picking up messages. This gave the Nazis the advantage that machines did not need to be set to different modes to transmit or receive signals, eliminating the possibility of someone using it incorrectly. The German command saw there would be potential for errors if the actual encoding process was too complicated. Most of the operators were not particularly well trained, and having different settings for the machine, depending on what it was doing at the time, would cause mistakes to take place, but it would also make the transmissions easier to unravel without this feature.

It was the task of the men and women at Bletchley Park to find a way to break the Enigma, and thus expose the Nazi forces' strategies. A designation system was put in place, each team dealing with a particular code. Welchman and his 'Hut 6' team dealt with the German air force codes, while the more complex naval codes were the responsibility of Turing in 'Hut 8'. The huts were supplied with intercepted data by 'Station X', the radio room in one of the towers of the mansion. Even Welchman himself acknowledged that Alan got the short straw; as he put it, this was 'a tougher problem than that facing Hut 6.'[21] The Luftwaffe codes were successfully deciphered by late 1940, but the naval codes were certain to prove more tricky.

Given the importance of the work he was involved in, and seeming to thrive on the challenge, Alan decided to defer his Cambridge research fellowship until the war was over. He dedicated himself to his work at Bletchley Park. Together he and Welchman developed the 'Enigma Bombe' a machine that would be able to get through the data retrieved much quicker, and seeing the value in such a device, the government backed the project with £100,000 of funding. The Bombes were built by the British Tabulating Company, using Turing's design. They were roughly 2m wide by 2m high, and a little under 1m deep. Each of them made its way through many different options until it found a match, slowly grinding down the code bit by bit. By March 1940, the first Bombes had gone into operation in Hut 11, and within a few months ten of them were in use.

Each morning, once Station X had captured enough signals, a menu could be constructed to run on the Bombe. This would determine the Enigma settings being used by the enemy on that particular day. It was then possible to use these settings to decipher further signals, and extract information on the Germans' battle plans. Hut 3 was placed in charge of sorting and prioritising messages. The staff worked around the clock in three shifts, to make sure no time was wasted in translating the coded information.

Once the menu had been worked out the Bombe would plod through the possible combinations. When a potential link was found, the machine would stop, and the position of the drums noted. After this was done, it was put through the Bombe's 'checking machine', to sort the 'false stops' from the 'good stops'. With this process completed, the information gleaned was utilised by the bank of 'Type X' machines to decipher the transmissions subsequently received.

The huts worked in pairs, Hut 8 would come up with the 'crib' needed for that day, Hut 4 then used this to decode all following messages. The decrypted information, known as 'Ultra' was then passed to a team of German translators, this in turn was conveyed onto the relevant intelligence departments, and finally into the hands of commanders in the field. The naval codes were sent straight to the admiralty. Churchill took a personal interest in the cryptographic process, and would receive daily updates on Bletchley's output.

The German naval codes were first compromised in Spring 1941. The Nazis, believing it was impossible to break the Enigma, instead suspected espionage was behind this breach. In early 1942, the navy hit back; a new version of Enigma was placed at its disposal. Greater complexity had been added to the machine, a fourth rotor being included, and this increased the number of substitutions immensely.

This was not the only problem Turing had to contend with. Since late Spring 1941, he had been close friends with Joan Clarke, a member of his team. The young girl had been one of Welchman's students at Cambridge, and joined the Bletchley team in mid-1940. The two spent their time together playing chess, cycling and taking weekend trips into the countryside. They began to see more and more of each other, and before long became engaged, though it was kept secret form their work colleagues.

The idea of a man not sexually attracted to women proposing to one seems a bit of a strange notion, but just like anybody else, Turing wanted a happy and contented home life, and it would not be for many decades that gay men or women would be able to openly share companionship with one another. Turing, in the same way that all of us are, was a made up of a great number of contradictions, but for the most part it was not him who was at fault; it was more to do with what society would allow, than any uncertainty about his sexual persuasion.

It appears that he admitted his true sexual persuasion to Joan, and she was willing to accept the situation. However, with time he realised it was not going to work out, and would be unfair on her to carry on in such a manner. In the end he decided to do the honourable thing and broke the engagement off. Though she continued to work at Bletchley, the two of them avoided being on the same shifts. She eventually married, becoming Mrs Joan Murray in the early 1950s, and died just a few years ago.

The second half of 1942 saw some small respite in the Allied losses. The figures dropped to 480 ships sunk compared with 530 in the previous six months. By this stage Bletchley had 200 Bombes in operation, and was finally starting to keep abreast of the Nazis' plans.

The next challenge was to try to decode the 'Fish' codes. These first saw use in 1940, and were based on the Baudot system, which made use of five digits to signify each letter, for example 00011 for A, and 11001 for B (showing some similarity to the code Wilkins proposed 300 years earlier). It was employed by German high-level command, and to break it would be another step towards victory. In order to do this a new device was designed, aptly called the 'Colossus'. The first of these machines was built at the Post Office Research Centre in Dollis Hill, under the guidance of Tommy Flowers, and installed at Bletchley in 1943. Each of the ten Colossi eventually put into operation contained close to 3,000 electromagnetic vacuum tubes (like those used by Tesla in radio communication), and could work its way through 25,000 characters per second.

Though the work of GC & CS was never about individuals, Turing's part in the proceedings was seen as something quite unique. 'He was easily the brightest chap in the place,' according to Peter Twinn, an Oxford mathematician amongst the first academics to join the Bletchley team. However, he recollected many times when Turing displayed a quite breathtaking ignorance of the real world. 'He had all kind of crackpot notions,' Twinn recalled, 'based on the fact that he didn't think the currency would stand up to a substantial war. He wanted to keep something of value, and put a lot of money into silver bars. Having extracted them from his bank, with the utmost difficulty, he went and buried them somewhere. He had an elaborate set of instructions for finding them after the war, but he never did.'[22]

Peter Calvocoressi, an RAF officer who served at Bletchley, said in his memoirs that Turing was someone that 'even Bletchley Park's most brilliant cryptographers put in a class of his own.'[23]

Jack Good, another Bletchley cryptoanalyst, once commented, 'I won't say that what Turing did made us win the war, but I daresay we might have lost it without him.'[24]

Alan was called upon to go once again to the USA in late 1942. He was given the role of Anglo-American liaison for the GC & CS, advising US Intelligence on decoding techniques. Hugh Alexander had effectively taken control of operations at Bletchley at this stage, and Turing's responsibilities in England were greatly reduced as a result. By then the Enigma codes were totally compromised, a German submarine (U-110) had been captured with its coding machine still intact and this gave the Allies the final missing pieces in the puzzle. By 1943, the British were decoding a thousand messages per day. In addition, other means of protecting its shipping became available. The development of the 'Huff-Duff' radio tracking system and the Liberator aircraft; which could travel much greater distances and thereby defend convoys for longer, had successfully closed up the 'Atlantic Gap' once and for all.

By Spring 1943, the number of ships lost had dropped by over forty percent and the tally of U-boats destroyed had risen sharply, with over a hundred put out of action in the first six months of the year. The Battle of the Atlantic was finally won, though close to 2,500 vessels had been destroyed by the time it was over. The breaking of the Enigma had proved invaluable in the land war, as well as upon the waves. It was instrumental in defeating Rommel at El Alamein in 1944. The empire that Hitler vowed would last for a millennium was over within just thirteen years. Victory in Europe would soon be attained, and Alan could look forward to returning to a normal life (well as normal as his life had ever been). Both he and Welchman were awarded OBEs in late 1945, though because of the top secret nature of their work, they were not allowed to talk about how they achieved such honours. Even Alan's family did not know the truth about their son's activities during the war. As far as they were concerned he had worked for the Foreign Office.

The Allies triumph was not to be without sacrifice on Turing's part. His prime had been stolen away from him. By the time he returned to research his powers were beginning to wane. Alan Turing's scientific career would be another casualty of war.

As pointed out in the Royal Society's Biographical Memoirs, Turing's war years were 'perhaps the happiest of his life, with full scope for his inventiveness, a mild routine to shape the day, and a congenial set of fellow-workers. But the loss to his scientific work of the years between the age of twenty-seven and thirty-three was a cruel one. Three remarkable papers written just before the war, on three diverse mathematical subjects, show the quality of the work that might have been produced if he had settled down to work on some big problem at that critical time.'[25]

He could not avoid the harsh reality that, for all the good he had done, he was not the same man he had been before he left Cambridge. Although it had given him a taste for something more rewarding and beneficial to the future of democracy, at the same time it had taken its pound of flesh from him. Like such varied individuals as Mohamed Ali, John Steinbeck, Wilfred Owen and Elvis Presley, Turing was cheated of his most productive years – stolen by the demands of their nation's military machine.

Though most of the previous chapters have pitched one person against another, in the struggle to be the first to achieve an allotted goal (whether it was the telephone, light bulb or radio communication) there is nothing to suggest that either Turing or von Neumann had anything short of mutual admiration for their counterpart. Though their backgrounds and personalities were poles apart, and they never worked in direct collaboration, each would benefit to some degree from the other's forethought.

Originally called Janos Neumann, the second individual that our attention will converge upon during this story, was born on 3 December 1903, in Budapest. He was the son of wealthy banker Maximilian 'Miksa' Neumann and his young wife Margrit Kann. He was their eldest child, two brothers, Michael and Nicholas, followed in 1907 and 1911 respectively. Miksa was a director of the powerful Magyar Jebzalog Hitelbank, and he and his family lived in a sizeable apartment on the more prosperous Pest side of the city.

Janos or 'Jansci', as he as known amongst his family, was exposed to an intellectually stimulating environment from an early age. His parents tried to give him a good grounding in history, literature, music, art and science. He was soon identified as a child prodigy, showing an uncanny skill for mental arithmetic. When he was only six he could converse with his father in Ancient Greek, and before reaching ten was capable of dividing two eight digit numbers in his head with consummate ease.

He learnt French, German and Latin in addition to his mother tongue of Hungarian. During his childhood both German and French governesses were employed at different stages, in order to improve his fluency in each of them. It was said of him in later life that he could talk as fast in any of seven languages as most people could in their native one.

He received private tutorage at home to begin with, then entered the Gymnasium Evangelikus when he was ten. Both Michael and Nicholas would follow their big brother there later. Miksa took a great deal of personal responsibility for his children's education too. At mealtimes he would conduct in-depth discussions on various subjects, and even gave them brief lectures on science and the arts. He also acquired a vast library, which Jancsi and his brothers took full advantage of whenever possible. Miksa had possessed a powerful intellect, with a convivial and entertaining nature, and these traits appear to have been passed on in abundance to his son.

The young lad had what can only be described as a photographic memory, able to absorb everything. As a party trick he could memorise a whole column of names and numbers from a phone book, and guests would pick ones to ask him about (i.e. what address or name they corresponded to), which he managed to answer almost instantly. His powers of recollection seemed virtually limitless. As he got older he could recite whole chapters from books with total accuracy. To the frustration of his teachers, Jancsi would turn up in school after not doing his revision for that day's assignment, then proceed to enter the discussion, or respond to questions, as if he had been studying all night.

His father received a title in 1913, as acknowledgement for the work he had done for the government's economic advisory committee, and thereafter Janos added 'von' to his name (the German way of signifying ennoblement). With Miksa's wealth growing, the family acquired a summer home just outside Budapest in the early 1920s.

Miksa was not a practicing Jew. When one night his son Michael asked him why they considered themselves Jewish if they did not follow it, not looking up from his reading matter, he replied 'tradition'. Janos would have a similar indifference when it came to religion. When he later married he had no qualms about converting to Catholicism, to improve relations with his wife's family.

Political unrest followed the end of the First World War. In March 1919, the liberal government, under Michael Karolgi, was ousted by the communists, led by Bela Kun. Hungary fell into a period of oppression, and the Neumanns fled, finding sanctuary in Austria. Luckily this self-imposed exile was relatively short-lived. The communists were overthrown before the end of the year, and Janos returned to complete his studies once a less extremist administration took power.

While Turing had not shown his greatness until much later in his life, Janos' grasp of the physical sciences impressed his teachers. They recognised soon enough that they were in the presence of a boy with extraordinary ability. His mathematics master at the Gymnasium, seeing the potential he possessed, recommended that his father got the child extra tuition. So to supplement his standard education, he received further coaching from Michael Fekete, of Budapest University's mathematics department. Before long it became more than a normal teacher-pupil relationship though. The wisdom beyond his years, that seemed so prevalent in von Neumann's personality, meant he began to be considered an equal rather than lowly apprentice. He and Fekete became co-workers, and by the time Janos was eighteen, they would publish a research paper together.

Later that year he enrolled at Budapest University to study mathematics, but his intention was not to stay in Hungary. He attended almost by proxy, just turning up for the exams. Despite the fact that he never went to a single lecture there, he still managed to obtain his diploma almost effortlessly. He was simultaneously taking a chemistry degree at Berlin University, where he spent most of his time. He was leading a double (if not triple) life during this period, and clearly did not want to run the risk of becoming bored, or having too much time on his hands. Once these courses where out of the way he decided to take more. In 1923, he entered the Eidgenoessische Technische Hochschule (ETH), in Zurich, where Einstein had studied. There he finished a chemical engineering degree in a little over two years. He received his doctorate in mathematics from Budapest University in 1926, but was still only twenty-two at the time. (Most students, myself included, would have just finished a single degree by this stage in their life, while von Neumann had obtained three and a Ph.D. to boot.)

He had condensed what would be by most standards an accomplished academic career, into the space of a little over five years, somehow managing to juggle several courses at a time. Just after turning twenty-three, he became the youngest person to lecture at Berlin University. That same year, with the help of a Rockefeller Foundation scholarship, he got a post-doctoral place at the University of Goettingen. It was there that the greatest minds in science and mathematics had over the years congregated. Its staff included the likes of the aforementioned Hilbert, Werner Heisenberg and Kurt Goedel.

It had gained a lot of attention for its part in the advancement of a new branch of physics that looked at the nature of sub-atomic particles; a subject which became known as quantum mechanics. The premise was that things at this level did not act in ways that we would consider normal. In fact their behaviour was downright odd. As mentioned when discussing the set-to between Hooke and Newton regarding whether light was a made up of waves or of particles, it was later shown by quantum mechanics that it was actually both at once. This was just the tip of the iceberg, however, things got far weirder. Until this stage every physicist would say if you did x and y then you knew z would be the result, but this was not guaranteed to be the case here, not by a long mark. Protons, neutron, electrons and photons were quirky little things, and never quite did what was expected of them.

To give an easy to follow example that illustrates how it works, we can apply it to everyday surroundings. For instance, if you lifted an object a metre off the ground and then let go, it would fall. What is more, the rate at which it fell could be predicted exactly, and if you did that same thing again a hundred times you would get precisely the same outcome. On the other hand if the world we are accustomed to worked on the principles of quantum mechanics this would no longer hold true. What might happen is say eighty times the object would drop just as normal, but then on another ten occasions it might drop at half the speed it had before, a further seven times it might actually go upwards hitting the ceiling, and three times it might hover where you left it.

Basically anything was possible, and it was just that some options were highly probable and others were not. To most people even today this is completely beyond all comprehension (and quite rightly too), luckily it need only concern scientific types. Back in the mid-1920s, however, even most of these guys could not get to grips with the concept either; there were perhaps five and six people in the world able to make sense of any of it.

One of the chosen few to be touched by the hand of this new philosophy was von Neumann, and not only did he understand it, he was possibly the only person around who could explain it to us lesser mortals. Using Hilbert's multi dimensional notation called 'Hilbert Space', he created a way of expressing the complexities of quantum mechanics in a more simplified way. Hilbert Space used a number of dimensions to signify different variables. Utilising this, he could indicate the energy states, velocities, directions and positions of bodies being investigated. This model would finally come to completion in his book *Foundations of Quantum Mechanics*, which was written in 1932.

During the 1920s von Neumann would do a great deal of work on the subject of mathematical logic, influenced once again by the work of Hilbert. His second paper *Toward the Introduction of Transfinite Ordinal Numbers* was published in early 1923, and this was followed two years later by his seminal work *An Axiomisation of Set Theory*. He was capable of making a great contribution to the evolution of both physics and mathematics, but would find his greatest opportunities would come from somewhere he was still not remotely familiar.

In 1927, he was given a research post at Berlin University, but keen to secure a professorship as soon as possible, he decided to transfer to Hamburg University in 1929. By this stage he had succeeded in placing himself at the forefront of mathematical development. He was seen as the young and vibrant trailblazer who would take up the mantle of his forebears and usher in a new epoch in this discipline. Now he had to look for another challenge (and possibly a chance to escape from Germanic cuisine too). Before long the 'New World' beckoned him.

As 1929 drew to a close, von Neumann received the offer of a part-time post in Princeton. The university was keen to establish itself as a bastion of scientific research, on a par with those of Europe. To do this it needed to recruit some of the scientific elite from across the Atlantic. Arriving there in February 1930, von Neumann took the position of guest lecturer in mathematics, and for the next three years split his time between teaching back in Germany, and the USA.

His transient existence had meant that he changed his first name almost as often as some people change cars, or hairstyles. He had been Jansci while in Budapest, Johan when his family had taken refuge in Austria and studying in Germany, before finally anglicising it completely upon arrival in Princeton. It was a time of new experiences for 'Johnny' as he now became known; a new job, new country, and new home life. He had only just married his childhood friend Marietta Kovesi days before leaving Europe. The couple had become close during the summer of 1929. Their parents had been acquainted for many years, and the two of them had effectively grown up together. They were wed on 1 January 1930, in Hamburg, before making their way to North America. She was just twenty at the time.

In January 1933, as the newly formed Institute of Advanced Study (IAS) opened its doors, von Neumann was one of the men considered for professorships. It offered him a permanent post in the United States, and given the atmosphere in Europe at that time, he was glad to have the option.

Oswald Veblen, Einstein and James Alexander were first to be confirmed, taking up their posts in February. von Neumann was installed in March, with Herman Weil and Marston Morse following soon after him. As professor of mathematics of the most high-profile academic institution of its era, von Neumann was now really viewed on the world stage.

In Spring 1934, he took part in a European exchange. Princeton was keen to gain the services of Cambridge's leading physicist Paul Dirac, for a term or so. In return von Neumann went over to England for a few months. It was here, as already related, that he met Turing for the first time. There is little doubt that von Neumann was everything the Englishman was not. Turing was the archetypal bumbling academic; unkempt, dishevelled, ill at ease around other people, and absent minded, while von Neumann was to an extent the very antithesis of this; gregarious, witty, and always immaculately dressed. Nevertheless they shared a common bond, an innate understanding of the complexities of the mathematical world, and more importantly the ability to stretch its parameters outside what had already been covered. For both of them it was this that would shape the course of the rest of their lives, and though they were never destined to work together, each benefited greatly from the other's wisdom.

Johnny appeared to have a happy enough lifestyle, though he was often too engrossed in his work to have enjoyed it as much as he should have. In 1936, the von Neumanns had a daughter, Marina, and later that year Johnny's brother Nicholas emigrated to the US, staying with them for a while. Following Miksa's death, his mother also came to live with him in North America.

Sadly his marriage to Marietta was completely over by mid-1936, and they divorced within a year. She had not received much attention from her husband, who was too wrapped up in the many different projects he had become involved with, and decided that life was passing her by. A mixture of the age gap, his workload, and their very different personalities all contributed to their relationship's demise. She finally ran off with another man, leaving von Neumann devastated.

It was in mid-1938, not long after he obtained American citizenship, that von Neumann met Klari Dan while back in Hungary visiting family. They were married that autumn. This time he tried to be more considerate of his partner's needs, and this appeared to work admirably.

One of the most interesting topics he was involved in during his pre-war years at Princeton was the development of 'Game Theory'. Basically this was a mathematical methodology that used statistical analysis to formulate tactics for two-player games. At the heart of this hypothesis was his 'Minimax' principle, which lends somewhat from the uncertainty he saw in quantum mechanics. Any game would have a huge number of possible outcomes, and it was a matter of examining the likelihood of each that allowed a player to formulate the best approach in order to triumph.

Take a game like chess. If the other player carries on in a normal way, while you followed a strategy in which each decision made during the course of the game statistically gives you the greatest gain, while at the same time offers the least amount of risk, then in principal you will win. To put it simply it is all about minimising your losses to the opponent and maximising your gains. In 1939, he published *Game Theory in Application to Economies*, and this was followed by *Theory of Games & Economic Behaviour* in 1943. Though he wrote extensively on the subject, and formulated a supposedly foolproof strategy for winning games, often using poker as an example, he was not actually that good at putting it into practice. By all accounts he was a poor card player.

Flying in the face of what was generally accepted, von Neumann considered economics to be a science. He once wrote that 'economics is much more difficult, much less understood, and undoubtedly in a much earlier stage of its evolution'[26] and because it was still only an emerging subject, it was unfair to expect its development to be comparable with the more antiquated scientific arts. In his opinion the 'arguments often heard, that because of the human element, or because there is – allegedly – no measurement of important factors, mathematics will find no application, can be dismissed as utterly mistaken.'[27] He pointed out that similar accusations could have been made in the past about other fields where an understanding of mathematics had now become vital, such as physics.

In late 1941, as the war began in earnest for the United States, von Neumann became a consultant to the National Defence Research Council (NDRC). Just as the British Government had wanted to benefit from Turing's incredible intellect, America was keen to make use of its adopted son's talents. He would be employed by the Ballistics Research Laboratory in Aberdeen, Maryland, as well as being involved in several other military projects. By 1942, becoming further embroiled in these matters, he and Klari were forced to move to Washington.

The following September, he was enlisted as a consultant to the Los Alamos Nuclear Research Laboratory, where work was already in process on the infamous 'Manhattan Project'. Roosevelt's advisers had estimated that a full-blown assault on the Japanese mainland could come at the price of half a million American lives. Invasion was clearly not an option, but if the war was to be brought to an expedient end then some sort of decisive move was needed. The only solution seemed to be development of nuclear arms, which could inflict huge losses on the enemy without endangering Allied troops. Research into the production of such weaponry had been underway for some time. Originally the US Government had been worried that Germany, with its huge lead in rocket science, might be able to create an atomic bomb before its own specialists could.

The group of scientists assembled for the task was, to put it mildly, impressive. So many great minds had not worked together on a single endeavour before; it was quite simply a 'Boffin Dream Team'. The list included Italy's Enrico Fermi, Denmark's Neils Bohr, Germany's Albert Einstein, and Hungary's Edward Teller, with American Robert Oppenheimer taking command (despite his known links with left-wing organisations) as director of operations. They would set about trying to put the work Bohr had done just before the war on the fission of uranium into practice.

The principle of nuclear fission is thus: if an atom of uranium-235 is hit by a neutron, then it envelops this within its nucleus. However, the atom rapidly becomes unstable and breaks apart. This creates two new smaller nuclei, gives off some excess energy (the sum masses of the component parts being less than the original mass, and by Einstein's legendary equation $E = MC^2$ the mass not accounted for being transformed into energy) and also emits two or three more neutrons in the process. Now supposing one of the neutrons given off hits another uranium nucleus, causing a further fission, and this in turn expels a neutron which hits another nucleus within the body of the substance. Each time this occurred a bit more energy would be given off. If it happened enough times then a huge amount of energy would be created. This is what ensues when a nuclear explosion takes place, a state of 'critical mass' is attained. Basically there are enough atoms of the uranium isotope that the odds are that at least one of the neutrons each fission gives off will hit another nucleus. With the average number of reactions above one (say statistically for every fission 1.2 of the neutrons produced initiates another reaction) then the whole process will rapidly escalate. So if enough uranium was packed together then it would result in the sudden release of a massive quantity of energy.

The issue was how to do this at will. It would be necessary for the uranium to be completely safe during the bomb's construction and trans-

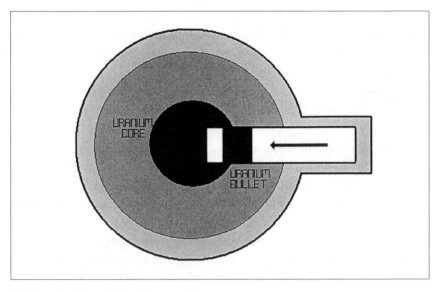

Uranium-based atomic bomb.

portation, but when the time came it would still be able to fulfil its onerous requirements. The staff at Los Alamos had to come up with a means of keeping apart two pieces of uranium-235, which on their own were too small to cause a chain reaction, but when slammed together would reach the level required to get the process started. The bomb they constructed for this purpose worked by firing a column of uranium into a cavity within a larger body of the substance, thus critical mass would be achieved and bingo! The column had to move very fast, so a quantity of cordite was exploded behind it, propelling it into the cavity.

For the next eighteen months this project was von Neumann's main concern. He and his colleagues faced countless problems and frustrations before the task was completed. Finally by Spring 1945 their labours were coming to an end. The test at the 'Trinity' site, out in the New Mexico desert on 16 July, proved successful. The device exploded, producing a crater over 300m wide. Following the test, the true extent of what they were about to do became apparent. Oppenheimer, in a moment of reflection, quoted a line from the Hindu scripture the Bhagavad-Gita. 'I am become Death, the destroyer of worlds.'[29] The time approached when it would be carried out for real. This was no longer some laboratory experiment, or theory scribbled upon a notepad, it was to be a physical act, one of stark and unadulterated violence.

Despite its nickname, 'little boy' weighed over four tons. The bomb was constructed in New Mexico and shipped to South East Asia in late July. Using reconnaissance, the US military made a list of viable targets, among them were the nation's capital city Tokyo, the former capital Kyoto, the principal port Yokohama, and the cities of Kokura, Hiroshima, and Nagasaki.

In the early hours of 6 August, at the US Air Force base on the island of Tinian in the South Pacific, just to the north of Guam, preparations were being made. The six-man crew boarded a B-29 bomber, which had been given the name *Enola Gay* only the day before, and were ready to get under-way. They took to the air just before three in the morning. By the time they had returned home later that day, over 60,000 people were dead.[28]

The 13 kiloton bomb (equivalent to 13,000 tons of dynamite) was dropped onto Hiroshima at 8.15 a.m. It exploded above the Shima Hospital, in the centre of the city, causing complete devastation. Hiroshima's popula-tion was close to 400,000. It is estimated that the blast, and the fire that engulfed the city afterwards, took 60,000 lives. The death toll as a direct result of the bombing rose to 100,000 by the time winter began, and a further 100,000 died within five years due to exposure to radiation.[29] The bomb had effectively killed half of the city.

The hypocentre of the explosion remains to this day exactly as it was sixty year ago. The ruined Industrial Promotion Hall, that stood a hundred metres away from the Shima Hospital, and the only building at the apex of the blast not completely flattened, still stands there as a chilling reminder of what humanity is capable of.

The USA had hoped that the horrors already witnessed would be enough to bring the war to an end, but the Japanese did not surrender. Newly sworn-in American President Harry Truman's warning of further bombings went unheeded. His threat to unleash a 'rain of ruin the like of which had never been seen before' was not enough to make them yield; the nightmare was not over yet.

The second bomb was to be made out of plutonium. The quantity of uranium-235 that could be got hold of was very small, it took a long time to extract enough of it from the element's natural form, and this would hold up the American assault for many months. It was hoped, of course, that Japan would have given up the fight after the first bombing, but if this was not the case, there had to be swift action or else the war would drag on.

By contrast plutonium was much easier to produce; a process had first been developed at the University of California, back in 1941, which facilitated its formation. All that was required was to bombard common uranium with deuterons (the nucleus of a deuterium atom – consisting of a neutron and a proton). The uranium and deuterons would merge to form a larger element (called neptunium), which had a very short half-life (i.e. it quickly decayed into another element). Within around two days half of it would have turned into the sought-after plutonium, two days after half of the remaining neptunium would have done so too, and so on. The relatively brief time frame required to create a substantial quantity of the material made it more suitable to use as the bomb's active ingredient. The only problem was that the same method of activation could not be employed as had worked with uranium, since plutonium atoms sometimes decayed by spontaneous fission, and this would make the device a little bit 'erratic' shall we say.

Seth Neddermeyer of the California Institute of Technology, had come up with the suggestion of using explosive charges to compress the plutonium into a state of critical mass. When explosives encasing it were detonated, the density of the plutonium increased. Squashed by the force exerted from each side it would reach a stage where the number of reactions taking place triggered the bomb.

Explosives on the outside of the plutonium core would be split into two layers; a highly reactive one on the outside, and a slower reacting one further in. The reason for this was to make sure the shockwaves produced would be synchronised together, and thus be more effective at crushing the plutonium. It would focus the blast just like a lens.

The 'implosion lens' appeared to be the perfect solution for initiating a plutonium-based nuclear explosion, but that in itself raised further problems. If the shockwaves did not act in unison the full force of the explosion would not be harnessed, and a chain reaction would not be instigated. It was impossible to run a series of tests to see what the outcome would be each time they tried something different. For obvious reasons, it was not something that could be worked on through simple trial and error. The only way to predict the outcome of different variations, and thereby find the correct orientation for the implosion lens, would be to make use of numerical modelling.

It was von Neumann who took on the challenge of seeking out a way to cope with the mind-boggling calculations needed to create the lens,

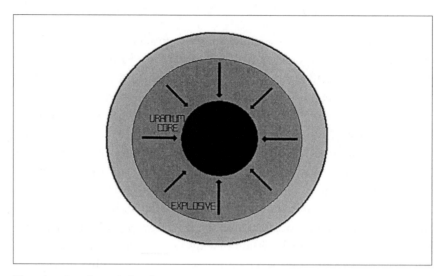

Plutonium-based atomic bomb.

and soon after began his quest for the equipment to complete the task. He looked at the electromechanical machines developed at Harvard by Aiken, and similar ones in use at Bell Laboratories, but they were not fast enough for such vast quantities of data. The immensity of the mathematical problem was truly overwhelming.

In the end a chance meeting would give him a way to complete the calculations, and additionally have a profound effect on the rest of his career. Through his involvement with the BRL he learnt of another venture that was underway, in which a group of academics were building an electronic calculating machine. The Electronic Numerical Integrator and Computer (ENIAC) covered 1,500sq.ft of floor space, and weighed over thirty tons. The project had begun in Spring 1943, at Pennsylvania University. The machine was designed by electrical engineer Presper Eckert and physicist John Mauchly, for the purpose of producing accurate firing tables for the army's long-range guns. It could calculate shells' trajectories, taking into account the firing angle of the gun, the wind's direction and its velocity, as well as humidity, temperature and atmospheric pressure. He could see that this would be able to deal equally well with all the variables in the implosion lens problem.

By hook or by crook, von Neumann managed to get the modelling calculations done, and thereby hold up his end of things. The lens system was finally judged to be satisfactory by late 1944. When the initial bombing of Hiroshima proved incapable of breaking Japanese spirits, it was this technique that would be employed to cause yet more devastation. The second bomb was known as the 'fat man' and lived up to its name far better than its predecessor. The spherical device was over 3m wide.

At eleven in the morning, on 9 August 1945, the US bomber *Bock's Car* unloaded its deadly payload over Nagasaki. This time an even more potent weapon was unleashed on the unsuspecting populace: the second bomb had the strength of 20 kilotons. Over 70,000 of the city's inhabitants were killed on the day or soon after, this again more than doubling within five years. There was finally no option; Emperor Hirohito issued a proclamation on 15 August confirming Japan's surrender. The Second World War had come to an end. In total the conflict had taken the lives of twenty million servicemen, and somewhere in the region of sixty to seventy million civilians.

The realisation of what their work had been responsible for weighed heavy on many of the scientists involved, not just Oppenheimer. Soon after Nagasaki, Einstein told a journalist that if he had known beforehand that his work would help generate such suffering, he would have become a watchmaker instead of a scientist.[30]

After the war ended, von Neumann was awarded the Medal of Merit by President Truman in recognition of his contribution to national defence, the highest civilian honour that an American can receive. His work on the first nuclear weapons had brought him into contact with the embryonic stages of electronic computing, and from this point on he devoted himself to the subject completely. But for some years before Johnny converted to this new faith, others had been putting in the groundwork, most notably two gentlemen by the names of Eckert and Mauchly.

Presper Eckert was born on 9 April 1919 into a well-to-do Philadelphia family. His father had made a very good living as a property developer, and this added to their already considerable ancestral wealth. The Eckerts had class oozing from them; they were an old money New England family. Presper joined Pennsylvania University in 1940 and was still to get his bachelor's degree when the first talk of ENIAC started to take place. John Mauchly was a physics doctorate who had come to Pennsylvania from John Hopkins University in mid-1941. He had been born in Cincinnati on 30 August 1907.

A little while before America had entered the war, Mauchly had visited his old pal Atanasoff (as mentioned earlier in this chapter), at Iowa State, and seen the rudimentary electronic calculating machine he had managed to design. The concept was completely new, and though it was still very rough around the edges, he could see there was something in it. He felt that a much more ambitious project should be embarked upon, to create a calculating machine capable of working on the most complicated of mathematical problems. He was personally interested in using such a machine to predict weather patterns, but could see there would be numerous other areas that could benefit from such a piece of hardware. In August that year, he put together a proposal to his department for the construction of an electronic computer. The problem was there simply was no money available. The project was well beyond the budget of even as reputable university as Pennsylvania. It was one of the most venerated academic establishments in the USA, tracing its origins back to the father of American science himself, Benjamin Franklin. Despite its impeccable heritage grants were scarce, especially in a country still recovering from the worst depression in its history, and with the ominous spectre of international conflict looming in the distance. What did not help either was the fact that Mauchly was not very well thought of as an academic, being seen by the higher echelons of the university as a bit of a dreamer.

At this time the university was helping the BRL out with some contract work on computation for its heavy artillery. When Mauchly learned about this, he saw an opportunity to take his design off the drawing board. He successfully sold his idea to the military, and the two projects were merged. In Spring 1943, 'Uncle Sam' confirmed it would back the ENIAC project, giving an initial payment of $150,000 to the university. Mauchly was put in charge of the operation, with Eckert as his number two, being given the title of chief engineer. Hermann Goldstine, a technically competent lieutenant, was made the BRL's liaison to the scheme, and took care of its funding needs. Work began in earnest that June. The ENIAC core team of Goldstine, Eckert and Mauchly was complemented by ten more engineering staff and several female operators.

ENIAC would be much faster than the Harvard machines, because it was electronic in origin. It relied on the movement of electrons, not shifting positions of mechanical components. Rather than relays, it made use of vacuum tubes. In total it contained 18,000 tubes and nearly 100,000 passive devices. It consumed a monstrous 170 kilowatts of power (a modern laptop, with many times more processing capacity, only expends four of five watts in comparison), and was over 25m long. It could store a set of twenty ten-digit numbers, but for all its advances it was still riddled with inadequacies. Reprogramming, for instance, would be a huge problem. It would require re-plugging thousands of switches, and take countless hours to complete. Even more important was the risk of failure of one of the countless tubes that were to be found in ENIAC's innards. The machine could easily burn out 2,000 tubes each month. If just one of them blew while a calculation was being run, the whole thing had to be started again.

Though Mauchly was the ideas man, it was Eckert's brilliance that really made the whole thing work. He excelled at practical matters, and soon started to tackle the issue of ensuring reliability. It was clear that the rate of tube replacement had to be minimised. Eckert noted that the tubes were more likely to blow under stress from being turned on. He therefore decided that ENIAC should be kept on all the time, as the power savings made by turning it off did not cover the cost of replacing tubes, or make up for the inconvenience caused. He also decided to run the tubes at the lowest possible voltage he could get away with, roughly a tenth of what they were specified to use, so as to not put any unnecessary strain on them.

At this point von Neumann was dividing his time between Aberdeen, Los Alamos and home. With the 'Manhattan Project' needing a way of getting through the immense amount of data that had to be analysed, and the mechanical-based machines on offer at the time sadly lacking when it came to speed, Johnny was desperate to find an alternative that was up to the challenge. A brief conversation at a train station was to change everything.

He met Goldstine while waiting on a railway platform in August 1944, and learnt of ENIAC. The lieutenant recognised von Neumann, and somewhat in awe of the well-known scientist, proceeded to discuss the project he was involved in. It was not really the height of discretion on his part. You would assume an army officer would display a little more caution, especially in wartime, but it turned out to be a good thing that he had been so talkative. The level of secrecy between different schemes meant useful partnerships and synergies were almost certainly overlooked on many occasions.

In principle ENIAC was a thousand times faster than the electromechanical dinosaurs to be found at Bell Labs and Harvard. Johnny visited Pennsylvania University later that month, and was impressed with what he saw. He was made a consultant to the project soon after.

Despite being the major reason for its construction, ENIAC would not be able to contribute fully to the war effort, as it was not completely operational until February 1946 (though it was used to calculate firing tables, and on the Manhattan project, even before being totally finished). At the end of that year, the machine was moved from its home in Pennsylvania to the BRL headquarters in Aberdeen. Here it continued to do computations for many years, until finally being shut down in 1955. By that stage it had completed 80,000 hours worth of calculations. In 1995, to celebrate the fiftieth anniversary of ENIAC, Pennsylvania University commissioned the design of a semiconductor chip duplicating the computer's entire functionality. It was just 3 x 1cm in size.

Johnny's involvement in ENIAC was really when things had reached the final stages. The design had been put together a long time before he joined the team. But plans were soon afoot to build a successor, and for this next endeavour he was determined to be a major part. It was here that he would lay down the foundations upon which every computer to come after would follow.

The Electronic Discrete Variable Automatic Computer (EDVAC) offered an opportunity to improve on all the errors interwoven into ENIAC's design that only became apparent when it was put into operation. It was also a chance to do things that there had not been time to finish on the first machine, and make up for some of the compromises the initial project had been burdened with. The venture started in October 1944, with von Neumann at the helm. Mauchly and Eckert came to resent the way their Hungarian colleague had managed to assume command of something that, in their eyes, had been getting along nicely without him. But this was just the beginning, tensions would be raised to new heights soon after.

EDVAC was planned to be the first stored program computer. The importance of a stored program was that it greatly increased programming flexibility. In many ways it was reminiscent of the infinitely long tape the 'Turing Machine' utilised, data could be used to set new tasks for the machine to do.

The 'First Draft of a Report on the EDVAC' was pieced together by von Neumann and Goldstine in Spring 1945. The hundred-page document was distributed later that summer. In the report he expressed the paramount importance of making better provision for data storage. He stated, 'the device requires a considerable memory. While it appeared, that various parts of this memory have to perform functions which differ somewhat in their nature and considerably in their purpose, it is nevertheless tempting to treat the entire memory as one organ.'[31] Eckert looked at the use of mercury delay lines to accommodate the voluminous memory required. Basically these used compression waves to store information.

The design which EDVAC followed would be the first example of what became widely known as the 'von Neumann Architecture'. It consisted of the following:

1. A Program Controller (Controller Unit)
2. A Processor (Arithmetic Unit)
3. A Memory
4. Input & Output Devices

The 'Processor' would execute the instructions set, or program. Each instruction gave the address of data required. The Processor took data from the 'Memory', performing one function at a time. There is not much effort needed to spot the similarity between this system and the one Babbage had proposed; the Mill replaced by the Processor, the Store translating into the Memory. The strict, step-by-step process EDVAC used would make the system easy to control with programming. The lynchpin of the system was the stored program. This allowed both instructions and data required to be held in the Memory, and thus the device did not need to be altered in order to perform each task (dispensing with the rewiring that machines like ENIAC required to change the program). This enabled the sort of machine which Turing had foreseen, or at least a close approximation of it, to become a reality. The tasks could be set without alteration to the machine itself. As the instructions were stored, the whole thing would be far quicker, there was no delay in getting data input into it by use of punched cards as this could all be entered before the calculation began. Also with certain operations that had to be repeated several times, the information could be taken directly from the Memory, rather than having to re-enter it again and again.

Von Neumann's paper owed greatly to Turing's 'On Computable Numbers' and Babbage's concept of an 'Analytical Engine', as well as the innovations Eckert and Mauchly had made with ENIAC. Turing was a pretty easygoing chap, so would not have been worried about his work being used in such a manner, and as Babbage had been dead for close to seventy years he was not going to kick up a fuss either. Conversely, Eckert and Mauchly both felt they had been defrauded. The pair became deeply annoyed by the report that von Neumann had penned. It implied that the ideas, which had been worked on as a joint effort (and some developed long before the Hungarian had even joined the team) all came from him. They believed that all von Neumann had really done was summarise the accomplishments of the whole group, but written it in such a way as to make it sound like it was simply his initiative. In an attempt to counter their disdain, von Neumann would claim that they only cared about the commercial possibilities for computers, and had purely selfish motives.

In August 1946, completely fed up with von Neumann's antics, they decided to hit out on their own. Over the next two years they fought a legal battle to claim the patent rights for their own, but without success. They went on to form the Electronic Control Corporation, and the company was triumphant in producing the first commercially available computer, UNIVAC (Universal Automatic Computer). The machine consumed eighty-one kilowatts of power, and had a thousand-word memory. It was held within a 20m.sq. water-cooled and air-conditioned room. Various large organisations bought one, mainly government departments and a couple of large enterprises. In total around a score of the machines were made. However, they never managed to create enough revenue to really put the business on a firm footing. Eventually the company was taken over by the Rand Corporation, although both kept consultancy posts there for some years. Other commercial computers were to follow later in the 1950s, the IBM 701 and the Lyons LEO 1 among them.

Mauchly died at his home in Abington, Pennsylvania, on 8 January 1980, following a brain aneurysm. Eckert lived for another fifteen years and was universally respected as one of the most important contributors to the development of the computer age. He was given the IEEE (Institute of Electrical & Electronic Engineers) Computer Society Award, in recognition of this, shortly before finally conceding defeat to cancer on 3 June 1995.

Once the war was over, everybody assumed Turing would return to Cambridge, and resume his research there, but this did not seem to whet his appetite any longer. His experiences with Britain's military intelligence community had broadened his horizons too far for him to follow the simple path of a bumbling college professor. He had gained a taste for the cutting edge and working under pressure. Now he developed the urge to seek out a new challenge.

Although a comfortable lecturing post awaited him, Alan resisted the temptation to cover old ground. In late 1945, J.R. Womersley, Superintendent of Mathematics at the National Physical Laboratory (NPL), approached him with a more attractive offer. He took up the post of Senior Scientific Officer there in October.

It was no longer just the Americans looking to develop computers, the Brits were keen to get in on the act too, and give the wealth of knowledge that people like Turing and his contemporaries had in this field, it was in a seemingly fortunate position. Work began at NPL on the Automatic Computing Engine (ACE) in early 1946.

Much as he had prior to the Second World War, Turing showed clear intentions to push back the boundaries of this area. In his paper entitled 'Intelligent Machinery' published soon after his arrival at the laboratory, he expressed the notion that civilisation would no longer need to employ 'an infinity of different machines doing different jobs. A single one will suffice. The engineering problem of producing various machines for various jobs is replaced by the office work of programming.'[32] This would take his 'Turing Machine' of the 1930s on to the next stage, a 'Universal Machine' capable of being programmed to do any task.

'A great positive reason,' he proposed, 'for believing in the possibility of making thinking machinery is the fact that it is possible to make machinery to imitate any small part of a man.'[33] Turing saw that, just like an infant, an intelligent machine would have to be educated, disciplined and allowed to develop its initiative. The Pilot ACE, a prototype for the final computing device, was to be developed first of all. It took up a little over 20sq.ft, and had a 4,000-word memory. The machine went into operation in May 1950, its first public demonstration taking place that December.

It was said that Turing had taken up running, toward the end of his time at Sherbourne, to avoid the ball in soccer games, but he clearly had a great propensity for it. Always a loner, he was considerably more comfortable with this activity than with team-orientated sports. He became a member of Walton Athletic Club, in Teddington, while at working at NPL, breaking its record for three miles, and later becoming its Vice President. He won both the three- and one-mile races at the laboratory's sports event in 1947, and later that year ran in the Amateur Athletics Association marathon, being placed fifth, and narrowly losing out on a place in the British Olympic team.

Ethel Turing described her son as 'broad, strongly built, tall, with a square determined jaw and unruly brown hair. His deep-set clear eyes were his most remarkable feature.' She also felt he had something of a 'child-like' appearance 'so much so that in his late thirties he was still at times mistaken for an undergraduate.'[34] Turing had a tendency to stutter, but several people pointed out that this was because he would get excited and want to get out what he was trying to say too quickly.

Turing was one of the few people outside the US to get a copy of the EDVAC report, which von Neumann had written. He used it to form the basis for a more comprehensive document, the 'ACE Report' for NPL. He noted that in the calculating machinery of previous eras, functionality had been limited to just 'four processes; addition, subtraction, multiplication, and division, together perhaps with sorting and interpolation, cover all that could be done.'[35] Turing was of the opinion that the type of 'electronic calculation now proposed should be different in that it will tackle whole problems.' He pointed out, much as von Neumann had, that frequently recurring pieces of any arithmetic process could be kept within the computer so that 'instead of repeated using human labour for taking material out of the machine and putting it back at the appropriate moment all this will be looked after by the machine itself.'[36] He saw that this offered certain key advantages. Firstly the speed of the machine in question would not be slowed down by the person operating it, secondly it removed the possibility of human error, and finally the processes to be done could be far more complex in form. As he put it, 'once the human brake is removed, the increase in speed is enormous.'[37] The components Turing identified in the structure of ACE followed a similar formula to that of EDVAC, but he took the basic ideas von Neumann had proposed and added more detail to them, considering a lot of other factors that his counterpart had not looked into. His design consisted of:

1. A series of erasable memory units (called the 'dynamic storage')
2. A set of quick reference temporary storage units
3. An input (or 'input organ' as he referred to it)
4. An output
5. A logic control unit; which assigned the order operations were carried out
6. A central arithmetic unit; which performed the calculating tasks
7. A clock; which synchronised the calculating process
8. A power supply
9. A binary/decimal converter; which translated machine code into a form that human operators could understand

In the same way von Neumann had thought of the component parts of EDVAC in terms of human organs, Turing's design easily lent itself to comparison with the body. The processor and controller were effectively the brain, the inputs the senses, the outputs the mouth, and the clock that ran it all acted as the heartbeat. The 'short code' computing language, he proposed to run the machine, was a way of dispensing with having to deal in machine code (i.e. binary numbers) which would soon become too cumbersome, as the complexity of programs increased. It would form the basis of a programming language, much like those used in PCs today.

Turing gave a lecture on ACE to the London Mathematical Society in February 1947. In it he contrasted the computer with the idealised system he had put forward in his 'Turing Machine'.

He was of the opinion that 'in general, the arrangement of the memory on an infinite tape is unsatisfactory in a practical machine.'[38] He could see that the machine would need to travel up and down the tape for incredibly long periods of time to find required pieces of information, and this made the whole point of using it completely futile. Such a computer would not give any greater level of efficiency. To offer an analogy he said that this 'difficulty presumably used to worry the Egyptians when their books were written on papyrus scrolls. It must have been slow work looking up references in them.'[39] Turing believed, much as von Neumann had, that the importance of adequate storage was 'the key to the digital computer' and if these machines were to be 'persuaded to show any sort of genuine intelligence, much larger capabilities than are yet available must be provided.'[40]

He stated that the problem of 'making a large memory available at short notice is much more important now than that of doing operations such as multiplication at high speed.'[41] ACE had 200 mercury delay lines in its memory, just like those Eckert had designed for EDVAC, as they were more accurate, and quicker. Each line was capable of holding over a thousand binary numbers. Turing predicted that ACE could replace a legion of 10,000 operators equipped with mechanical desk calculators.

He began to consider the new horizons this technology would be capable of reaching. In the past computing machines were only meant to carry out 'processes that they are instructed to do' and if they did 'something other than what they were instructed then they have made some mistake. It is also true that the intention in constructing these machines in the first place is to treat them as slaves, giving them only jobs which have been thought out in detail, jobs such that the user of the machine fully understands what in principle is going on all of the time.'[42] Basically they would just take care of tasks that had been worked out beforehand by the user, and there would be no room for additional input or 'thought' by the computer itself.

Even though until this stage 'machines have only been used in this way', he reasoned that it was not 'necessary that they should always be used in such a manner?'[43] Turing's perspective was far more sweeping than those of his peers. He did not think of computers simply as number crunchers, but envisioned a much broader range of possibilities for these machines, and was concerned not to put limitations on what they were capable of. He could see that as operations got more and more complicated, the machine would find means of performing them that had greater efficiency, and the manner in which these things were actually achieved would be beyond the intellectual grasp of its operators. He likened it to how 'a pupil who had learnt much from his master, but had added much more by his own work. When this happens I feel that one is obliged to regard the machine as showing intelligence.'[44] Turing predicted that a point would be arrived where the machine would effectively 'learn' from its experience.

Alan became preoccupied with this idea that development of more complex computers was going to result in machines capable of independent thought. They were no longer going to be just glorified calculators, they were rapidly approaching a stage when they would show signs of what we would describe as intelligence.

Turing was the first to put forth the concept of 'Artificial Intelligence'. He thought the commonly held rule 'that machines cannot give rise to surprises is due, I believe, to a fallacy to which philosophers and mathematicians are particularly subject. This is the assumption that as soon as a fact is presented to a mind all consequences of the fact spring into the mind simultaneously with it.'[45] This would therefore lead to the assumption 'that there is no virtue in the mere working out of consequences from data and general principles.'[46] He pointed out that if a machine can do something that was not considered in its original program, and in effect surprise the programmer, then it must be able to perform what he called a 'creative mental act'.

He defended what he felt some might think was a 'fundamental contradiction in the idea of a machine with intelligence, it is certainly true that "acting like a machine" has become synonymous with a lack of adaptability. But the reason for this is obvious. Machines in the past have had very little storage, and there has been no question of machines having any discretion.'[47] Turing went on to suggest a way to nurture this artificial intelligence was to allow the computer in question to 'have contact with human beings in order that it may adapt itself to their standards. The game of chess may perhaps be rather suitable for this purpose, as the moves of the machine's opponent will automatically provide this contact.'[48] He conjectured that it 'would probably be quite easy to find instruction tables which would enable ACE to win against an average player.'[49]

In May 1997 Deep Blue, IBM's supercomputer, beat World Chess Champion Gary Kasparov. The machine took two games to the Russian's one. Computers were no longer just capable of 'outthinking' the regular person (most of us at some stage having a computer chess defeat inflicted upon us), they now had the ability to take on the best human in a given field and still come out as the superior. Turing, who was a keen chess player himself, would have no doubt been in a state of euphoria if he had lived to see it.

In an interview published in the *Times* in 1949, Turing described the research he was doing to find the 'degree of intellectual activity of which a machine was capable, and to what extent it could think for itself.'[50] It would be something that he would return to many times over the following years.

The engineering assistance on ACE that was expected from the Post Office department at Dollis Hill (which had been so helpful in the building of equipment for Turing and his Bletchley Park chums during the war) was not forthcoming this time around. At the best of times ACE had three to four men working on it, which was ludicrous in comparison to the projects underway in North America at this point.

ACE's completion did not arrive until the mid-1950s, and it would have little if any influence on the development of computer science, the industry had moved on far too quickly. By this stage Turing was long gone. In Autumn 1947, he wrapped up his involvement with NPL. The organisation had been too belated with its efforts to make a serious contribution, and this was not to his liking. The US Government had put over $100,000 into the initial development of ENIAC, almost without breaking a sweat, and by the time the project was completed it had cost over $400,000. NPL was, by comparison, more than a little bit miserly in its outlook. Turing was scaling the highest peaks of computing conception, but the engineering staff charged with the task of supporting him remained rooted to terra firma amongst the foothills. He had a sabbatical for twelve months in late 1946, and given the lack of progress made when he returned, lost all motivation. He could see there was not likely to be any change in the situation.

In 1948, Alan took up a position at Manchester University. His old pal Newman, who was the Professor of Mathematics there at the time, persuaded him that he might be better fortuned there. Just after the war Newman had managed to secure a large grant for the development of a computer system, Alan was given the title of Deputy Director of the Computing Machine Laboratory.

For von Neumann, any opportunity to lighten his workload when the war finally came to an end soon disappeared. Even once peace resumed, he continued his work into nuclear weapons research. The Los Alamos laboratory in New Mexico was joined by the Lawrence-Livermore laboratory in California, and Johnny would be expected to regularly show his face at both establishments. Along with the Pole Stanislaw Ulam, he worked on the H–Bomb during the early fifties. His strong anti-communist stance meant he was keen to play a part in the protection of the United States from the growing threat posed by Stalin and the increasingly powerful Soviet Union.

As well as his government work, von Neumann began to make ties with big business. He was far from an academic in its purest form, becoming a consultant for a series of commercial ventures. Perhaps his family background in the financial world had made him a little less idealistic than some of his contemporaries.

Following his first work on ENIAC, his wife recalled, 'from then until his last conscious hours, he remained interested and intrigued by the still unexplored aspects and possibilities of the fast-growing use of automata.'[51] He would spend every waking moment trying 'to convince others in many varied fields that numerical calculations done on fast electronic computing devices would substantially facilitate the solution of many difficult unsolved scientific problems.'[52]

By Autumn 1945, the IAS also wanted to start investigating the possibilities for computing. Nobody else in the organisation had even remotely the same level of experience as von Neumann in this particular field, so he was the obvious choice to lead the endeavour. Leaving the EDVAC project to plod along by itself, Johnny concentrated his efforts on this new challenge. The 'IAS Machine' would not be able to command the same sort of budget that had been afforded to ENIAC, but taking into account the great strides that had been made in this arena since then, and given von Neumann's remarkable abilities, this was not seen as an issue.

It required only 2,000 tubes, much less than ENIAC or EDVAC, but was capable of performing much more demanding calculations. Cathode ray bulbs (like those used in standard television sets) formed its memory bank. This made the system a great deal quicker and more accurate too. To supplement the funding from the university he used his ambassadorial skills to gain more financial backing here and there. A little bit from the military, and a bit more from the private sector. Both RCA and IBM played their part in the project, in return for access to the completed design, so that they could use it in their own ventures.

He bought one of his ENIAC minions onboard almost straight away. Goldstine became his deputy in early 1946. Johnny was also keen to procure Eckert's services, hoping he would join the team as his chief engineer, but he wanted nothing to do with von Neumann by this stage.

The IAS Machine papers von Neumann published in 1946, while it was being built, allowed others to produce similar computers. It would result in the creation of a full generation of machines, rather than having one or two very different projects going on in complete isolation, in other parts of the world. The IAS Machine would be cloned by various military research establishments, universities and business enterprises, such as the laboratories at Los Alamos, the universities of Munich, Sydney, and Stockholm, as well as the offices of IBM and Rand Corporation, to name just a few. It was a reference design used for other machines all across the globe. This had great significance to the future of computing as it allowed a standard plan of construction to be followed, encouraging greater proliferation. The machine first went into operation in mid-1951, and was utilised for many different tasks, including calculations for the H-Bomb, and meteorological projects.

Meanwhile Turing was beginning his new life in Northern England. He bought a house in the quaint town of Wilmslow, just on the outskirts of Manchester. His involvement in the university's project would mainly concern the development of software, while Freddie Williams and Tom Kilburn took responsibility for the engineering side of things. Before Alan had arrived they had managed to construct Manchester's Small Scale Experimental Machine or 'Baby'. It was the first fully programmable computing device, as unlike its US equivalents it required no mechanical alterations to be done if new tasks were embarked upon.

The 'Baby' was followed by a second, far larger computer, the Manchester Mark I. Here Turing was afforded the opportunity to play a major role. The system would use cathode ray tubes (like those von Neumann was employing in the IAS machine) to store the data. It was put into operation in October 1949, with Alan Turing producing the 'Programmers Handbook' to accompany it soon after.

Though things seemed to move at a far quicker pace than at NPL, Turing felt that he was not given as much creative license as he wished for. Williams once said of him that he knew 'a lot about computers and substantially nothing about electronics.'[53] In his opinion the practicalities of producing something that was capable of aspiring to Turing's vision were beyond the scope of the technology available to the team. A personality clash arose between them over who had overriding control, and this would impede the fulfilment of many of the ideas that Alan hoped to incorporate within the machine. The attempts he had made to take his designs off the drawing-board, both at the NPL and in Manchester, were not to be successful. Turing theories on computing would find better fortune at the hands of other protagonists than they did with their originator.

After the war Alan continued to work with British Intelligence as a consultant. Following the conflict the GC & CS became the Government Communications Headquarters (GCHQ), its operations centred in Cheltenham.

Turing's last great work on the development of computers, 'Computing Machinery and Intelligence', was published in the magazine *Mind* in 1950. In this he further pursued the concept of 'Artificial Intelligence'. He proposed a test to help decide whether a computer could display the ability to think. The 'Turing Test', as it became known, goes something like this. You have an interrogator, a computer and a human. The interrogator's job is to ask questions to the others, who are hidden from him, in order to affirm which one of them is the computer. The questions and answers are transferred between the different parties via keyboards and monitors, so nobody can see each other during the test. It is in the human's interest to tell the truth during the test and try to help the interrogator to make the right decision, it is in the computer's interest however to lie.

The basic aim of the interrogator's questions would be, as Turing described it, to 'discover whether one really understands something or has learnt it parrot fashion.'[54] It could be assumed that standard answers to questions could be programmed into the machine (i.e. What is your name? Who is the Prime Minister? Which is your favourite member of Westlife?), and given enough storage capacity, the quantity of data at the computer's disposal would be huge. Therefore, these questions would serve no purpose when it came to revealing the true identity of each participant. What would be required was a way of judging whether they could actually reason.

Turing was elected a Fellow of the Royal Society in March 1951, an honour bestowed on his great grandfather, George Stoney, some ninety years before. In his paper 'Celestial Basis of Morphogenesis' presented to the Society later that year, he examined how mathematical modelling could be used in relation to the form that living creatures took.[55] Again this showed the huge versatility of Alan's mind; he could create groundbreaking works in mathematics, computing and even natural sciences.

In early 1952, Turing's world fell apart. This man, who had spent his whole life not trying to harm anybody, was to become the victim of a vindictive, overbearing society, still effectively living in Victorian times. Homosexuality was deemed a serious offence in Britain until the late 1960s, and Alan was to fall foul of this archaic law.

He had been having an affair with a young man, and became ensnared in an awkward situation as a result of this. One of his lover's associates burgled Turing's house and furthermore, it is believed, tried to extract hush money. The police investigation that ensued uncovered the nature of their relationship. It is quite amazing how homosexuality could be considered such a grave violation of decency, but robbery and blackmail were judged of only secondary importance. What made this whole fiasco even more ridiculous is that Turing had actually gone to the police himself. It seems as if the 'Establishment' of 1950s England had not moved since it sentenced Oscar Wilde to years of hard labour in the 1890s.

In the blink of an eye, Turing had been turned from hero into villain. Although he had been decorated by the king for his aid in bringing the war to an end, and had helped save countless lives, he was now considered the lowest of the low. On 31 March, he was found guilty of homosexual acts. His punishment was either to go to prison, or alternatively take a twelve-month course of oestrogen injections. The administering of the female sex hormone was designed to abate the libido of sex offenders. Turing decided to take the second option. However, the shame and humiliation was not to stop there. He was now a convicted criminal, which made him a security risk. After many years helping to protect the safety of the realm, His Majesty's Government withdrew Turing's security clearance. Perhaps this was the final indignity for him.

His housekeeper came in on the morning of 7 June 1954, just following the Whitsun bank holiday, to find Turing's lifeless body strewn on the bed.

The light on his bedroom was still on, the cutlery from the meal he had the night before was left out, clothing he had just bought and theatre tickets were to be found upon his desk. It was discovered that the cause of death had been traces of potassium cyanide on an apple he had eaten. He was only forty-two years old at the time of his passing.

In the years following his demise there has been some debate as to whether Alan had meant to take his life or not. He did keep chemicals in his home, for research he was doing in electrolysis, so it was possible that he had contaminated the apple he consumed by accident (some residue on his hands being transferred onto it). Others have suggested that he intended to kill himself, but he wanted to make it look as though it was an accident, so as not to upset his mother. Regardless, the coroner's verdict was one of suicide, the poison 'being self-administered while the balance of his mind was disturbed'.

It is now recognised that the drastic measures the authorities had taken in giving oestrogen to convicted homosexuals were unwise. The injections appear to have upset the psychological make up of their unfortunate recipients. This resulted in acute alteration of their mood, and often led to depression. It does not take much of a mental leap to suggest that the hormones could have had an effect on his state of mind, and been a contributing factor to the tragic end of his life.

Following Turing's death, his mother Ethel endowed an annual award, the 'Alan Turing Prize for Science' to be given to the most promising pupil in scientific disciplines at Sherbourne, where he had studied as a boy. On 23 June 2001 a memorial to him was placed in Sackville Park, Manchester. It states 'Alan Mathison Turing (1912–1954), Father of Computer Science, Mathematician, Logician. Wartime Codebreaker, Victim of Prejudice'. To commemorate the fiftieth anniversary of his death, in June 2004 a plaque was unveiled at his former home in Winslow, and later that summer Manchester University established the Alan Turing Institute. Its brief is to build collaborations between industry and academia in the field of mathematical research.

Back across the Atlantic the Americans were not going to be accused of undervaluing von Neumann the way that the British had done with Turing. During the 1950s he became a technical adviser to IBM, which would require him to spend a few days a month in New York. This was yet another commitment which left little time for him to carry on with his academic research. Much of the last few years of his career were to be relinquished to his obligations to both state and industry. In total, he would be involved with more than twenty different government departments and industrial combines during this period.

He accepted President Dwight Eisenhower's appointment to the Atomic Energy Commission (AEC), officially joining the body in March 1955. His participation in so many government programs persuaded him to move back to Washington DC, he and Klari setting up their permanent home there that spring.

In the mid-1950s he was given the Enrico Fermi Award for his contribution to computing, and the Albert Einstein Commemorative Award for Science. He planned to take a post at the University of California, once his term with the Commission was concluded, but his hopes of realising this dream were dashed following a medical check up.

In August 1955, Johnny went to Bethesda Military Hospital after suffering acute pain in his left shoulder for some time. When exploratory surgery was performed it became apparent that he had a malignant growth. Klari recalled that 'the pattern of our active and exciting life, centred around my husband's indefatigable and astounding mind, came to an abrupt stop.'[56]

His wife later stated, 'The ensuing months were of alternating hope and despair; sometimes confident that the lesion in the shoulder was a single manifestation of the dread disease.'[57] Sadly this was not to be the case, with time it was confirmed that the cancer had spread, and was completely inoperable. By November, the tumours had overrun his body, some developing on his spine. As a result of this he began to have difficulty walking, and by January was confined to a wheelchair. Despite his frail condition he was still engaged in AEC affairs, attending its meetings as often as possible. By the time April arrived he was too ill to be kept at home. He was admitted into Washington's Walter Reed Hospital. Klari and his brother Nicholas kept him company. That spring he received the 'Freedom Medal' from President Eisenhower, in a ceremony conducted at the Whitehouse, it was his last public appearance.

Even when his condition meant he was almost completely debilitated and could no longer leave the hospital, Johnny kept on carrying out his duties. He had a hotline installed between his room and the government offices, and worked from his hospital bed. Finally it was just not possible for him to go on any longer, he became simply too weak. According to his wife 'Johnny's exceptional mind could not overcome the weariness of his body.'[58]

Johnny died on 8 February 1957, he was only fifty-three at the time. He was buried in Princeton, the place which had given him the one and only constant that he had through his varied and complex life. His brother Nicholas said of him, 'his incredible diversity and multiplicity of topics in many branches of mathematics, physics, economics, and other disciplines, covered sufficient contributions in each class alone to fill a lifetime career for any great scientist.'[59] His wife Klari said, 'his capacity for work was practically unlimited.'[60]

The *Times* obituary read, 'for a man to whom complicated mathematics presented no difficulty, he could explain his conclusions to the uninitiated with amazing lucidity.'[61] He once said that people did not believe that mathematics was simple, and this was only because they did not realise how complicated life was.

His book *The Computer & the Brain*, which he had worked on while in hospital, would be published posthumously in 1958. His devoted spouse Klari died, by drowning, in November 1963. It is assumed she committed suicide, unable to cope without the presence of her soulmate.

The contribution von Neumann could provide to computing would be limited to an interval of just over a decade, making his achievements all the more exasperating. Tragically he was not able to devote a greater period to this field. In 1990, the IEEE established the John von Neumann Medal, an annual award 'for outstanding achievements in computer-related science and technology'.

The skill for mental arithmetic that he displayed in his youth was maintained as he grew older. He still had the ability to stun people whenever he chose. His breadth of versatility almost knew no limit. He could apply himself with equal ease to quantum mechanics, computing, atomic physics or economics. The reason for this was his understanding of the underlying cohesive stem all of them shared; namely mathematics. It offered a basic commonality on which each of them relied; in the same way that grasping the principles of Latin will help you learn French, Italian or Spanish.

It was a tendency of von Neumann's to analogise human physiology with the design of computers, often referring to different machine parts as 'organs' or comparing the inputs to human senses. Others, like Eckert, tried to avoid this, feeling it might be misconstrued, though as things have moved on these approximations became more and more accurate. Johnny proposed the concept of a self-reproducing machine that, like living cells, would be able to create exact copies of itself, and even contemplated use of DNA to form computer memory elements, foreseeing the advent of genetic computing, something that is currently being developed. As he put it, 'chromosomes and their constituent genes are clearly memory elements which by their state effect, and to a certain extent determine, the functioning of the entire system.'[62]

As already shown, even the inventions that seem relatively simple in conception still drew upon the work of several individuals. It is therefore no huge surprise to see that so many people contributed to the central theme of this chapter. The early computing devices required the support of many different minds, proficient in a wide variety of disciplines.

Equally there is no single machine that can easily be defined as the first computer. It is a grey area, just as it would be hard to say where exactly the line that divided men from apes was crossed. The Difference Engine, Colossus, ENIAC, EDVAC, and many others, all have attributes that helped to make the modern computer. Thus Babbage, von Neumann, Aiken, Eckert and Turing, to name just a few, all had integral parts to play.

I have chosen not to include Babbage in the final selection for the 'Nearly Men', but this is very obviously a judgement call. My reasoning is thus: although he was clearly a precursor for what would become the computer age, and the work that people like von Neumann and Turing did a century later owed a great deal to his foresightedness, it was still only theoretical. He never managed to create his 'Analytical Engine', as the sheer cost and complexity made it unfeasible back then. The 'Difference Engine' is sometimes described as the first computer, but using the modern definition of this statement is in fact untrue. The machine was capable of incredible feats given the context of the era in which it was constructed, but it was still only a mathematical calculator based on the use of simple addition. A real computer, whether mechanical, electronic or even organic in nature, would have to have the capacity to be programmed to perform different functions.

Babbage could not have made a true computer in the period of history he occupied, since without the help of electronics that level of intricacy would have been impossible to realise, and though he managed to identify the basic form such a machine would take, it would not be practicable until well into the twentieth century.

Now by this proviso, Turing could be said to deserve the same consideration, as nearly all of his work was just done on paper. This is true, but was more down to geographical barriers rather than technological ones. The means to make the 'Turing Machine' flesh were there, but they were unfortunately not at the disposal of a British academic. They were all on the other side of the ocean.

The Harvard Mark 1 could not be said to have any real case to contest when it comes to resolving what deserves the title of first computer. It was basically a calculating machine, just like those Leibnitz and Pascal had constructed, but on a much grander scale. It had virtually no commonality with computers used today, so we can discount its claim. An abacus could be considered a form of computer if you took this way of thinking to its extreme.

The Colossus, which the GC & CS had developed during the war, is another pretender to the throne, but in reality this was not a true computer either, as it was only capable of a very specific task. A real computing device would be able to perform a whole series of different functions, depending on what it was programmed to do. ENIAC, which is more widely considered as the first computer, has the same problem. Again, it was not in all honesty general purpose, and could only really perform certain predetermined actions. Any change in its operation required mechanical alterations and rewiring. It did not have the ability to use other programs directly.

In principle there are many different ways of judging which was the first computer. Depending upon your perspective, several different creations could present a reasonable argument in their favour. But the real issue is not which one crossed the line first, but which had lasting influence? Which machine made future generations of computers possible? The fact is none of those just mentioned have anything remotely in common with how all the computers that followed were formed, whereas EDVAC acted as the basic blueprint for all of them.

So it is the design for EDVAC that probably deserves the title of first real computer, when judged by modern standards, but it does have to be conceded that it was just the design. The construction of the machine itself was greatly delayed, and the layout was applied to other computers long before the one it had been meant for was finished.

It can be argued that though von Neumann took innovations from Turing, Eckert and others, then subsequently palmed them off as his own, he was only doing so for the benefit of civilisation. It is probably fair to say that von Neumann acted to get things done as quickly as possible, and for this assignment he was the ideal candidate. Throughout his life he made use of ideas that other people put forward, but it is probably harsh to accuse him of outright thievery. In general he took underdeveloped concepts and by adding his own thoughts made something far more substantial, bringing extra clarity and depth to them. Nevertheless, in retrospect, it would appear that he was given more of the commendation than he deserved, and other vital contributors were overshadowed by his presence, now it seems only fair the credit be more accurately apportioned.

His involvement on the engineering front was almost certainly minimal. This was Eckert's forte. Johnny was more of a frontman; the acceptable face of the project, rather than some scruffy egghead, as the US Army would have seen Mauchly or Eckert. He acted as the visionary who would give the whole thing direction and creative drive. Whether von Neumann was key to the EDVAC venture itself is hard to be certain. It is quite possible that Eckert and Mauchly would have been able to complete its design without him, though it would probably be fair to say that his incredible intellect must have been a plus point.

What cannot be questioned is that it was von Neumann that encouraged the widespread use of the designs of both EDVAC and the IAS Machine, and that allowed far quicker progression in this field. The other two thought it was better for the discoveries the team made to remain in their possession, so they could try to reap the commercial benefits. For von Neumann money was not the most important thing, he just wanted to see more widespread use of computers. Even if his technical contribution was completely redundant, the design that he told the world about would endure for many decades, and is still utilised by the vast majority of computers even today. His EDVAC report formed the basis for all the machines that followed, it is only in relatively recent times that parallel processing has been employed to improve system speeds, and has started to supplant this. This technique is only in its infancy however, and the 'common or garden' PC is still based firmly on von Neumann's work. He effectively laid the foundation stone upon which the whole computing age was constructed.

Both Turing and von Neumann were of the opinion that progression of computer technology should be for the benefit of all. In this respect at least, they were straight down the line academics, looking to develop computers for the aid and protection of humanity. Therefore, they wrote papers and published articles telling the whole world about what they were doing, and tried to encourage others to get involved. Mauchly and Eckert on the other hand were keen to keep the technical details of ENIAC and EDVAC secret, and patent the technology, which would have held things back.

It is quite likely that von Neumann was aware of Turing's work 'On Computable Numbers' from their time in Princeton together, and used a lot of the Englishman's thoughts on the subject to help him create EDVAC. Though even if he did, it is still hard to count this as plagiarism. Turing had the intellectual power, but not the support to take his vision and create something of real substance with it, he would have just been happy to see his work not going to waste.

The irony is that although Turing is the guy who effectively lost out, he is still, in most circles, more famous than the guy who actually won. Maybe this is because of his untimely death, his persecution at the hands of his own people, or his contribution to the defeat of the Nazis. Has the fact that Turing died so young had an effect on how he is perceived? Just like certain actors and rock stars (Monroe, Dean, Morrison, Cobain, etc.) whose lives ended at the peak of their powers, and did not have to suffer the aging of their bodies, or crumbling of their creative abilities, has he been given some sort of halo of immortality?

At a stage in history when commercial interest in computing was still limited, it needed collaboration with the armed forces to help it gain traction, and they were not going to give away the money they had spent so much time taxing people out of to just anybody. ENIAC, EDVAC and the IAS machine all benefited from von Neumann's credibility, he bridged the gap between the cutting-edge engineers who knew it all, but had no scientific reputation, and the respected but clearly ignorant older generation.

Of course it is no surprise to anybody that our hunger for technology has been to an extent driven by the need to kill. Humanity's deep-engrained instinct to destroy one another has proved the most compelling of motivations for our creativity, resulting in a quite bizarre dichotomy. For centuries it has fanned the flames of innovation. Perhaps the primal force propelling our scientific advancement has been that of warfare. The widespread use of radio was encouraged by the navies of Japan, the United States and Britain employing it to help orchestrate attacks on their enemies. The Internet was initially developed by the Defence Advanced Research Projects Agency (DARPA) to aid communication between different departments, and improve the efficiency of projects being embarked upon. At the turn of the last century the aeroplane was just a toy for the over-adventurous with too much money and too little sense. It was not until the First World War that Britain and Germany began to exploit its potential in battle, and generally speaking today's commercial aircraft owe their development to the military ones that went before them. The whole basis of the 'space race', the Russians' success in putting a man into orbit and the American's triumph in placing one on the surface of the moon, stem directly from the expertise both sides managed to snatch from the production of V2 rockets following Germany's surrender in the Second World War.

Many of the greatest minds of the twentieth century, von Neumann among them, had some sort of involvement in the creation of the first atomic bomb. The war acted as a catalyst for the advancement of computing on both sides of the Atlantic. Firstly it freed up cash to fund the projects, and meant that machines like ENIAC and Colossus could be built. On top of this, it broke down the barriers that academics tend to build up around their work. Things could no longer be done in isolation, they had to pool their resources together and exchange ideas for the greater good.

It is of course notable that the team that developed the 'A' Bomb was made up of many German and Eastern European Jews who had been forced to leave their homelands. One can only think what would have been the consequences if Hitler had waited to capitalise on the richness of scientific talent at his disposal to develop such weapons for himself, before persecuting the Jewish race.

But enough politics, let's return to practical issues. Fifty years ago the 'Universal Machine' that Turing predicted must have been viewed with a lot of scepticism, today each of us sees one at close quarters on a regular basis. The PCs that we use in our workplace or at home are just that. They can act as typewriters, games consoles, CD players, televisions, photo processing tools, address books, calculators, or diaries. They can remind us not to miss our mother-in-law's birthday, allow us to view movies, or fulfil all manner of creative urges we may have; whether musical, artistic or literary based. They offer a gateway to the web; a virtually infinite resource of information (some of it actually mildly accurate), with the use of Voice-over-IP they are capable of acting as a telephone system for cheaper long-distance calls, they can even facilitate the purchase of goods without having to haul our carcasses down to the shops. We would not be able to be half as lazy if it was not for the computer.

The Manchester 'Baby', ENIAC and EDVAC all required huge budgets to build, operate and maintain. Each needed hundreds of thousands of dollars worth of financing. By contrast, today's run-of-the-mill PC that you can get from any high street store, has astronomically greater processing power and is far more reliable than any of these machines, but will probably come with a price tag of less than £500. However, all this needed a paradigm shift to achieve.

Big corporations all assumed there would be a very limited amount of places that could utilise computers, and were thus cautious about entering whole-heartedly into their development. What they did not foresee was that as computers became smaller and cheaper, the number of tasks they could be employed for would be greatly compounded.

By the 1970s the middle management in these corporations became so infuriated by the length of time they had to wait for a data processing department to come back with results, that they would start demanding accessibility to computers of their own. Thus, the market for personal computers was created. At the same time the hobbyists around the world were gaining interest in putting together their own machines, and so companies like Apple were able to exploit this need, which would eventually give rise to the use of computers in the home. All this would be made possible by the engineers discussed in the next chapter, they would find a way to put the functionality of gargantuan machines, like the Manchester Mark 1 or ENIAC, onto pieces of material no bigger than a postage stamp. Vacuum tubes would be replaced by microprocessors. Delay lines got substituted for non-volatile memory chips. Punch cards were ousted by floppy disks, and have now been supplanted by CD ROMs and memory sticks. All this required the development of semiconductor technology, the final theme this book will address.

CHAPTER 7
Small Wonders

It is hard to imagine in an era of huge multinational corporations which boast tens of thousands of employees that a single man could have such an influence on the world as to be responsible for creating an entire industry, and one that brings in an annual revenue measured in hundred billion dollars to boot. However, this is basically what the subject of this next chapter managed to do, almost single-handed.

The last episode of this story brings us right up into the modern age, dealing with the semiconductor revolution, something which allowed many of the inventions we have already looked at to take huge leaps in terms of what they were capable of doing. Telephones would no longer be fixed to one position, they could come along with you when you went out of the house. Radios did not have to be a great hulking piece of equipment that weighed fifteen kilos and took twenty minutes to warm up, but were something you could carry around, and would not lose reception whatever your location was. Computers would not take up acres of space anymore, they would be compact and affordable to the ordinary man on the street. Semiconductors, for better or worse, changed the world completely, and in the following pages we will look at the innovations that allowed this to happen.

In the previous chapters we have travelled from the depths of the Amazonian rainforest to the mountains of the Balkans, from the beauteous wonder of islands of the Galapagos to the squalor-ridden streets of revolutionary Paris, from the sight of blazing buildings along the Thames as medieval London was destroyed and the smouldering rumble of Hiroshima to the creation of modern New York. This final instalment, however, for the most part, is centred on just one tiny part of the world. It is a 40km stretch of land that separates Los Gatos from Palo Alto (containing Santa Clara, Milpitas and San Jose) bordered to the north by San Francisco Bay, and wedged between the Santa Cruz Mountains to the east, and the Coastal Range to the west. Until the end of the Second World War the region was best known for prunes and apricots, its inhabitants just simple fruit farmers. But its rustic origins are now all but forgotten, today it has more Ph.Ds per square kilometre than any other locale on the face of the planet. Time and again it has been the crucible in which the world's technological advancements have formed. This region of Northern California has become known as 'Silicon Valley'. Though the name was not used to describe it until a journalist with *Electronics News* (a magazine which I have had occasion to write for, from time to time), called Don Hoeffler, coined the phrase in the early 1970s, by then the entity had in fact been in existence for quite a while.

A great deal of the thanks for the creation of the environment that would eventually spawn 'Silicon Valley' must go to the philanthropic efforts of Leland and Jane Stanford, who established a university in Palo Alto in the 1890s, to compete with the Ivy League colleges on the East Coast. It would gain a reputation that was second to none in the domain of science and technology. Its alumni include the likes of Oracle CEO Larry Ellison, Cisco founders Leonard Bozak and Sandy Lerver, Yahoo's Jerry Yang and David Filo, plus Intel CEO Craig Barrett.

Back in the 1950s, Stanford University set in motion an ambitious enlargement plan, but soon found that its financial reserves were not sufficient to carry it through to completion. Luckily Fred Terman, an engineering professor, came up with a way of raising the money needed. The university owned a considerable expanse of land, over 8,000 acres in fact. Terman suggested leasing out part of it to technology-based firms, and encouraged some of his brighter students to start up their own businesses.

Two Stanford graduates who took advantage of the scheme were Bill Hewlett and Dave Packard. They had started a rather humble little business together in the late 1930s, working part-time out of Packard's garage in Palo Alto. With the onset of war their electronic components found use in the US military's sonar and radar equipment. By the time the university started its business development project, their company had gone from a two-man cottage industry with annual revenue of just a few thousand dollars, to a complement of some twenty people, and a turnover above the five million mark. Hewlett-Packard would be joined by other firms, Eastman Kodak and General Electric soon becoming its neighbours. Over the years that followed the factories and offices continued to sprawl out further and further, eventually taking up every piece of available space.

Though this will be the battlefield on which this chapter is played out, its principle characters did not originate from there. Both men, whose lives we will in turn focus on, came from provincial towns in America's Mid-West.

The integrated circuit, or 'microchip', as it is more commonly known, is to be found in nearly everything that you can think of in modern life. Your car, your washing machine, your mobile phone, your TV, your iPod, your pacemaker perhaps, even your bank card, you name it – its got a chip in there somewhere. But while most people associate Darwin with evolution, Newton with gravity, Edison with the light bulb, and Marconi with radio, it is unlikely that more than a fraction of you would have heard of either of the people most prominent in its conception. Yet it could be argued that what they created has greater influence on our lives than any of the others.

John St Clair Kilby, was born in Jefferson City, Missouri, on 8 November 1923, the only son of Hubert and Vina Kilby. He and his family moved to Kansas not long after he was born, so most of his formative years were spent in the town of Great Bend, on the Arkansas River. Even as a young boy he knew that he wanted to work in electronics. His father ran a small power company, serving rural Kansas. The company was 'scattered across the west side of the state', and Kilby recalled on one occasion 'they had a big ice storm that took down all the telephone and many of the power lines, so he began to work with amateur radio operators to provide some communications, and that was the beginning of my interest.'[1] Back then the young boy decided, 'this field was something I wanted to pursue'[2]

He entered the University of Illinois in 1945, studying electrical engineering. At the time there were not really any proper courses devoted to electronics. It was still a very new subject, and there were just a few people around who knew enough about it to attempt teaching it. Although it was not to be a core part of his curriculum, Kilby managed to take a few extra classes in the field of vacuum tube engineering, as he thought that this might be of use to him in the future. Tubes, as already mentioned, were another child of the inventiveness of our old friend Thomas Edison (there is hardly a chapter in this entire book that does not have some mention of his work).

In 1947, Jack got married. He and his wife, Barbara, would have two children; Ann and Janet. He also graduated in that year, and with a wife to support, and a family planned, he found employment soon after. Kilby joined Centralab, based in Milwaukee, which made printed circuit boards. It was a fairly compact operation there, and this meant a certain amount of multi-skilling was required. Kilby would be involved in a number of different tasks, taking products from the design phase right through to production. The company's most complex device was a three-tube hearing aid, for which 'tubes had to be soldered in place or plugged into very small sockets individually.'[3] Each circuit within the device was completely different, there was no repetition whatsoever. He knew this meant the manufacturing would thus be far less efficient, and the products were more expensive as a result.

Back then, even the most intricate electronic equipment, like a television set, had perhaps only thirty circuits in it, of which no two would be alike. 'Had this continued to be the case,' Kilby later stated, 'there probably would not be a market for the integrated circuit today.'[4] At this stage the technology was able to cope with the demands being put on it, but things would not stay that way for long.

He could see that in the future there would need to be a change in the situation. The public would at last begin to visualise electronic equipment which was 'much more complex than anything that had been realised. And if this equipment had been built with existing technology, it would have been too big and too heavy, too expensive, and use too much power to be useful. That was collectively described as the tyranny of numbers. The number of parts that was required was just prohibitive.'[5]

The vacuum tubes that had been used in computers up till this stage (like ENIAC and its brethren) would be totally deficient when it came to supporting any significant evolution of these machines. Engineers were reaching a point where they could contemplate the design of much more ambitious systems, but if they were to utilise the prevailing technology they would be 'too big, too heavy, consume too much power and simply got too hot to work'[6] as Kilby put it.

'A huge step forward came in 1948, when Bell Labs unveiled the transistor'[7] he recalled. The device basically used layers of positively and negatively orientated material to form a gate through which current could be passed, or closed off as required. It offered a way of achieving the same functions previously done by vacuum tubes, but without all the inconvenience inherent in these older electronic components. As mentioned in the previous chapter, tubes were very delicate, they were prone to damage while being transported (being made out of glass), and likely to burn out while being used. They were also somewhat bulky, normally 7 or 8cm long by 3cm in diameter. The invention of the transistor meant that such devices could finally be dispensed with. They offered a more reliable, compact alternative and had far faster response times.

The transistor was developed following the war at Bell Laboratories. This centre for technological research was founded back in 1925, and had been responsible for a vast array of important innovations. Within its workshops the laser, the image sensor, and the first telecommunication satellites were all developed, its technicians were also the first to detect the residual radiation left from when the universe began, proving the 'Big Bang' theory. Initially transistors were made from germanium, but the substance was not particularly robust, and lacked resistance to heat. Later versions used silicon, which could handle higher operating temperatures, and was not so power hungry.

Kilby later described how 'the transistor pointed the way to the future.'[8] With the creation of this 'solid-state' technology, the reliance on tubes was coming to an end. This meant 'the race for miniaturised electronic circuits intensified.'[9] While at Centralab he had the opportunity to do a two-week training course at the Bell Labs headquarters in New Jersey. Here he gained first-hand experience of advanced transistor technology.

He started to work towards a master's degree at Wisconsin University in his spare time. Jack felt that 'working and going to school at the same time presents challenges, but it can be done and it was well worth the effort.'[10] He finally received his diploma in late 1950.

By the time 1958 rolled around, Kilby had decided Centralab was just too conservative in its mindset for him to remain there for the rest of his career. He picked up a good familiarity with cutting-edge technology, thanks to his evening classes and his visits to Bell Labs, and thus his demands had begun to outgrow those that could be satisfied if he remained in a small provincial firm. He started to look for another job. Kilby was searching for somewhere a bit more forward thinking, where he would be able to pursue his ideas on miniaturising electronics. He got offers from a number of corporations. Motorola was keen, and willing to allow him to concentrate on his research on a part-time basis, but Kilby suspected this would not be enough for him to make serious progress. Texas Instruments (TI), however, was a bit more enthusiastic, and thereby won the day. The Dallas-based firm had started out making electronic equipment for use in geophysical research, helping locate oil reserves, then during the war it had moved into components for use in radar. The company offered Kilby a position that would allow him to work on miniaturisation full time (and I had always thought everything was bigger in Texas!). He started there that May.

The summer break was coming up, but as a new recruit who had only joined the company just a few weeks before, he had not accrued any leave days. Over ninety percent of the workforce was away during what was known as the 'mass vacation', and although it probably made Kilby feel a little lonely at times, it afforded him a great opportunity to make some headway on his miniaturisation work. 'I was left with my thoughts and imagination'[11] he said later.

As he saw it, his 'goals were simple: to lower cost, simplify the assembly, and make things smaller and more reliable.'[12] Kilby realised that 'the cost structure within TI was very different'[13] to what he had experienced at his previous working location. He felt that 'probably the only thing they could make cost effectively were semiconductor products. This triggered the thought that maybe you could make everything from semiconductors.'[14] Kilby could clearly see that the major problem with all the existing approaches to miniaturisation was that they involved different materials and manufacturing processes. This made the whole circuit much more complicated to create, and therefore cost more money. It became clear that if the only thing that TI could make effectively was semiconductors, then he had to find a way to create everything that was required within the circuit in exactly that manner. He realised that maybe 'semiconductors were really all you needed.'[15]

If he could make all the basic parts that constituted a normal electronic circuit; the resistors, transistors, capacitors, diodes, and so on, out of just one single material, and manage to connect them all together afterwards, then in principle he could get rid of all these different components and integrate everything into a single device. On 24 July 1958, Kilby first wrote in his lab notebook a short description of the concept he had come up with. It became known as the 'Monolithic Idea'.

In principle all the various bits needed to build an electrical circuit could be made from semiconductors, instead of the usual materials required. However, the real problem was they would not 'perform as well as the best ones made with more conventional techniques and materials.'[16] Nevertheless, he felt that if the advantages of size and manufacturing cost outweighed this, any other deficiencies would not really matter.

During the hot, sticky Texan summer, this rookie employee, barely two months with the company, tried to come up with a way of making an entire circuit using just one material. Once the holidays were over, and everybody returned to work, Kilby showed the design to his supervisor, Willis Adcock. He was somewhat enthused by what this new member of his team had done, but was still sceptical about the practicalities. He wanted to see proof that circuits made entirely from semiconductors could work. Kilby busied himself building up a circuit made up of just silicon components. He formed resistors 'by cutting small bars of silicon and etching them to the required values'[17] while 'capacitors were cut from diffused silicon power transistor wafers.'[18] Each of them was crudely fashioned, and wires soldered between them to connect the whole thing together.

He managed to demonstrate the device to Adcock at the end of August, and to both men's surprise it actually worked. The test had proved that 'all circuit elements could be built of semiconductor materials.'[19] Although the circuit's components were made out of one material, this could not yet be described as integrated, since the pieces were all separate. Kilby reasoned that if all the components could be made with a single kind of stuff, then it should be possible to consider making them all within a solitary piece of it. This clearly offered the benefits he was looking for when it came to cost effectiveness of fabrication, and reduced circuit size. It was similar to how Ford had transformed the motor industry five decades before. Early automobiles were built one by one, to meet each customer's specific requirements. This meant they took a lot longer to construct, and made them very expensive. By creating production lines, his company could churn out a standard 'one size fits all' product that was cheap and easy to produce. If it worked for cars, why shouldn't it work for semiconductors? Over the following weeks he spent his time trying to fit a whole circuit onto one tiny slither of germanium.

On 12 September, Kilby gave a demonstration of his 'integrated circuit' to a group of TI's top brass. By his own admission, Kilby found these executives' response as 'somewhere between moderate and restrained'. There were four basic objections. The first was a belief that production yields would always be too low to be profitable, i.e. the number of actual working devices turned out in comparison to the junk they had to throw away. During this early stage in the evolution of the semiconductor industry, the ratio of good to bad products was notoriously low. On average less than ten percent of all transistors manufactured actually worked properly, if you were trying to put

everything on one piece, it would only make matters worse, as there would
be more things that could go wrong. The second problem Kilby faced was
the fact that most people felt it did not make very good use of materials,
since the best resistors and capacitors were not produced from semiconductor substances, like germanium or silicon, thus the products' quality would
be, to say the least, dubious. Thirdly the engineers were against denigrating
the products they had created, as Kilby put it they 'didn't want to see their
elegant devices messed up with all the other stuff on the chip.'[20] Finally and
perhaps most importantly of all, many people were worried that if semiconductor technology could replace a lot of the electronics that made up
the televisions and radios, or the pieces of industrial machinery being built,
then all the circuit designers needed previously would suddenly become
superfluous, and lose their jobs. Back then people had no clue that the use
of such technology would open the door for people to create more electronic equipment, rather than encroaching on the business that was already
there. Kilby recalled 'the thought that everybody might have a personal
computer at their desk or their home was certainly not on the mainstream
of anybody's activity at that time.'[21]

He had to admit that at this stage 'these were difficult arguments to counter, because they were basically all true.'[22] It was certainly not going to be
easy. The issue was not one of creating a product that worked properly, it
was convincing people that they really needed it in the first place. Kilby had
to find a way of making people believe.

TI first publicised the integrated circuit in January 1959. Kilby felt
that following the announcement, he and the small band of like-minded
engineers who had faith in this technology 'provided the technical entertainment at professional meetings for the next five years.'[23] To many people
Kilby's concept of squeezing an entire circuit onto a single chip equated to
something similar to the Loch Ness monster, or the lost city of Atlantis: it
was sheer fantasy.

The company filed for a patent on Kilby's 'Miniaturised Electronic
Circuits' on 6 May that year (US Patent 3,138,743), though it was still unsure
if it had any long-term potential. The patent described the production of
'integrated electronic circuits fabricated from semiconductor material', and
proposed that 'miniaturisation can best be obtained by use of as few materials and operations as possible.' It went on to claim that it would be possible
'to achieve component densities of greater than 30 million'[24] units within a
single cubic foot of the substance.

The first commercial available products were released in 1961, and cost roughly $500 each. Now at the time a competing non-integrated module would only set you back say fifty bucks. It meant that for most applications they were just priced out of the market. Kilby and his compardres were forced to accept this was not going to be an instant success. They had to be in for the long term.

By the time 1964 arrived TI's 'Solid Circuits' were capable of carrying the equivalent of seventy different components on a single piece of silicon. According to Kilby, by then 'a few adventurous companies had begun docking integrated circuits into their commercial equipment'[25], but universal interest in the invention was still some way off.

When I was given the chance to interview the octogenarian Kilby in Spring 2005, he told me it had been a very slow process. 'We had to take it one bit at a time, they weren't going to get widespread use overnight,' he noted, 'we chose the applications to target very carefully. Aerospace and military were obvious places to try, and offered an important starting point for us.'[26]

There would be two key developments that forced a change of thinking within the electronics world, and both of them would be fuelled by rising tensions between the United States and the Soviet Union. Firstly there was the 'Cold War'. Occurrences such as the Cuban Missile Crisis had taken the threat of nuclear attack from being a remote possibility to a strong likelihood, and so 'Uncle Sam' embarked on a program to construct intercontinental ballistic missiles. Although the Americans had developed the first nuclear devices, the Soviets had quickly caught up, and by the late 1950s were considered to occupy a superior position. The Minutemen, as the US missiles became known, were designed to close the lead the Red Army had gained, but they required far more complex electronics than had been used in the past. Each missile had a computer system made up of twenty-two different circuit types. It would be impossible to use vacuum tubes and big discrete components in such a design. The need for a more integrated approach now became apparent. The US Air Force first started to look at TI's integrated circuits in mid-1961; and over the following decade would utilise them in increasing numbers.

The second factor was that of the 'Space Race'. The NASA Mercury and Gemini programs had been unable to keep pace with Soviet space efforts. The USSR was first in placing a satellite into orbit (Sputnik 1, in October 1957) and sending a living creature into space (the dog Laika, in November

1956). It had been responsible for putting both the first man (Yuri Gagarin, in April 1961) and woman (Valentina Tereshkova, in June 1963) into space, as well as completing the first space walk (executed by Aleksei Leonovinm, in early 1965). In the world's eyes the Eastern Bloc was wiping the floor with the capitalist West when it came to cosmic exploration. Edward Teller, one of the scientists who had worked at Los Alamos on the early atomic bombs, said that the US had effectively lost a battle of greater importance than Pearl Harbour.[27] However, the most momentous prize of all was still up for grabs; sending a human being to our nearest neighbour in the heavens – the Moon.

In early 1961, President John F. Kennedy had pledged that the USA would put a man on the lunar surface before 1970 had begun. The goal of the Apollo programme was to do just that, but it certainly was not going to be easy. The Apollo 8, scheduled for late 1968, would attempt to orbit the Moon; if the mission successful the next stage would be to try to get a spacecraft to land there. The Apollo Guidance Computer (ACG), which the vehicle needed to perform this marvel, required a processing system that was tough, small, frugal with power, and very fast. The invention of the microchip had come just at the right time, such an endeavour would have been impossible with out it. The AGC would use over 4,000 chips.

The turning point had come, Kilby's foresightedness had paid off, and by the late 1960s 'most of the engineers had accepted that integrated circuits were here to stay.'[28] He had not been alone in his thinking, though, for on the other side of the country others had been at work on a similar idea. What is more, it would transpire that much of the spoils were to head in their direction, rather than Kilby's.

Born in Burlington, Iowa, on 12 December 1927, Robert Norton Noyce was an average kid from an average neighbourhood, in an average Mid-Western town. His father, Ralph, was a Protestant minister and his mother, Harriett, a typical American housewife. He was the third of four boys; Donald was the eldest, followed by Gaylord, and Ralph Jnr the youngest. Just like Kilby he grew up in America's 'heartland', lots of wide open spaces, miles of cornfields, and precious little to do to relieve the boredom. From an early stage Robert knew he did not want to end up stuck in Iowa for the rest of his life, he had great expectations. In school he was seen as a model student, and showed signs of extraordinary ability, though whether he would get to use it to full effect was uncertain. In 1941, still in his early teens, he almost ended his life story before it had even had a chance to begin. He and his brothers had read in a scientific magazine about a box kite capable of carrying a person. They decided to try to create one of their own. The boys put a frame together, using a similar design to the one described in the article, then set about giving it a trial run. It was down to the intrepid Robert to act as test pilot. Possibly it was because he was smaller, and thereby lighter than his older brothers, that he was picked. Or it could have been the fact that as he was smaller, and they just simply did not give him much choice in the matter. Taking the kite they had constructed with him, he clambered up onto the roof, and with a huge tug on the rope the others tried to launch him into the air. Their first endeavours into the field of aviation were not that impressive. Robert did not so much sail into the stratosphere as plummet headfirst toward the ground. Luckily he suffered no serious injury, but had been fortunate not to break his neck.

Noyce began to study physics at nearby Grinnell College. There he was tutored by a young professor by the name of Grant Gale. Gale had read electrical engineering at the University of Wisconsin with John Bardeen, one of the team that went on to develop the transistor, and had also worked for a couple of years at Bell Labs himself, before taking up teaching in the 1930s. His connection with the laboratories allowed him to get hold of few of the very earliest transistors, and he brought them into one of his lectures to show Noyce and the rest of the class. Noyce was fascinated by the technology, and soon installed himself as Gale's star pupil. He was not your normal tech geek though. He was a charming, incredibly popular, gregarious character, who enjoyed playing tennis and going horse riding. He was on the school diving team, winning the Midwest Conference Championship in 1947, as well as being an accomplished musician and an actor (gaining a leading role in a soap opera broadcast on local radio).

Despite his quite strict religious upbringing (being involved in church youth groups and Sunday schools) Noyce knew how to have a good time. He dated local girls, and partied regularly. Unfortunately one night his work-hard/play-hard attitude would go a little too far, and land him in a whole heap of trouble. In July 1948, he and a couple of college friends stole a pig for a barbecue. The next day, once the haze of a punch-induced hangover had cleared from his head, Noyce decided, thanks to his strong Christian principles, that he should do the right thing. He paid a visit on the farmer whose animal he had taken, acknowledged his guilt, and offered to reimburse him for the trouble caused. The farmer was clearly impressed with the young lad's sense of responsibility, and his willingness to make amends for his prior stupidity. So to show that he could make an equally magnanimous gesture, and display his kind and merciful nature, he immediately called the cops – so much for forgiveness.

Noyce was living in a small farming community with a highly conservative populace. Incidents like this, though they sound ludicrously minor in a modern context, were not treated as adolescent pranks. Stealing was stealing, end of story. He was looking at almost certain expulsion, and for a potential high-flyer this could have done irrevocable damage to his career. The top universities would not touch him with a 20ft barge pole with this on his record.

His older brothers had already done very well for themselves; Donald had achieved a glittering academic career, receiving a Ph.D. from Columbia University, then taken a post doctoral position at Berkley, while Gaylord had also excelled in college, and just secured a teaching post in Turkey. Robert did not want to become known as the 'no good loser' of the Noyce clan.

Gale knew that his young protégé had the ability to really be someone, and so he made it his mission to ensure that Noyce was not going to blow it all over a stupid drunken lark. If he had any permanent record of criminal activity to blacken his name, the chances of him getting into a decent university would be close to zero. After negotiations with the school authorities, a compromise was reached; Noyce was suspended for three months. It would give him enough of a rap across the knuckles to make him think twice about getting involved in similar stunts in the future, but would not impede his potential for greatness in the long term. He graduated with honours, his file unblemished, and gained a place at Massachusetts Institute of Technology (MIT), starting his studies there in Autumn 1949.

MIT was among the most well-respected seats of learning in the whole United States, it had been founded back in the 1840s, and was America's first university to concentrate on science and engineering. Former students included George Eastman (founder of Eastman Kodak) Robert Oppenheimer (of atomic bomb fame) and William Coolidge (inventor of the tungsten light bulb), to name but a few. While at MIT, Noyce was involved in the amateur dramatics society, and here he met a young girl, Elisabeth 'Betty' Bottomley, whom he married in late 1953. They had two children; William was born in 1954, and Penny the following year.

By the time he left MIT, in possession of a doctorate in physics, there was
no shortage of companies interested in hiring Noyce. The likes of Bell Labs,
RCA and IBM all made him attractive offers. But he decided to work for a
smaller, less well-known firm, based out of Philadelphia, called Philco. His
reasoning for this was he felt picking a company that was only just starting
out in the semiconductor market would offer a better chance to quickly
move up the ladder. He would not stay there long, opportunities soon presented themselves elsewhere.

William Shockley was a graduate from Stanford University, and had also
been at MIT, receiving his doctorate there. In 1947, while working at Bell
Labs, Shockley had led the team that created the transistor. In 1955, he had
decided to make a new life for himself. He headed back west, and chose
his old hometown of Palo Alto as the location for his business venture.
Like Eastman, Hewlett, Packard and others had done, he took advantage
of Stanford's offer of cheap real estate, and with financial backing from
Beckman Instruments, Shockley Semiconductor was born.

The cream of the world's electronics talent was pulled towards Shockley
and his new enterprise, his reputation as the man leading the way towards
a new era of technology drew the engineering elite to him like moths to a
flame. Robert had missed out on the opportunity to work with Shockley
when he chose to accept Philco's offer instead of Bell Labs, but he was not
going to do the same thing twice. In Spring 1956, Noyce packed in his job
in Philadelphia. The story goes that he, with quite amazing levels of self-
belief, left the East Coast with his whole family in tow, arrived in California,
bought a house that morning, then after lunch went to see Shockley for a
job, in precisely that order.

Despite his limited previous experience and relatively young age, Noyce
became Shockley's Director of Research. However, the honeymoon period
was not to last long, some harsh realities soon became apparent. Shockley
had an incredible analytic mind, but he was not a 'people person' by any
stretch of the imagination. He treated the management of his employees
in the same way he would try to solve an arithmetical problem. Though he
knew his way around an electrical circuit like no other man in the world,
the complexities of the human psyche were a total mystery to him.

The working practices that Shockley engaged proved completely
abhorrent to all involved. Firstly he was suspicious of everybody. He was
concerned that employees might be spies for competing firms, and constantly worried that projects were under threat from sabotage. He wanted to

make his workers have polygraph lie detector tests, to put his mind at rest. Some of the staff started to refer to the company's address by the nickname '391 Paranoid Place' rather than its true location of '391 San Antonio'. His egocentric nature was also a problem. He undervalued his engineers' efforts in comparison to his own, and this bred an atmosphere of resentment.

Further stunts that did little to promote team spirit included his 'peer ratings' scheme, in which employees would be publicly assessed by their colleagues. He even put up a list of all the staff members' salaries on the noticeboard (cringe!). There was a growing tide of anger in the lower ranks against the boss, and before long it was certain to end in a full-scale mutiny. Noyce grew increasingly dissatisfied with the situation, and he was not the only one. A group was forming that were determined to find a better alternative to this sort of treatment.

In November 1956, Shockley received the Nobel Prize for Physics for the 'discovery of the transistor effect', which offered some respite from the pressures that Noyce, and the rest of the workforce were feeling, but it would not hold off the deluge of dissention for long. At first the rebels tried to oust Shockley; going over his head they approached Arnold Beckman, the company's financial backer, asking him to act before it was too late. Unfortunately, Beckman was not having any of it, unwisely putting his support firmly behind Shockley.

In mid-1957, Noyce, along with Gordon Moore, a former California Institute of Technology post-doctorate, and Jean Hoerni a Swiss-born engineer who had gained doctorates from the universities of both Geneva and Cambridge, persuaded another five of Shockley's staff to walk out. He had managed to get Fairchild Camera & Instrument, based in New York, to back his own semiconductor venture. The chairman, Sherman Fairchild, agreed to front up enough money to get the firm on its feet. Shockley, always one for dramatics, branded them the 'traitorous eight'.

Fairchild Semiconductor opened its plant in nearby Santa Clara in mid-1959. Noyce's plan was to create a vibrant, ambitious firm which could really take this technology on to the next stage. The mother company would have the option to buy the founders out for the sum of $3 million, if the enterprise turned out to be a success, but other than that it gave them free reign.

The industry was young and dynamic, with an abundance of ideas, and by now Shockley was just too old and cranky to be part of it. Despite all this, the importance of his role in the forming of the high–tech armada should not be underestimated. Firstly he had been the real pioneer of the semiconductor industry thanks to the creation of the transistor, but on top of that he would also be the man who sustained it, although this was achieved more indirectly. He brought together the men (and subsequently drove them away) who would take this still very niche-orientated market into a new epoch where it became almost completely ubiquitous.

Deserted by his top executives and most valuable engineers, Shockley tried to carry on the business as before, but made little headway. He was finally bought out in 1960, and returned to safer ground back in the world of academia, taking a professorship at Stanford. From this point on the less we learn about the deeds of William Shockley the better. While back in university he developed rather erroneous theories about the connection between racial type and levels of intelligence. He argued that black people usually had lower IQs than Caucasians, and tended to procreate more. He believed that these 'intellectually inferior' people, as he described them, were diluting the gene pool, and as a result mankind was starting to regress. His solution to stop this was to pay black males $30,000 a time to be sterilised.

He was not the first to suggest such extreme measures, the idea of 'eugenics' had been around for some time. Over sixty years before, the messiah of healthy living, breakfast cereal inventor and renowned nutcase, John Harvey Kellogg, had been spouting this rubbish too. One thing to bear in mind here is that for all his engineering talents, Shockley had no in-depth knowledge of genetics or human biology, he would just churn out this stuff, but lacked scientific proof to corroborate it.

He would gain a great deal of attention for his rather distasteful views, and even stood for the American Senate. Luckily he came a poor seventh in the poll. Interestingly, though, some rather important people chose to tolerate his views. During the 1970s, he served on a government taskforce charged with looking into methods of conserving the Earth's diminishing resources. The senator who headed the commission, and had recruited Shockley, was none other than future US President George Bush Snr. Shockley finally died of cancer in 1989, and was probably not that greatly missed.

In many ways Noyce was the anti-Shockley; he had a warm and trusting nature, he was generous, and had the ability to inspire and motivate his team. His management technique was fast and loose. Noyce gave his employees the opportunity to do things the way they wanted to do them, take chances, and develop greater levels of aptitude. He did not want to rule with a rod of iron; firstly it would not suit his image as everybody's pal, and secondly it would not be beneficial to the long-term prospects of the company. Cramping his people's creativity would just breed resentment like it had under Shockley. Noyce wanted to provide a fertile and favourable environment that allowed staff members to realise their potential, not just use up their hard work and goodwill, and take all the glory for it.

Noyce's path to the integrated circuit was slightly different to his Dallas-based counterpart. Though Kilby's concept was sound enough, translating it from a bunch of hand-carved semiconductor components crudely connected together into a device which could be mass-produced was the real challenge. Having to use separate wires to connect all the components on the chip defeated the whole object. There had to be a better way to unite the working parts of the circuit together.

Noyce and Hoerni came up with the idea of putting the interconnections on to the chips without all this hassle. Basically the Swiss engineer had developed a new manufacturing technique that would allow complete transistors to be mass produced in big blocks, with a protective oxide layer attached. These could then be broken up into individual devices, rather than having to add the protective layer after they had been separated. This made the components far sturdier, and more reliable. Noyce began to think to himself, where was the sense in separating them if they had to be connected up again when they were placed in the customer's circuit? Why not sell them in little blocks of ten or twelve transistors, all connected together beforehand.

They had found a way to deposit an insulating layer of silicon oxide over the top of each device, almost like a cake factory would place frosting onto its donuts. This would isolate each transistor from the others, so that no charge could pass between neighbouring devices and short the circuit out. All the interconnections could then be placed on the wafers quickly and easily, instead of having to endure the precarious and time-consuming task of wiring up each piece by hand. What is more, the printing process was ubiquitous, it did not matter if you were making ten chips or 10,000, the time and effort involved would be just the same. Thus the more chips being made, the cheaper the process would effectively be. Fairchild put forward a patent for the 'planar process' in July 1959. It was granted on 25 April 1961 (US Patent 2,981,877).

Just as it had in TI, the integrated circuit polarised the staff at Fairchild. The company's management was divided about using this process to make its transistors, feeling that it presented a great risk. The amount of money needed to spend on research was enormous, and thus it would transpire that Fairchild had to scrimp on development of the standard devices it was already making. This essentially meant the firm would put all it eggs in one basket. Noyce was effectively gambling with the entire company's future, but he knew that Fairchild was still a minnow in this particular pond, and it would only be by innovation, and occasionally taking risks, that it would ever get any bigger. He decided to implement the planar process for all its forthcoming products.

Noyce was a visionary; he could see the way the wind was turning. The manufacturing cost of discrete devices would only go up, while in time integrated semiconductors would become cheaper to make. To keep up with the big discrete suppliers would need a massive expansion in production facilities, which obviously posed difficulties for a small firm like his to implement, and would eventually only take them down a dead end street anyway. Sooner or later the industry would need to embrace a whole new ethos. If they did it now, as a small flexible company, they would get the jump on the larger, stodgy firms which dominated the sector at that time.

As well as the technical advance that allowed the microchip's use to become more widespread, Noyce pulled a marketing masterstroke. He figured that if they priced these devices a little lower than the complete bill of materials of all the standard discrete components needed, then customers could no longer ignore them. Of course this meant that for a long time, the company would lose money on every chip it made, but once the volume really started to ramp up, as he assumed it would, then the cost of producing each chip would drop considerably. It was a big gamble, but Noyce had backed a winner.

The first generation of integrated circuits were far from impressive. The principle was there, but the ends did not justify the means yet. The level of functionality and reduction in manufacturing cost, compared to standard components were at best slight, and more likely to be non-existent, because these early chips only contained a handful of transistors. What was needed was something that became known as a Large Scale Integration (LSI); by putting many devices on a single chip the benefits became much more apparent. As volumes increased, manufacturing costs per chip were reduced, and this in turn made them more accessible to other application areas. As these new areas began to employ chips, it pushed up the number being produced still further, and in turn the economies of scale meant prices would be reduced by an even greater factor. The whole thing kept on spiralling upwards, and this has pretty much continued ever since.

Gordon Moore, even from an early stage, could see a trend beginning to take shape. In an article published by *Electronics* magazine in Spring 1965, he tried to create a mathematical model of how this movement would play itself out, and the pattern by which the semiconductor business would grow. It became known as 'Moore's Law' and for anyone who has any involvement in the electronics sphere it is virtually the first thing you learn. It serves as the driving principle for the entire industry. The basic precept of the law is as follows: the number of transistors (and hence the functional capacity) of a semiconductor device doubles every year and a half. Okay big deal. Ah, but at the same time this means that the cost to make a chip effectively halves. So what you have is an exponential increase in the processing power of devices, and an equally dramatic reduction in price. The prediction meant that as time went on the number of places you would find a microchip just kept on growing, not just by a little bit but by a factor of two, every eighteen months. The law has stayed true right up till the present day, and will probably continue to do so for several decades yet.

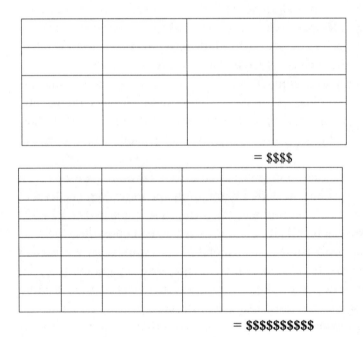

= \$\$\$\$

= \$\$\$\$\$\$\$\$\$\$

Costs have kept on being pushed further and further down. Firstly because the manufacturing yields keep on improving, more good are chips produced. Also wafers continue to get larger, which means that more chips can be turned out per production run. The chips have also become more complex, and thus deliver higher performance, as the structures on each chip have kept on getting smaller and smaller. Alternately it is possible to make chips that can do the same job as the previous generation, but cover substantially less silicon. This means that with each wafer produced you can fit more chips on to it, again reducing the expense involved in creating each one.

The huge proliferation of electronics that the development of integrated circuits allowed meant that manufacturing equipment was better utilised. New machines could be used to form products by the million rather than the thousand. These savings could be passed on to the customer, and the huge price erosion meant the devices were not expensive luxuries any-more. This formula described what Kilby's instincts had told him; if volume demand could be found the costs would drop, this in turn would open up new markets, and the whole thing would snowball from there. He noted that back in the late 1950s 'a single silicon transistor sold for about $10. Today, $10 will buy you something with twenty million transistors'[29]

While back in the late fifties the ownership of the idea that had all sparked this off was not something that anybody was in a rush to admit to, as the demand for microchips gained greater fortitude, the debate as to who was really its inventor also intensified. Noyce would claim he had not heard of what Kilby was up to down in Texas, and his and Hoerni's work had been completely independent, but this is highly unlikely. The industry was still very small, and somewhat incestuous, at this stage. It is almost certain that word would have made its way along the semiconductor grapevine. The legal battle that ensued would continue for over a decade, but despite the fact that Noyce's patent had been applied for after Kilby's, it encompassed not just the basic concept, but the manufacturing technique needed and this put him in stronger position. Eventually TI and Fairchild managed to work out a deal that everybody could live with. The firms cross-licensed each other's technology, the Texans getting a share of the royalties.

By 1967, Fairchild had grown from its original eight founders into a multi-million dollar firm with over 10,000 employees. It was close to becoming the largest semiconductor manufacturer in the entire world. At this stage it was the most successful part of the entire Fairchild organisation, in fact it was the only division actually making a profit. Despite Noyce's clear ability to manage a successful business venture, the multinational was not interested in letting him progress within its own hierarchy. Also as the company had taken up the option to buy the founders out a couple of years after its inception, it was now fully under Fairchild's jurisdiction. Noyce, who had previously been allowed to carry on virtually independently, was now expected to toe the party line. He was certainly peeved by the fact that, regardless of what he had achieved, he was not taken seriously by the board of directors over in New York. The 'Noyce Administration' was nothing like what the boys back east were used to. They found his free-and-easy management style way too outlandish.

Once again Noyce was ready to up and leave. He had managed to realise his dream of building a ground-breaking young firm to an extent, but the paymasters that were to be found behind its façade were just the same as those he had tried to get away from, and this would always limit how far he could actually take things. If he was really to fulfil his vision in its entirety, he needed to go out on his own.

Arthur 'Art' Rock, a San Francisco-based venture capitalist who had helped to instigate the deal with Fairchild Camera, was asked to find backers for a new project. The reputation that Noyce and Moore had managed to make for themselves held them in good stead. Rock was able to raise $2 million almost instantaneously. Legend has it that it took just one afternoon.

They were at first unsure what to call the start-up; they knew it needed to be something with a bit of punch to it. Moore suggested 'Intelligent Electronics', which seemed to fit. Soon afterwards, they decided to contract this to 'Intel'. The company was incorporated in mid-July 1968, its mission, as Moore saw it, was to address the 'product areas that none of the other manufacturers'[30] were covering.

Intel could certainly count on attracting the industry's hottest properties, the status of the men behind it had already attained assured that. The problem was what were they going to pay them with? All the best engineers were on good wages with other firms, but this start-up's budget would not stretch to the sums of money required to persuade others to jump ship. When you decide to join another company, you expect to get a bit more money for your trouble, otherwise what would be the point in going there in the first place? However, the primary batch of Intel employees were not to be blessed with such luxuries, they would have to wait until the company had gained some traction in the market before salaries could be raised. The management was just hoping this would not take too long. At best his staff's pockets were almost as full as they had been with their previous companies, and in most cases decidedly lighter, so Noyce had to find a way of cushioning the blow. He did this by offering the new arrivals shares in the firm.

He had tried to use similar incentives at Fairchild, but the parent company, with its rather conformist East Coast outlook, would not agree to this. Giving employees stock options not only helped Noyce to get away with capping their wages, but gave everybody a sense of purpose. It boosted their productivity, because they were not just making somebody else rich with their efforts; if the company did well, then so would they. It would be the strategy that nearly every semiconductor manufacturer would adhere to from this point on. Noyce had started a whole new business philosophy.

The company would grow at a phenomenal rate. Much as he had tried to do at Fairchild, Noyce created an atmosphere that harboured greater inventiveness, and encouraged his team to have passion for what they were doing. He was not looking to control the company like a dictator, subjugation just did not suit his personality. As Art Rock put it, Noyce liked 'to help people make their own decisions'.[31]

Though his new business was coming on leaps and bounds, his home life suffered as a result. Relations between him and his wife Betty became increasingly cold, and the countless hours he devoted to Intel meant that he overlooked her needs too often. They divorced in 1974, and she went to live back east. Surprisingly it did not take long for Noyce to recover, he married again soon after, but was clearly thinking that it would be more convenient to be with someone that he could spend a greater amount of time with. His second wife was Ann Bowers, Intel's human resources manager, who was a good ten years his junior.

By this stage Noyce was contented with what he had achieved, and was sensible enough not to want to waste the rest of his life on his career. He pulled back from the crazy circus that the electronics world was rapidly becoming. From this point on his involvement with Intel would be on a much more relaxed footing. As its chairman, he would act as an ambassador for the company, rather than a boss. His high repute, and the sense of authority emanating from him, made for an invaluable figurehead, but he did not want to concern himself with the trivialities of hands-on management. He left the day-to-day running of the firm to the over-zealous Andy Grove, and the strategic decisions to the ever-thoughtful Moore. Through the rest of the 1970s he acted as just the front man for the company, rarely getting involved in other matters.

Despite that fact he was now in his fifth decade, Noyce had barely gained a pound since his teens. He had maintained an athletic physique, and decided to make a little more time for sporting activities. He took up hand-gliding and scuba diving, as well as becoming a proficient skier. He also started taking flying lessons, and before long bought his own plane.

He used the wealth he had obtained to aid new companies within the high-tech arena, investing in several firms that would eventually become major players. In 1977, he helped found the Semiconductor Industry Association (SIA), and served as its first chairman. Three years later he was awarded the National Medal of Science by President Ronald Reagan in a ceremony conducted at the White House. He was inducted into the National Inventors Hall of Fame in 1983.

Noyce died of a heart attack while swimming in his pool on 3 June 1990. He was just sixty-two, and had appeared to be in incredibly good shape for a man of his age. Later that year the SIA inaugurated a special prize to honour his memory. The Robert Noyce Award is the highest accolade that the organisation gives.

Throughout his life he had been a person who possessed great charm and vitality, boasting a highly magnanimous nature. His management technique was very laid back. He believed that the best way to get one hundred percent from your people, and to encourage them to put in the extra hours, was to make sure they liked the environment they worked in, and they felt their efforts were appreciated. It has been the template used by nearly every other high-tech company in Silicon Valley to this day. He has been described by many writers as being blessed with a 'halo effect', instilling the seeds of inspiration in the people around him.

Noyce effectively wrote the manual on how to be a semiconductor chief executive, nearly every top manager that followed stuck to the pattern he had forged. He was an inspiration, both on technical and entrepreneurial fronts, to the whole of the electronics industry, and is certain to be remembered as one of it most influential individuals.

Though most electronics historians speak of the wrangling between Texas and California, as to who was behind the first integrated circuit, very few tell of the Englishman who proposed the whole idea several years earlier. This is the hapless story of Geoffrey Dummer, a radar engineer with the British Ministry of Defence. If ever the old adage about being in the wrong place at the wrong time was true about anyone, he would have to be in the running.

Dummer, who is sometimes described as the 'Prophet of the Integrated Circuit', was born in the town of Lacock situated in Wiltshire, back in 1909. He was educated at Regent Polytechnic in London, before going to study in electrical engineering at Manchester College of Technology in 1931. He married his childhood sweetheart Dorothy in 1933, and they had a son, called Stephen, two years later. After he received his degree he joined the Royal Radar Establishment (RRE), based in the English border town of Malvern.

Dummer worked on the development of automatic gain-to-range control for the Royal Air Force's radar systems, thus enhancing the span of how far they could track aircraft. He also created the plane position indicator, which was used to observe the first ever ground-controlled aircraft interception, and built simulators that gave air crews greater experience without the huge expense of actual flying time. During the Second World War he was made Group Leader of the Trainer Design Group, and under his guidance some 2,000 training machines were constructed. This was the equivalent of 4,000 aircraft doing two million miles of manoeuvres, and saved the British Government over £50 million.

For his assistance in helping secure the Allied victory, Dummer received the MBE in 1945. The following year he became a Fellow of the IEE, it would have happened sooner but his defence work prevented him from being part of such an organisation during wartime.

The first published reference to the possibility of constructing integrated circuits was made by Dummer in May 1952. At the Symposium on Progress in Quality Electronic Components, held in Washington DC, he told the audience, 'with the advent of the transistor and the work in semiconductors generally, it seems now possible to envisage electronic equipment in a solid block with no connecting wires.'[32] What he foresaw was a piece of semiconductor fabric that would 'consist of layers of insulating, conducting, rectifying, and amplifying materials. The electrical functions being connected directly by cutting out areas of various layers.'[33]

Before the end of that year, he bought together a plan to create a fully integrated device, like the one he had described while in the USA. He proposed putting a complete circuit on to a piece of silicon roughly a centimetre square, and started building a basic prototype. Over the next few years he attempted to perfect his model, and construct a whole electronic circuit on a single chip. He managed to get his department to back the scheme at first, but it soon lost interest. He was to fall foul of the same arguments that would be used to attack Kilby's ideas. It was considered just too costly to produce such things, and they were not as good as the separate components they looked to replace, so what was the point? With typically poor foresightedness on the part of the British Government, the project was finally axed. All Dummer's efforts had come to nothing, with funding officially stopped in late 1957.

Dummer was just a few years off with his timing. Had he voiced his ideas a little further down the line, when the need for integration was starting to make sense, and he had been in a more forward-thinking environment, he probably would have been on to a winner. He was to suffer a similar fate to that which befell Turing and Babbage. It is the same old story; Britain's potential for greatness being scuppered by lack of financial support. Just as the UK had been responsible for creating the sports (almost without exception) played around the globe, yet we are still totally awful at nearly all of them; our nation's ability to reap the rewards of its citizen's inventiveness has gone awry on many occasions. To continue the sporting analogy a little further, Dummer had given Britain a five-year head start in developing the integrated circuit, but we still managed to come in un-placed.

After the horse had bolted, the RRE renewed its efforts in the research of integrated circuits, and it was there that in 1960 the first devices of this kind in Europe were to be produced, but that was little consolation for Dummer. Nevertheless he did become the continent's most esteemed spokesperson on semiconductor development, chairing many important conferences, and publishing a series of books.

Dummer finally retired from the RRE in 1966, and began a new life as a consultant. He worked into his eighties until a stroke finally prevented him from continuing. His first wife died in 1992, but Geoffrey remarried not long after. His second wife, June, cared from him until he finally passed on in February 2002, at the age of ninety-three.

Even Kilby himself acknowledged that 'Dummer suggested that it was possible to foresee a time when all electronics would be made as a single block.'[34] Though he pointed out that the Englishman had not been able to transform his vision into a working device. When I was fortunate enough to talk with him he told me 'Dummer was able to propose the idea very early on, but he didn't really explain how it was going to be realised, and that was what counted.'[35]

For over three decades the United States ruled the semiconductor world, and even in modern times, when it has had to share power with SE Asia and

Japan, it is still one of the most important countries in this sector. Britain, which could have pre-empted the whole technology movement had it listened to Dummer, lost out on all this, as it had with so many other things.

The success that Kilby had inspired began to have a profound effect on his own career. He was made the Manager of Engineering of TI's ever more important semiconductor division in 1966, and in February 1970 rose to the title of Director of Engineering and Technology. Later that year he decided to take some time out. The previous twelve years had been filled with a great deal of challenge and excitement, but the pressure had taken its toll. Kilby started a leave of absence that began in November 1970, to do some independent research of his own. The projects that he undertook were varied, but among them was the use of semiconductors to produce electrical power from sunlight. He was among the first to make serious forays into this area.

When I asked him if he felt that the microchip had been a hard act to follow, and made it more difficult to start anew with other innovations, he responded, 'No, I suspect if anything it helped, people are more willing to listen to you once you have something like that under your belt. Solar power was the biggest project that I worked on, but there were a couple of sizeable things that I attempted in telecoms.'[36]

In 1978, Kilby took up a professorship at Texas A&M University, though he confessed that 'I didn't do much "professing". However, I did have a rewarding time doing research and working with students and faculty on various projects.'[37] He continued to hold this post until 1984. By then he was well into his sixties, and felt that he had earned a bit of a rest. He went into retirement, though still kept an involvement with TI in an unofficial capacity.

Over the years, the IEEE presented him with a string of medals, making him its most highly commended member. Back in 1965 he had been given the Outstanding Achievement Award, in 1978 he was the recipient of its Cledo Brunetti Award, and finally in 1986 he received the Medal of Honour. Kilby's contribution was acknowledged elsewhere too. He was inducted into the Engineering & Science Hall of Fame in 1988, and later that year was given an honorary doctorate from the University of Illinois, where he had studied all those years before.

The National Medal of Technology was awarded to him in 1990, and he was given the Kyoto Prize for Advanced Technology three years later. In September 1997, TI paid its own tribute to the achievements of its illustrious former employee, by opening the Kilby Centre, a $150 million research and development complex, based at its North Dallas campus. The company's then CEO, Tom Engibous, said 'not only is Jack Kilby's invention one of the most significant of our time, it is one of the most significant of all time. Jack did more than invent the integrated circuit on that day ... he invented the future.'[38]

Kilby had won the most impressive collection of tributes that his industry could bestow on him, but a greater honour was still to materialise. In Autumn 1999, it would become apparent that he would receive the Nobel Prize in Physics. It came like a bolt out of the blue. Jack had no idea that he was even being considered for the accolade.

Supposedly he heard the news in the most bizarre of circumstances. It was not until a reporter started knocking at his front door to get his reaction to the announcement, that he knew anything about it. Normally the Nobel Foundation would notify laureates with a phone call before any formal statement is made. But as Jack wore a hearing aid, and tended to take it out during the evenings, he was often hard to get hold of. 'It was completely unexpected, I really had no idea that this would happen,' as he told me, 'as I recall I was informed in October, and the award ceremony took place two months later.'[39] On 10 December 2000, at a ceremony held in Stockholm, Kilby was presented with the Nobel Prize 'for his part in the invention of the integrated circuit'. He received $450,000.

Though the award was presented to him, Kilby was generous to say the least, acknowledging the efforts of his rival Noyce in ensuring the future of microelectronic technology. He stated that even though 'Robert and I followed our own paths, we worked hard together to achieve commercial acceptance for integrated circuits. If he were still living, I have no doubt he would have shared this prize.'[40] Kilby's view was that any sins against him had been adequately atoned for. Though 'there was some litigation' as far as he was concerned 'in the end it was all resolved amicably. Even though we never worked together, and were effectively on different sides we were to become friends.'[41] In one interview he expressed his feeling that 'there's beginning to be some general agreement on that now, and it's not as controversial as it was a few years ago. Today people tend to credit me with having the original idea and having made the first circuits. They tend to credit Noyce with having made a major improvement with the use of planar technology, which tremendously expanded the field and contributed to cost reduction.'[42]

Moore's take on things is a little different however. He feels that though 'Noyce and Kilby are often given credit as co-inventors of the integrated circuit,' what they 'contributed was dramatically different: Kilby made a laboratory model by hook and crook, Noyce took the planar technology and extended it so you could make a complete structure using the material processing operations we had developed. So you could cover a whole wafer with identical structures again and cut them apart and package them individually,'[43] which was an important consideration. Just creating an integrated circuit would not cut the mustard, it had to be formed in a way that was cheap and easy to produce. In his opinion what Kilby had produced 'was very much a laboratory device' but 'the technology we had at Fairchild was the path to make the practical integrated circuit.'[44]

Texas Instruments, which was without question the dominant force in the semiconductor arena up till the early eighties, is still one of the top five manufacturers in the sector today, and boasts annual revenues of close to $8 billion. And though this is thanks to the work of many gifted and hardworking people, at its bedrock is the outcome of one man's epiphany.

Sadly on 20 June 2005, just a few weeks after I had the opportunity to speak with him – and in what transpired to be the last interview he would ever give – Jack's life finally came to an end. He had been battling with cancer for a brief period. The whole electronics world paid tribute to the man who had set it all into motion. In the eulogy that TI's Tom Engibous gave upon the announcement of Kilby's demise, he was compared to other leading figures of American industrial history, such as Edison and Ford. He had collected over sixty patents during his life.

Kilby never lost his passion for technological advance. Right up till his death, he continued to keep an avid interest in how the industry unfolded. He told me that 'the tremendous growth and change that has characterised electronics has been extremely exciting to watch, and I am very pleased to be a part of it.'[45] Kilby further remarked that 'the field is still growing rapidly, and the opportunities ahead are at least as great as they were when I graduated from college.'[46]

The Luddite reaction that he and his colleagues received was ill-founded. Engineers had feared that the creation of the microchip would bring an end to people's jobs, as electronic circuits would be a lot easier to build and so fewer designers would be needed. In fact this is exactly the opposite of what happened. Thanks to integration semiconductors became affordable enough to fit in everywhere. A staggering propagation took place, and a huge industry employing hundreds of thousands of people was born.

The situation is possibly a little different to the other affairs which have been investigated in the previous chapters. There can be little doubt in this occasion that both parties contributed to naissance of the microchip. Even Kilby and Noyce themselves would have had to acknowledge the other's importance in its early development. But in this case it is not that one man got the fame at his counterpart's expense, rather it is the fact neither of them got it. This phenomenon does not appear to be unusual in this era, in fact if you think of any great invention in the last sixty years or so, it is unlikely that you can say who was responsible for it. Innovation has become an increasingly anonymous affair.

This may not be a big problem when you think of the throwaway consumer rubbish and 'techno bling' that fills our television ad breaks, there is only so much importance that can be placed on the Easy-Mop or the self-assembly wardrobe, after all. But when we talk of something so primal to everything of note that humanity has done in the time following its inception, then it is hard to argue that the microchip is a special case, and therefore those responsible for bringing it to the world deserve more consideration. It would not be an over-exaggeration to say that Kilby and Noyce affected all our lives, though the true extent is not likely to ever be fully appreciated.

Conclusion

'In science the credit goes to the man who convinces the world, not to the man to whom the idea first comes.'[1]

Sir Francis Darwin

Having read the previous chapters, it is likely that most of you will agree that this statement holds huge validity for some if not all of the cases described. It is particularly true for Meucci, who may have conceptualised the telephone, but seemed unable to get anybody to take more than a passing interest in his creation. It needed someone like Bell, who could use his contacts in the financial world to get the project off the ground. Similarly, Tesla may have planted the seeds that brought forth radio communication, however he lacked the faculties required to convert it from a scientific curiosity into a commercially viable invention. Marconi did not possess anything like the same level of technical skill, but this was easily compensated by his proficiency in the art of public relations. He could use his ability to create media hype, and take advantage of family connections to make sure the powers that be sat up and took notice.

While Kilby had come up with the concept of the microchip, it required Noyce to find a way of constructing these devices so they could actually be mass-produced. In the case of Swan and Edison, it has to be admitted that though Swan clearly had a brilliant mind, he was far too naive to really make the electric light business his own. Despite the fact that Edison is often portrayed as a monster who would use any means at his disposal to swindle someone out of their ideas, possibly more than any other person in history he combined great technical proficiency with a head for commerce, and this enabled him to succeed while his counterpart only found dejection.

In reality there are no true 'Eureka' moments. The path of innovation is far too complex to be thought of in such a manner. It would be more apt to describe it in the form of a relay race. One person's achievements being the starting point for another to take it a little further, to pass the baton on to the next, or as we have seen in several examples already described, to have it unceremoniously snatched from them. History has sadly tended to concentrate its attention on those who ran the final stretch, and subsequently crossed the finish line. Unfortunately, when it comes to invention all too often it is not in fact the person who takes the anchor leg that can be said to have been pivotal.

The telephone was a fusion of the work that Meucci, Gray, Bell and several others had done. With regard to the theory of evolution, Darwin's must be considered part of an ensemble cast in which Chambers, Lamarck and Wallace all played supporting roles. Once again, in the development of the light bulb Edison, Swan and Starr had all been significant contributors. For radio communication Popov, Lodge, Tesla and Marconi all deserve thanks. In computing, Eckert, Mauchly, Turing, von Neumann, Babbage and others too numerous to mention were all involved, and it certainly seems unfair to give any single person the praise in its entirety.

Some important individuals in this book did not make the final cut to join the 'Nearly Men'. Among these were Eckert, Babbage and Gray. This is only my appraisal and you may feel that some of those mentioned actually had better cases to present. It is of course your prerogative as to where you feel the merit is best deserved. The point is that by looking at the story as a whole, rather than concentrating on the deeds of just one person, we can see that rarely is there a situation where someone took things all the way from first principles through to the finished item, whether it was a scientific theory like evolution, or a piece of gadgetry such as the light bulb. How much of the thanks should go to each contributor is something that everybody has to decide for themselves, based on how they judge the evidence, but it must be accepted that these are not solo performances.

By way of one final illustration, take American James Watson and Englishman Francis Crick of Cambridge's Cavendish Laboratories, who are credited with formulating the structure for DNA, the basic code that all life is dependent upon. Little is said of the fact that they managed to put the whole thing together not by their own research, but by making use of another scientist's work. Rosalind Franklin, of Kings College London, had been trying to uncover the form that these nucleic acids took for some time, but in historical terms her efforts would count for nothing. Thanks to a rather deceitful friend of theirs at Kings, Watson and Crick gained access to the results of X-ray photography she had done, without her consent. Using it, along with information obtained elsewhere, they put together their own model of its molecular arrangement. In 1962, they received a Nobel Prize for the discovery, but by then Franklin had died, the victim of cancer at the age of just thirty-seven. As the prize cannot be given posthumously, she was even denied this acknowledgement for her part in the discovery.

Of course almost without exception, anyone who claims to have made some great advance is going to have a bunch of wannabes following them, saying 'I did it first'. So it is clearly a matter of sorting those with some credibility from the chancers just out to capitalise on the situation. For example, there was talk just a few years ago of how failed US presidential candidate Al Gore had claimed he had actually been responsible for the Internet. Huh! Where did that come from? Suffice to say Tim Bernards Lee and Robert Taylor probably did not take this too seriously. (This proved to be a bit of a sham in fact. Gore had actually said that during his service in the US Congress he took 'the initiative in creating the Internet'. What he really meant was he helped promote its widespread use. He was apparently not trying to claim that he invented it, although his choice of words did invite that conclusion. The interview in which he said it was with supposedly unbiased, but in most people's opinion Republican supporting, CNN, so the rumour quickly spread. Thank God for reputable journalism eh?!)

With so many opportunists looking to get in on the act, the waters are often muddied by erroneous claims, and it is hard to come to a real verdict as to who has a valid case. Also it must be said that sometimes there can be only a thin line partitioning what could be regarded prodigious genius from the ravings of a lunatic. Newton's incredible mind had discovered the mechanism by which the planets move, but it also drove him into dealings with the occult and alchemy. Likewise Wallace's willingness to challenge widely held views allowed him to uncover the system that makes all living things evolve, however his open-mindedness meant he was prepared to believe in the far less substantiated practises of spiritualism too. Tesla's boundless imagination gave him the ability to empower a new era of civilisation, bringing it a convenient way of producing electricity, but carried him into a fantasy world of alien beings and other such nonsense.

Timing is another important factor obviously. Lamarck's ideas were just too outlandish to be accepted by the laypeople of his particular period in history. A similar appraisal could be given to Turing. His notion of how a computer would go from just simply adding up numbers to being an artificial being capable of thought was just not possible to make a reality in 1940s England. This was partially due to technology and funding, but also his colleagues' lack of foresightedness played a part. Often innovation has to wait until people's minds are ready to cope with the concepts involved.

In addition we must realise the truth can be an expandable thing. The outcome of who is adjudicated responsible for an invention or discovery is often dependent on the cultural and geographical perspective of those judging. As already seen, the French appear to prefer the idea that local boy Lamarck was the instigator of evolutionary theory, rather than Darwin, and Reis is considered inventor of the telephone in Germanic eyes. Likewise Britain regards the television to have been completely down to John Logie Baird, when in fact Japan's Kenjiro Takayanagi and Russia's Vladimir Zworykin had also offered important assistance to its development. Cuba has more recently got in on this act, in 2004 issuing a series of phone cards which paid homage to one time resident Meucci, and the invention he created on Cuban soil.

I remember reading an article in the *Guardian* some years back, which pointed out how different editions of a certain encyclopaedia stated conflicting information on who had been responsible for particular inventions. For example, the British version claims Scottish inventor Alexander Graham Bell came up with the first telephone. The US version favours Bell too, but refers to him as American (which he did eventually become, but long after patenting the telephone), while an Italian copy, as you might expect, chooses to cite Antonio Meucci. Likewise, when it comes to the light bulb, the British version acknowledges Joseph Swan with its invention, but the US adaptation credits Thomas Edison. What proportion of the basis of each case is down to knowledge of historical fact, and what is just simply patriotism, I cannot say.

It probably has not escaped some of the more perceptive of you that residents of one relatively small country appear more than any other. You guessed! A far greater percentage of Britons seem to have lost out in the stories discussed over the preceding pages. Too frequently we have been the ones who took up the initiative, only to be beaten in the end. As well as 'Nearly Men' Swan, Turing, Hooke and Wallace, there are also the more minor characters to consider, such as Dummer and Babbage. So what is it about us limeys? Are the population of this 'sceptred isle' intrinsically easy to rip-off, do we have some sort of metaphysical 'L' stamped on our forehead at birth? Well possibly. It has to be said that we do tend to still live up to the old stereotypical characteristics. I think it is great that we keep a strong sense of fair play, and try to do the honourable thing, but there are times when it certainly does us no favours.

Neither should the importance of class be underestimated. Marconi took full advantage of his aristocratic influence to expedite his rise to prominence, and though Bell had no place in the landed gentry, he did have powerful and affluent allies to aid him in his work. As already discussed, Darwin was born into a life of comfort. He came from a relatively wealthy family and married into money (in the form of the heiress to the Wedgwood fortune). This gave him the opportunity to devote himself wholeheartedly to his research, and meant the establishment were more willing to hear what he had to say. John von Neumann was also the product of a well-off family, and though he had quite extraordinary talents, would he have achieved so much if he had been born a pauper?

Higher social standing or just better financial security would have certainly furnished some members of the 'Nearly Men' with greater opportunity to tell the world of their discoveries. In Meucci's case it might have given him a platform with which to present his work to a much broader cross-section of American society. It could have meant that Hooke was considered an equal by his colleagues at the Royal Society rather than being trapped in the role of their servant. With Lamarck it would have allowed him to prioritise his own research rather than having to earn a meagre wage to save his family from starvation. This in turn may have meant that he would have got further with his theorising, perhaps spotted the errors in his view of the evolutionary process, and reformulated it correctly.

For me the concept of genius is too often bandied about. In particular, when the likes of Newton or Tesla are discussed, this word nearly always surfaces. I have tried not to bring it up more than a couple of times during the course of this book, and even on these occasions did so with an element of cautiousness. The idea of some god-given ability seems to be completely baseless. What part of an inventor's cerebral make-up is cogent, and what is not is very hard to say. I personally side more easily with Edison's view, that 'genius is 1% inspiration and 99% perspiration.'[2]

There is no question that the people discussed within this book all had incredible talents, but how much of this was down to nature and what part was simply nurture is impossible to estimate. The use of genius to describe an individual's intellectual powers is quite simply poetic fancy, it does not really have any rational foundation. Yes, it cannot be questioned that people like Tesla and Newton made some truly superhuman achievements, but at what cost? Their feats were not down to some mystical quality, but the product of sheer hard work and almost complete neglect of every other aspect of their lives. In order to succeed, they sacrificed virtually every comfort and enjoyment that the rest of us treasure so deeply. They effectively became living caricatures, so consumed by their work that they were no longer able to relate to other people, and devoid of any understanding of conventional life.

So what then is the key to being a true innovator? Is it to explore completely uncharted territory and try against the odds to make a mark, like Kilby or Babbage? Is it to build upon the research of others that went before in order to make something of practical use, as Marconi or Bell did? Should you play the foresighted maverick using intuitive guesswork, like Hooke and Tesla, or take the role of the methodical plodder investing vast quantities of time and energy into the allotted goal, like Newton and Edison?

It was Edison who once said, 'Society is never prepared to receive any invention. Every new thing is resisted, and it takes years for the inventor to get people to listen to him.'[3] Kilby, Meucci and Babbage would all have to concur with this. With human nature so adverse to change, perhaps persistence is the most primary attribute that needs to be found in the soul of every person looking to follow this calling. It was this which allowed Marconi to succeed where Tesla could not, and meant that Newton tasted the sweetness of monumental accomplishment, while poor old Hooke feasted only on resentment.

It is also worth noting that the great inventors all managed to find ways of isolating themselves from the everyday distractions that we less remarkable examples of humanity have to contend with. As already mentioned, Edison to this end benefited from his deafness. Kilby took advantage of the summer exodus from the company's plant to do much of his most significant research, while Bell worked through the night so as to avoid any disruptive influences. Similarly Newton relocated to the countryside, away from all the trappings of civilisation. The chance to follow a more plebeian lifestyle separated him from the commotion or intrusions of the outside world, and resulted in his most prolific period. Darwin's illness gave him an excuse to withdraw from public life, and this opened up far more time for him to concentrate on developing his theory of evolution. Tesla made a point of not trying to build any close relationships, staying focussed on his investigations at the expense of any real private life.

Hooke, Newton, Turing and Tesla never married. Turing's reasons were more to do with sexual preference than devotion to his work. He did of course consider the possibility and was briefly engaged, but realised he would be living a lie if he continued down that path. Hooke never wed, but he was in a long-term relationship, of sorts, with his niece. It may have been that this was more out of convenience than deep love. It offered him companionship and sexual gratification, without the complications of marriage, children and so forth, thereby allowing him to continue with his scientific research with the minimum of hindrance. Both Newton and Tesla seem to have taken to leading lives of chastity, becoming totally fanatical about their endeavours. This has raised speculation, in both cases, that they may have also been homosexual. But that is all it is. There seems to be no substantial evidence that either of them had such tendencies, and in these enlightened times it really does not matter anyway.

Another telling difference that seems to separate the 'haves' and the 'have nots' when we look into the matter further, is those who succeeded almost always managed to direct all their efforts on to one sole objective. Meucci, Hooke and Tesla would try to cope with working on many unrelated areas of research simultaneously, and to an extent this was probably instrumental in their downfall. Whereas those who were successful, like Edison, Bell, Newton and Marconi, focussed completely on the matter in hand until their respective goals were achieved.

'The progressive development of man is vitally dependent on invention. It is the most important product of his brain. Its ultimate purpose is the complete mastery of mind over the material world, the harnessing of the forces of nature to human needs. This is the difficult task of the inventor who is often misunderstood and unrewarded'[4] was how Nikola Tesla felt about the trade he plied. It is possible that his own luckless experience helped to form this view, but maybe there is some truth to be found with in it. Perhaps some inventors have been willing to accept that their struggles would not be fully appreciated. For somebody like Tesla, who had such important influence on the way we lead our lives today, maybe it was achieving this, rather than necessarily getting the acknowledgement for it, that really mattered. This could certainly be said for Alfred Russel Wallace who seemed content with a life of almost total anonymity, but knowing personally that he played a key role in changing natural science forever. Artists like Vincent Van Gogh or Paul Cezanne were not to acquire any serious wealth or fame for their work during their lifetimes, but this was not what motivated them. The passion for what they did was of far greater importance than any worldly gains they might receive.

So maybe our 'Nearly Men' should be satisfied with the spiritual rewards, happy in the knowledge their interventions in some way helped mould a brave new world, even if only a few knew the extent of their involvement. Wealth and notoriety are just transitory things after all, they cannot be said to have equal value to aiding the sustained growth of civilisation.

Though all of them lost out, each has in some way been compensated for the pain they suffered. Even when someone has been deemed the loser it does not necessarily mean that things cannot change for the better with time. History can on occasion redress the balance. Meucci received vindication from the US Congress, and has been made an idol of sorts by the Italian American community. The Supreme Court eventually overturned its decision regarding Tesla's radio patents (although he never lived to see it), and he has developed a cult following to boot. Likewise, Hooke has had something of a renaissance in recent years, with a series of books, television documentaries and exhibitions all helping to remind the world what a remarkable person he really was. Similarly, Kilby was destined to lose out to Noyce when it came to capitalising on the microchip, but this eventually went full circle. He was to receive a Nobel Prize, which meant his final years were not filled with bitterness about the squabbles surrounding the integrated circuit's invention.

In fact the way 'winners' are judged in historical terms is again very much open to debate. For example, take the race to reach the South Pole. The Norwegian Roald Amundsen was victorious, but the world remembers Englishman Captain Robert Scott's noble failure with far greater affection.

We should not look to take part in the character assassination of those who were successful. This would be unfair. It is difficult to say that they are completely without merit. Whether Marconi, Bell, Edison or any of the others regarded victors did really come up with the original idea or not shouldn't be the only consideration, they certainly put the hours of hard graft in, and so it can be argued that they earned their place in history through their tireless labour. That opens up a whole new debate; is the person that happens to be visited by serendipitous inspiration and perhaps does not take it any further really deserving of all the consideration, or alternatively the one whom may not have been blessed with the initial vision but by their own industriousness and determination made something of it. Reis, Dummer, Lodge, Hertz and others were all given a sneak preview of the shape of things to come, but did not manage to exploit this good fortune. Though fate smiled kindly upon them, they still weren't prompted into taking what could be considered significant action.

Advisor to the Florentine dynasty of the Borgias, and all-round nasty piece of work, Niccolo Machiavelli wrote, 'Men nearly always follow the track made by others and proceed in their affairs by imitation.'[5] Perhaps surprisingly, this sentiment is echoed in the words of Joseph Swan. He was of the opinion that 'an inventor is an opportunist who takes occasion by the hand, and having seen where some want exists, successfully applies the right means to attain the desired end; the means may be largely, even wholly something already known, or there may be certain originalities or discovery in the means employed, but in every case the inventor uses the work of others.'[6] He equated him to 'a man who essays the conquest of some virgin alp. At the onset he scans the beaten track, and as he progresses in the ascent, he uses the steps used by those who proceeded him'[7] and it would only be 'after the last footprints have died out that he takes the ice axe in his hands and cuts the remaining steps.'[8]

Newton, described it as standing on the shoulders of giants, while Edison was not so lyrical, and basically said it was just unadulterated thievery. The indisputable fact is all of them knew that they had to draw upon the toil their predecessors had gone through to aid them in reaching their objective.

So has the purpose of invention been distorted? Victorian writer H.G. Wells concluded his novel the *Time Machine* with the following lines. The central character, after observing the future, is filled with distress by the path humankind has followed. In contemplating what he had witnessed he 'thought but cheerlessly of the Advancement of Mankind, and saw in the growing pile of civilisation only a foolish heaping that must inevitably fall back upon and destroy its makers in the end.'[9] Maybe ten years ago we would have taken these lines to infer the threat of nuclear war and the impending doom associated with it. In these supposedly more secure times, it is not this that menaces the future of civilisation. It is not enemy attack that we should be scared of, but something far closer to home.

Our relentless pursuit of technological advancement, with little thought for what the long-term effects would be, may not make the Earth a better place to live in, but will rather result in humankind reverting back to barbarism as Wells prophesised in his book. Some will say that this is already happening, and it is just the increments are so small that most of us do not see it. There certainly seems little doubt that the Western populace is getting fatter, stupider, and lazier. Technology does not appear to be helping things, but conversely looks more likely to make matters worse.

Invention is no longer aimed at meeting any well-defined requirement, but now only serves to fulfil our ever-growing consumerist appetites. Marketing is creating needs for things that we managed to get along without just fine beforehand. Innovation is not helping narrow the Third World debt, solve the energy crisis, irrigate the Earth's barren regions, or save children from hunger and disease. Instead it is allowing us to send photos through our mobile phones, put TV screens into the back of our petrol-guzzling SUVs, and create a more ignorant and selfish world. If our culture is not to risk stagnating completely we must look at how to realign the objectives being set.

It is hard to say which of the many individuals we have looked at should be given the final words, but I think nobody could argue that Newton is at least as fitting as any of the others. In probably his most famous quotation, he compared his scientific discoveries to a small boy playing on the seashore finding shells while the ocean of truth lay unknown before him. Maybe this is as true now as it was back then; though our eruditeness can keep on pushing towards the horizon, there is forever a new horizon emerging in the distance.

Timeline

Here is a brief summary of key scientific events that took place over the period which this book covers. It makes specific reference to the discoveries and inventions that our leading protagonists were responsible for, but also includes other important deeds that played a part in our technological advance.

1543 (Frauenburg, Poland) – Copernicus manages to publish *De Revolutionibus* just before his death. The book is the first in Christendom to propose that the Earth goes around the Sun, and not vice versa.

1633 (Rome, Italy) – Galileo is sentenced by the Holy Inquisition to life imprisonment for openly supporting Copernicus' theories. His punishment is later commuted to house arrest by the Pope.

1665 (London, England) – Hooke's *Micrographia* goes on sale. It is the first serious investigation of the subject of microscopy.

1687 (Cambridge, England) – Newton publishes *Principia Mathematica*, which signifies the beginning of the new era of scientific investigation. Within it he sets down the laws of motion.

1752 (Pennsylvania, USA) – Franklin's kite experiment shows that lightning is actually static electricity.

1769 (Glasgow, Scotland) – James Watt patents the first steam engine

1771 (Bologna, Italy) – Galvani investigates the effect of electricity on muscles, paving the way for research into neurology.

1800 (Pavia, Italy) – Volta invents the acid battery.

1809 (Paris, France) – Lamarck pens *Zoological Philosophy*, in which he suggests that plant and animal species are not fixed but have in fact evolved.

1831 (London, England) – Faraday discovers the induction of electrical currents, thereby inventing the first dynamo.

1833 (London, England) – Babbage begins to conceptualise the 'Analytical Engine' a forerunner to the modern computer, but never manages to build a device that lives up to his design.

1859 (London, England) – Darwin's *Origin of the Species* goes into print, bringing the theory of evolution by natural selection to the public eye.

1865 (Brünn, Czech Republic) – Mendel discovers the secrets of genetics, but his work is ignored completely until long after his death.

1876 (Boston, USA) – Bell patents the telephone, though he has to face a series of court cases before his priority is secured.

1878 (Newcastle, England) – Swan gives first public demonstration of the carbon filament incandescent lamp.

1880 (New Jersey, USA) – Edison receives patents in the US and UK for a lamp with very similar attributes to Swan's.

1891 (Budapest, Hungary) – Tesla works out a method to generate AC electricity more efficiently than had previously been considered possible. His 'polyphase system' becomes the basis of modern electrical distribution.

1895 (Wurzburg, Germany) – Roentgen discovers X-Rays.

1897 (Paris, France) – Curie discovers radium, and sets down the basic principles of radioactivity, though she dies of cancer as a direct result of her research.

1901 (St John's, Canada) – Marconi defies his critics and manages to send a radio signal across the Atlantic Ocean.

1907 (Chicago, USA) – de Forest develops the 'Audion' tube, the earliest form of electronic device.

1916 (Berlin, Germany) – Einstein publishes his *General Theory of Relativity*. The underlying principles of physical science, set down by Galileo and Newton, are shown to no longer hold true under certain circumstances.

1936 (Cambridge, England) – Turing puts forth the concepts that form the foundation of the modern computer, with his paper 'On Computable Numbers'.

1945 (New Mexico, USA) – The team at Los Alamos finish the construction of the first atomic bomb.

1946 (Pennsylvania, USA) – The ENIAC computer goes into full operation.

1946 (Pennsylvania, USA) – The basic architecture that will be used by all computers to follow is put onto paper by von Neumann, while designing the EDVAC.

1948 (New Jersey, USA) – Shockley and his team at Bell Laboratories produce the first transistor.

1953 (Cambridge, England) – Watson and Crick give a detailed description of the structure of DNA.

1958 (Dallas, Texas) – Kilby creates first integrated circuit.

End Notes

CHAPTER 1 – CROSSED LINES

1. Letter from Antonio Meucci to Teresta Garibaldi, 31st October 1887.
2. US Congressional Report: Investigations of Telephone Industry, 1939.
3. Figures from report by the Gartner Group, December 2003.
4. Meucci's Affidavit, 15th December 1885.
5. Meucci's Affidavit, 15th December 1885.
6. Meucci's Affidavit, 15th December 1885.
7. Antonio Meucci's Affidavit, 9th October 1885.
8. *New York Times*, 31st July, 1871.
9. *New York Times*, 31st July, 1871.
10. *The Harper's Weekly*, 12th August, 1871.
11. *The Harper's Weekly*, 12th August, 1871.
12. *The Harper's Weekly*, 12th August, 1871.
13. *The Harper's Weekly*, 12th August, 1871.
14. US Patent Caveat 3335 'Sound Telegraph', p.2, 28th December 1871.
15. US Patent Caveat 3335 'Sound Telegraph', p.3, 28th December 1871.
16. US Patent Caveat 3335 'Sound Telegraph', p.3, 28th December 1871.
17. US Patent Caveat 3335 'Sound Telegraph', p.3, 28th December 1871.
18. Letter from A.Meucci to the L'Eco d'Italia, October 1865.
19. Letter from A.Meucci to the L'Eco d'Italia, October 1865.
20. *New York World*, 2nd October, 1886.
21. *New York Herald*, 19th October, 1889.
22. Letter from Antonio Meucci to Teresta Garibaldi, 31st October 1887.
23. US Patent Caveat 3335, 14th February 1876.
24. US Patent Caveat 3335, 14th February 1876.
25. US Patent Caveat 3335, 14th February 1876.
26. *The Papers of Thomas A. Edison*, by P. Israel, K. Nier, & L. Carcat, p.185, John Hopkins University Press, 1998.
27. *The Papers of Thomas A. Edison*, by P. Israel, K. Nier, & L. Carcat, p.185, John Hopkins University Press, 1998.
28. *The History of the Telephone*, by H. Casson, p.34, Kessinger Publishing, 2004.
29. US Congress Press Release, 11 June 2002.
30. Figures from report by the Worldwatch Institute, September 1998.

CHAPTER 2 – THERE WAS A CROOKED MAN

1. *Aubrey's Brief Lives*, by J. Aubrey, p.164, edited by D. Lawson, Seckler & Warburg, 1949.
2. *Life of Robert Hooke*, by R. Waller, p.i, London, 1705.
3. *Life of Robert Hooke*, by R. Waller, p.xxvi, London, 1705.
4. *Life of Robert Hooke*, by R. Waller, p.v, London, 1705.
5. *The History of the Royal Society* of London, by T. Birch, Volume I, Royal Society, 1756.
6. Council Minutes of the Royal Society (1663-1682), Volume I, p.11, Royal Society, 1682.
7. Council Minutes of the Royal Society (1663-1682), Volume I, p.35, Royal Society, 1682.
8. *The Correspondence of Henry Oldenburg*, Voulme II, (1663-1665), p.297, Hall & Hall, 1966.
9. *The Illustrated Pepys*, p.106, edited by R. Latham, Penguin Books, 2000.
10. *The Diary of Samuel Pepys*, p.463, edited by C. Tomalin, Penguin Books, 2003.
11. *Philosophical Transactions of the Royal Society (1665-1678)*, Volume I, p.27, Royal Society. 1963.
12. *Micrographia*, by R. Hooke, preface, Royal Society, 1665.
13. *Philosophical Transactions of the Royal Society (1665-1678)*, Volume I, p.27, Royal Society. 1963.
14. *Micrographia*, by R. Hooke, page 210, Royal Society, 1665.
15. *Hooke's Micrographia (1665-1965)*, by A. Hall, p.14, University of London Press, 1966.
16. *Hooke's Micrographia (1665-1965)*, by A. Hall, p.26, University of London Press, 1966.
17. Extracts from Hooke's *Micrographia*, by R. Hooke, p.55, Oxford, 1926.
18. *Philosophical Transactions of the Royal Society (1665-1678)*, Volume I, p.3, Royal Society. 1963.
19. *The Illustrated Pepys*, p.108, edited by R. Latham, Penguin Books, 2000.
20. *The History of the Royal Society*, by T. Birch, p.523, Volume IV, 1756.
21. Lectures & Discourses on Earthquakes & Subterraneous Eruptions, by R. Hooke, p.293, Arno Press, 1978.
22. *The Illustrated Pepys*, p.163, edited by R. Latham, Penguin Books, 2000.
23. *The Illustrated Pepys*, p.165, edited by R. Latham, Penguin Books, 2000.
24. The Diary of Robert Hooke, edited by H. W. Robinson & W. Adams, p235, Taylor & Francis, 1935.
25. *The Illustrated Pepys*, p.258, edited by R. Latham, Penguin Books, 2000.
26. *Life of Robert Hooke*, by R. Waller, p.ix, London, 1705.
27. *Philosophical Transactions of the Royal Society (1665-1666)*, Volume I, p.386, Royal Society, 1963.
28. *The History of the Royal Society*, by T. Birch, p.180, Volume I, 1756.
29. *Isaac Newton's Papers & Letters on Natural Philosophy*, edited by B. Cohen, p.111, Harvard University Press, 1978.
30. *Isaac Newton's Papers & Letters on Natural Philosophy*, edited by B. Cohen, p.111, Harvard University Press, 1978.
31. *The Wordsworth Dictionary of Quotations*, edited by C. Robertson, p.412, Wordsworth Reference, 1997.
32. *Philosophical Transactions of the Royal Society (1665-1678)*, Volume VIII, p.6092, Royal Society, 1963.
33. *Newton: Texts, Background, Commentaries*, edited by B. Cohen, & R. Westfall, p.25, Norton, 1995.

34. *Newton: Texts, Background, Commentaries*, edited by B. Cohen, & R. Westfall, p.13, Norton, 1995.

35. *Philosophical Transactions of the Royal Society (1665-1678)*, Volume VII, p.4004, Royal Society, 1963.

36. *Philosophical Transactions of the Royal Society (1665-1678)*, Volume VII, p.4004, Royal Society, 1963.

37. *Philosophical Transactions of the Royal Society (1665-1678)*, Volume VI, pp.3075-76, Royal Society, 1963.

38. *The Optical Papers of Isaac Newton*, Volume I, edited by A.Shapiro, p.583, Cambridge University Press, 1984.

39. Newton's Notebook, p.15. Portsmouth Collection.

40. *The Optical Papers of Isaac Newton*, Volume I, edited by A. Shapiro, p.583, Cambridge University Press, 1984.

41. *The Correspondence of Isaac Newton*, Volume I, p.412, edited by H. Turnbull, Cambridge University   Press, 1959.

42. *The Correspondence of Isaac Newton*, edited by H. Turnbull, Volume I, pp.416-417, Cambridge University Press, 1961.

43. *The Diaries of Robert Hooke*, by R. Nichols, p.29, Book Guild, 1994.

44. *The Diaries of Robert Hooke*, by R. Nichols, p.33, Book Guild, 1994.

45. *The Diaries of Robert Hooke*, by R. Nichols, p.33, Book Guild, 1994.

46. *The Diaries of Robert Hooke*, by R. Nichols, p.33, Book Guild, 1994.

47. *The Diary of Robert Hooke (1672-80)*, p.114, edited by H. Robinson, Taylor & Francis, 1935.

48. *The Diary of Robert Hooke (1672-80)*, p.11, edited by H. Robinson, Taylor & Francis, 1935.

49. *The Diary of Robert Hooke (1672-80)*, p.15, edited by H. Robinson, Taylor & Francis, 1935.

50. *The Diary of Robert Hooke (1672-80)*, p.91, edited by H. Robinson, Taylor & Francis, 1935.

51. *The Diary of Robert Hooke (1672-80)*, p.114, edited by H. Robinson, Taylor & Francis, 1935.

52. *The Diary of Robert Hooke (1672-80)*, p.237, edited by H. Robinson, Taylor & Francis, 1935.

53. *The Diary of Robert Hooke (1672-80)*, p.306, edited by H. Robinson, Taylor & Francis, 1935.

54. *The Diaries of Robert Hooke*, by R. Nichols, p.151, Book Guild, 1994.

55. *The Diaries of Robert Hooke*, by R. Nichols, p.151, Book Guild, 1994.

56. *The Diaries of Robert Hooke*, by R. Nichols, p.152, Book Guild, 1994.

57. *The Diaries of Robert Hooke*, by R. Nichols, p.151, Book Guild, 1994.

58. *The Diary of Robert Hooke (1672-80)*, p.418, edited by H. Robinson, Taylor & Francis, 1935.

59. *The Diary of Robert Hooke (1672-80)*, p.419 edited by H. Robinson, Taylor & Francis, 1935.

60. *The Diary of Robert Hooke (1672-80)*, pp.422-23, edited by H. Robinson, Taylor & Francis, 1935.

61. *The Diary of Robert Hooke (1672-80)*, p. 347, edited by H. Robinson, Taylor & Francis, 1935.

62. *The History of the Royal Society*, by T. Birch, p.179, Volume I, 1756.

63. *Philosophical Transactions of the Royal Society (1674-1678)*, Volume IX-X, p.13, Royal Society, 1963.

64. *The Diary of Robert Hooke (1672-80)*, p.314, edited by H. Robinson, Taylor & Francis, 1935.

65. *The Correspondence of Isaac Newton*, edited by H. Turnbull, Volume II p.297, Cambridge University Press, 1961.

66. *The Correspondence of Isaac Newton*, edited by H. Turnbull, Volume II p.297, Cambridge University Press, 1961.

67. *The Correspondence of Isaac Newton*, edited by H. Turnbull, Volume II p.297, Cambridge University Press, 1961.

68. *The Correspondence of Isaac Newton*, edited by H. Turnbull, Volume II p.297, Cambridge University Press, 1961.

69. *The Correspondence of Isaac Newton*, edited by H. Turnbull, Volume II p.300, Cambridge University Press, 1961.

70. *The Correspondence of Isaac Newton*, edited by H. Turnbull, Volume II p.300, Cambridge University Press, 1961.

71. *The History of the Royal Society*, by T. Birch, p.519, Volume III, 1756.

72. *The Correspondence of Sir Isaac Newton*, edited by J. Edleston, p.264, Frank Cass & Co., 1969.

73. *The Correspondence of Isaac Newton*, edited by H. Turnbull, Volume II p.309, Cambridge University Press, 1961.

74. *The Correspondence of Isaac Newton*, edited by H. Turnbull, Volume II, p.313, Cambridge University Press, 1961.

75. *The Principia*, by I. Newton, translated by I. Cohen & A. Whitman, University of California Press, 1999.

76. *The Correspondence of Isaac Newton*, edited by H. Turnbull, Volume II, p.431, Cambridge University Press, 1961.

77. *The Correspondence of Isaac Newton*, edited by H. Turnbull, Volume II, p.431, Cambridge University Press, 1961.

78. *The Correspondence of Isaac Newton*, edited by H. Turnbull, Volume II, p.433, Cambridge University Press, 1961.

79. *The Correspondence of Isaac Newton*, edited by H. Turnbull, Volume II, p.437, Cambridge University Press, 1961.

80. *The Correspondence of Isaac Newton*, edited by H. Turnbull, Volume II, p.435, Cambridge University Press, 1961.

81. *The Correspondence of Isaac Newton*, edited by H. Turnbull, Volume II, p.442, Cambridge University Press, 1961.

82. *The Correspondence of Isaac Newton*, edited by H. Turnbull, Volume II, p.442, Cambridge University Press, 1961.

83. *The Correspondence of Isaac Newton*, edited by H. Turnbull, Volume II, p.438, Cambridge University Press, 1961.

84. *Newton: Texts, Background, Commentaries*, edited by B. Cohen, & R. Westfall, 305, Norton, 1995.

85. *Newton: Texts, Background, Commentaries*, edited by B. Cohen, & R. Westfall, Norton, 1995.

86. *Life of Robert Hooke*, by R. Waller, p.xxvi, London, 1705.

87. *Life of Robert Hooke*, by R. Waller, p.xxvi, London, 1705.

88. *Life of Robert Hooke*, by R. Waller, p.xxvi, London, 1705.

89. *Aubrey's Brief Lives*, p.165, edited by D. Lawson, Seckler & Warburg, 1949.

90. *Aubrey's Brief Lives*, p.165, edited by D. Lawson, Seckler & Warburg, 1949.

91. *Life of Robert Hooke*, by R. Waller, p.xxvi, London, 1705.

92. *Life of Robert Hooke*, by R. Waller, p.xxvi, London, 1705.
93. *Life of Robert Hooke*, by R. Waller, p.vii, London, 1705.
94. *Aubrey's Brief Lives*, p.167, edited by D. Lawson, Seckler & Warburg, 1949.
95. *The Wordsworth Dictionary of Quotations*, edited by Connie Robertson, p.30, Wordsworth Reference, 1997.

CHAPTER 3 – MAKING WAVES

1. *My Inventions*, by N. Tesla, p.27, edited by B.Johnston, Hart Bros, 1982.
2. *My Inventions*, by N. Tesla, p.8, Skolska Knjiga, 1977.
3. *My Inventions*, by N. Tesla, p.9, Skolska Knjiga, 1977.
4. *My Inventions*, by N. Tesla, p.10, Skolska Knjiga, 1977.
5. *My Inventions*, by N. Tesla, p.12, Skolska Knjiga, 1977.
6. *My Inventions*, by N. Tesla, p.27, edited by B.Johnston, Hart Bros, 1982.
7. *My Inventions*, by N. Tesla, p.37, Skolska Knjiga, 1977.
8. *My Inventions*, by N. Tesla, p.37, Skolska Knjiga, 1977.
9. *My Inventions*, by N. Tesla, p.40, Skolska Knjiga, 1977.
10. *My Inventions*, by N. Tesla, p.41, Skolska Knjiga, 1977.
11. *My Inventions*, by N. Tesla, p.41, Skolska Knjiga, 1977.
12. *My Inventions*, by N. Tesla, p.40, Skolska Knjiga, 1977.
13. *My Inventions*, by N. Tesla, p.40, Skolska Knjiga, 1977
14. *My Inventions*, by N. Tesla, p.40, Skolska Knjiga, 1977.
15. *My Inventions*, by N. Tesla, p.44, Skolska Knjiga, 1977
16. *My Inventions*, by N. Tesla, p.45, Skolska Knjiga, 1977.
17. *My Inventions*, by N. Tesla, p.46, Skolska Knjiga, 1977.
18. *My Inventions*, by N. Tesla, p.8, Skolska Knjiga, 1977.
19. *My Inventions*, by N. Tesla, p.51, Skolska Knjiga, 1977.
20. *Nikola Tesla: Lectures, Patents, Articles*, edited by V.Popovic, p.6, Tesla Museum Publishing, 2004.
21. *Nikola Tesla: Lectures, Patents, Articles*, edited by V.Popovic, p.6, Tesla Museum Publishing, 2004.
22. *My Inventions*, by N. Tesla, p.20, Skolska Knjiga, 1977.
23. Article in the *Electrical Review*, August, 1897.
24. Article in the *Electrical Review*, August 1897.
25. U.S. Patent 645576, March 1900.
26. U.S. Patent 645576, March 1900.
27. *Inventions, Writings, & Research of Nikola Tesla*, p.359, edited by T. Commerford-Martin, Barnes & Noble, 1995.
28. U.S. Patent 645576, March 1900.
29. U.S. Patent 645576, March 1900.
30. U.S. Patent 645576, March 1900.
31. U.S. Patent 645576, March 1900.
32. U.S. Patent 645576, March 1900.
33. *Scientific American*, June, 1897.
34. *My Father, Marconi*, by D. Marconi, p.8, Frederick Muller, 1962.
35. *Nobel Lectures*, p.196, Elsevier Publishing, 1967.
36. *Proving the Practicality of Wireless Telegraphy*, by G. Marconi, p.11, McLure, 1926.
37. *My Father, Marconi*, by D.Marconi, p.112, Frederick Muller, 1962.

38. *Proving the Practicality of Wireless Telegraphy*, by G. Marconi, p.10, McLure, 1926.
39. Interview with Guglielmo Marconi, conducted by H.J. Dam, *Strand Magazine*, March 1897.
40. *My Father, Marconi*, by D.Marconi, p.8, Frederick Muller, 1962.
41. *Developments in Wireless Telegraphy*, by G. Marconi, pp. 3-4, Birmingham & Midland Institute, 1921.
42. *Nobel Lectures*, p.201, Elsevier Publishing, 1967.
43. *Inventions, Writings, & Research of Nikola Tesla*, p.371, edited by T. Commerford-Martin, Barnes & Noble, 1995.
44. *Developments in Wireless Telegraphy*, by G. Marconi, pp.3-4, Birmingham & Midland Institute, 1921.
45. *My Father, Marconi*, by D. Marconi, p.103, Frederick Muller, 1962.
46. *My Father, Marconi*, by D. Marconi, p.103, Frederick Muller, 1962.
47. *My Father, Marconi*, by D. Marconi, p.103, Frederick Muller, 1962.
48. *New York Times*, 15th December, 1901.
49. *Daily Telegraph*, 15th December, 1901.
50. *Talking with the Planets*, by N. Tesla, *Colliers Weekly*, February 1901.
51. *Talking with the Planets*, by N. Tesla, *Colliers Weekly*, February 1901.
52. *New York Times*, 19th December, 1901.
53. *The Merchant of Venice*, by W. Shakespeare, Act.3, Scene.5.
54. *The Electrical Experimenter*, p.293, September, 1917.
55. Letter from N. Tesla to the American Institute of Electrical Engineers, 8th November 1919.
56. *New York Times*, 11th July, 1934.
57. *Songs of Freedom*, by R. Johnson, p.42, Century, 1897.
58. *New York Times*, 8th January 1943.
59. *New York Times*, 8th January 1943.
60. *My Beloved Marconi*, by M. Marconi, p.19, Dante University Press, 1999.
61. *My Beloved Marconi*, by M. Marconi, p.19, Dante University Press, 1999.
62. Article in the *Daily Mail*, 31st July, 1934.
63. Article in the *Daily Mail*, 31st July, 1934.
64. *My Beloved Marconi*, by M. Marconi, p.250, Dante University Press, 1999.
65. *My Beloved Marconi*, by M. Marconi, p.254, Dante University Press, 1999.
66. *My Beloved Marconi*, by M. Marconi, p.254, Dante University Press, 1999.
67. *My Beloved Marconi*, by M. Marconi, p.301, Dante University Press, 1999.
68. The *Times*, 22nd July 1937.
69. The *Times*, 22nd July 1937.
70. *My Father, Marconi*, by D. Marconi, p.3, Frederick Muller, 1962.
71. *My Father, Marconi*, by D. Marconi, p.3, Frederick Muller, 1962.
72. *My Beloved Marconi*, by M. Marconi, Dante University Press, 1999.
73. *My Inventions*, by N. Tesla, p.27, edited by B. Johnston, Hart Bros, 1982.
74. *My Inventions*, by N. Tesla, p.27, edited by B. Johnston, Hart Bros, 1982.
75. Interview with Guglielmo Marconi, conducted by H.J. Dam, *Strand Magazine*, March 1897.
76. Report from Research Machines, 1999.

CHAPTER 4 – ROUGE ELEMENTS

1. *Sir Joseph Wilson Swan F.R.S*, by J. Swan, p.17, Ernst Benn, 1929.
2. *Sir Joseph Wilson Swan F.R.S*, by J. Swan, p.19, Ernst Benn, 1929.
3. *Sir Joseph Wilson Swan F.R.S*, by J. Swan, p.19, Ernst Benn, 1929.
4. *Sir Joseph Wilson Swan F.R.S*, by J. Swan, p.19, Ernst Benn, 1929.
5. Swan United Lamp Catalogue, 1883.
6. *Wordsworth Dictionary of Quotations*, edited by C. Robertson, Wordsworth Reference, 1997.
7. *Edison: The Man Who Made the Future*, by R. Clark, p.12, MacDonald & James, 1977.
8. The Diary & Sundry Observations of Thomas Alva Edison, edited by D. Runes, p.44, Philosophical Library, 1948.
9. *Edison: The Man Who Made the Future*, by R. Clark, p.12, MacDonald & James, 1977.
10. *Edison: The Man Who Made the Future*, by R. Clark, p.13, MacDonald & James, 1977.
11. *Edison: The Man Who Made the Future*, by R. Clark, p.13, MacDonald & James, 1977.
12. *The Papers of Thomas A. Edison*, by P. Israel, K. Nier, & L. Carcat, p.473, John Hopkins University Press, 1998.
13. *The Papers of Thomas A. Edison*, by P. Israel, K. Nier, & L. Carcat, p.473, John Hopkins University Press, 1998.
14. *The Papers of Thomas A. Edison*, by P. Israel, K. Nier, & L. Carcat, p.473, John Hopkins University Press, 1998.
15. *New York Herald*, 27th April 1879.
16. *Sir Joseph Wilson Swan F.R.S*, by J.W. Swan, p.90, Ernst Benn, 1929.
17. The *Times*, 28th May, 1914.
18. Sir Joseph Swan & the Invention of the Incandecsent Lamp, by K. Swan, p.50, Longmans, 1948.
19. Sir Joseph Swan & the Invention of the Incandecsent Lamp, by K. Swan, p.50, Longmans, 1948.
20. *Wordsworth Dictionary of Quotations*, edited by C. Robertson, Wordsworth Reference, 1997.
21. The Diary & Sundry Observations of Thomas Alva Edison, edited by D. Runes, p.54, Philosophical Library, 1948.
22. *New York Times*, 19th October, 1931.
23. *New York Times*, 19th October, 1931.
24. *New York Times*, 19th October, 1931.
25. *New York Times*, 19th October, 1931.
26. *New York Times*, 19th October, 1931.
27. *New York Times*, 19th October, 1931.
28. *New York Times*, 19th October, 1931.
29. Interview conducted by R. Sherard, *Pall Mall Gazette*, 19th August 1889.
30. *Edison: The Man Who Made the Future*, by R. Clark, p.67, MacDonald & James, 1977.
31. The Diary & Sundry Observations of Thomas Alva Edison, edited by D. Runes, p.43, Philosophical Library, 1948.
32. *Edison: The Man Who Made the Future*, by R. Clark, p.10, MacDonald & James, 1977.
33. The Diary & Sundry Observations of Thomas Alva Edison, edited by D. Runes, p.169, Philosophical Library, 1948.
34. The *Times*, 28th May, 1914.
35. Figures from report by the Mintel International, November 2003.

CHAPTER 5 – A DESIGN FOR LIFE

1. *The Lamarck Manuscripts at Havard*, p.1, edited by W. Morton-Wheeler, & T. Barbour, Havard University Press, 1984.
2. *Lamarck: His Life & Work*, by A.S.Packard, p.12, Longmans, 1901.
3. *Lamarck: His Life & Work*, by A.S.Packard, p.13, Longmans, 1901.
4. *Zoological Philosophy*, by J. Lamarck, p.8, Hafner Publishing 1963.
5. *Zoological Philosophy*, by J. Lamarck, pp.9-10, translated by Hugh Elliot, University of Chicago Press, 1984.
6. *Philosophie Zoologique*, by J. Lamarck, p.34, translated by I.Johnston, University of Malaspina, 2000.
7. *Zoological Philosophy*, by J. Lamarck, p.37, translated by Hugh Elliot, University of Chicago Press, 1984.
8. *Philosophie Zoologique*, by J. Lamarck, p.34, translated by I.Johnston, University of Malaspina, 2000.
9. *Zoological Philosophy*, by J. Lamarck, p.36, translated by Hugh Elliot, University of Chicago Press, 1984.
10. *Philosophie Zoologique*, by J. Lamarck, p.40, translated by I.Johnston, University of Malaspina, 2000.
11. *Philosophie Zoologique*, by J. Lamarck, p.36, translated by I.Johnston, University of Malaspina, 2000.
12. *Philosophie Zoologique*, by J. Lamarck, p.36, translated by I.Johnston, University of Malaspina, 2000.
13. *Zoological Philosophy*, by J. Lamarck, p.36, translated by Hugh Elliot, University of Chicago Press, 1984.
14. *The Lamarck Manuscripts at Harvard*, p.xix, edited by W. Morton-Wheeler, & T. Barbour, Harvard University Press, 1984.
15. *The Lamarck Manuscripts at Harvard*, pp.103-4, edited by W. Morton-Wheeler, & T. Barbour, Havard University Press, 1984.
16. *The Age of Lamarck*, by P. Corsi, University of Press, 1988.
17. *Lamarck: The Mythical Precursor*, by M.Barthelemy-Madaule, p.109, MIT Press, 1982.
18. *Zoonomia*, by E. Darwin, Volume I, Preface, Byrne & Jones, 1794.
19. *Zoonomia*, by E. Darwin, Volume I, Preface, Byrne & Jones, 1794.
20. *Zoonomia*, by E. Darwin, Volume I, p.546, Byrne & Jones, 1794.
21. *Zoonomia*, by E. Darwin, Volume I, p.524, Byrne & Jones, 1794.
22. *Zoonomia*, by E. Darwin, Volume I, p.524, Byrne & Jones, 1794.
23. *Zoonomia*, by E. Darwin, Volume I, p.550, Byrne & Jones, 1794.
24. *Zoonomia*, by E. Darwin, Volume I, p.556, Byrne & Jones, 1794.
25. *Zoonomia*, by E. Darwin, Volume I, p.558, Byrne & Jones, 1794.
26. *Zoonomia*, by E. Darwin, Volume I, p.558, Byrne & Jones, 1794.
27. *Zoonomia*, by E. Darwin, Volume I, p.558, Byrne & Jones, 1794.
28. *Zoonomia*, by E. Darwin, Volume I, pp.558-559, Byrne & Jones, 1794.
29. *The Autobiography of Charles Darwin*, p.2, by C. Darwin, Icon Books, 2003.
30. *The Autobiography of Charles Darwin*, p.7, by C. Darwin, Icon Books, 2003.
31. *The Autobiography of Charles Darwin*, p.7, by C. Darwin, Icon Books, 2003.
32. *The Autobiography of Charles Darwin*, p.7, by C. Darwin, Icon Books, 2003.
33. *The Autobiography of Charles Darwin*, p.7, by C. Darwin, Icon Books, 2003.
34. *The Autobiography of Charles Darwin*, p.9, by C. Darwin, Icon Books, 2003.
35. *The Autobiography of Charles Darwin*, p.12, by C. Darwin, Icon Books, 2003.

36. *The Autobiography of Charles Darwin*, pp.11-12, by C. Darwin, Icon Books, 2003.
37. *The Autobiography of Charles Darwin*, pp.11-12, by C. Darwin, Icon Books, 2003.
38. *The Autobiography of Charles Darwin*, p.19, by C. Darwin, Icon Books, 2003.
39. *The Autobiography of Charles Darwin*, p.19, by C. Darwin, Icon Books, 2003.
40. *The Autobiography of Charles Darwin*, p.23, by C. Darwin, Icon Books, 2003.
41. *The Autobiography of Charles Darwin*, p.25, by C. Darwin, Icon Books, 2003
42. *The Autobiography of Charles Darwin*, p.25, by C. Darwin, Icon Books, 2003
43. *The Autobiography of Charles Darwin*, p.27, by C. Darwin, Icon Books, 2003
44. *The Autobiography of Charles Darwin*, p.27, by C. Darwin, Icon Books, 2003
45. *The Autobiography of Charles Darwin*, p.30, by C. Darwin, Icon Books, 2003
46. *The Autobiography of Charles Darwin*, p.30, by C. Darwin, Icon Books, 2003.
47. *The Autobiography of Charles Darwin*, p.33, by C. Darwin, Icon Books, 2003.
48. *Beagle Diary*, by C. Darwin, edited by R. Darwin-Keynes, p.3, Cambridge University Press, 1988.
49. *The Autobiography of Charles Darwin*, p.36, by C. Darwin, Icon Books, 2003.
50. *Beagle Diary*, by C. Darwin, edited by R. Darwin-Keynes, p.17, Cambridge University Press, 1988.
51. *Beagle Diary*, by C. Darwin, edited by R. Darwin-Keynes, p.17, Cambridge University Press, 1988.
52. *Beagle Diary*, by C. Darwin, edited by R. Darwin-Keynes, p.3, Cambridge University Press, 1988.
53. *The Autobiography of Charles Darwin*, p.34, by C.Darwin, Icon Books, 2003.
54. *Voyage of the Beagle*, by C. Darwin, p.63, Penguin Books, 1989.
55. *Voyage of the Beagle*, by C. Darwin, p.82, Penguin Books, 1989.
56. *The Autobiography of Charles Darwin*, p.35, by C. Darwin, Icon Books, 2003.
57. *Voyage of the Beagle*, p.269. by C. Darwin, 1989, Penguin Books.
58. *Voyage of the Beagle*, p.279. by C. Darwin, 1989, Penguin Books.
59. *The Voyage of Charles Darwin: His Autobiographical Writings*, edited by C. Ralling, Ariel Books, 1982.
60. *Voyage of the Beagle*, p.286. by C. Darwin, 1989, Penguin Books.
61. *Beagle Diary*, by C. Darwin, edited by R. Darwin-Keynes, p.3, Cambridge University Press, 1988.
62. *The Autobiography of Charles Darwin*, p.41, by C. Darwin, Icon Books, 2003.
63. *The Autobiography of Charles Darwin*, p.56, by C. Darwin, Icon Books, 2003
64. *The Autobiography of Charles Darwin*, p.53, by C. Darwin, Icon Books, 2003.
65. *The Autobiography of Charles Darwin*, p.57, by C. Darwin, Icon Books, 2003
66. *The Autobiography of Charles Darwin*, p.52, by C. Darwin, Icon Books, 2003.
67. *The Autobiography of Charles Darwin*, p.53, by C. Darwin, Icon Books, 2003.
68. *The Autobiography of Charles Darwin*, p.53, by C. Darwin, Icon Books, 2003.
69. *The Autobiography of Charles Darwin*, p.54, by C. Darwin, Icon Books, 2003.
70. Vestiges of the Natural History of Creation, by R. Chambers, p.112, University of Chicago Press, 1994.
71. *Life & Letters of C. Darwin*, Volume II, edited by F. Darwin, p.187, J.Murray, 1887.
72. *The Origin of the Species*, p.107. by C. Darwin, 1985, Penguin Books, 1985.
73. *The Origin of the Species*, p.117. by C. Darwin, 1985, Penguin Books, 1985.
74. *The Origin of the Species*, p.448. by C. Darwin, 1985, Penguin Books, 1985.
75. *The Origin of the Species*, p.444. by C. Darwin, 1985, Penguin Books, 1985.
76. *The Origin of the Species*, p.117. by C. Darwin, 1985, Penguin Books, 1985.
77. *Life & Letters of C. Darwin*, Volume II, edited by F. Darwin, p.380, J. Murray, 1887.

78. *Life & Letters of T.H. Huxley*, Volume I, edited by L. Huxley, p.175, MacMillan & Co., 1900.

79. *Life & Letters of T.H. Huxley*, Volume I, edited by L. Huxley, p.176, MacMillan & Co., 1900.

80. *The Autobiography of Charles Darwin*, p.58, by C. Darwin, Icon Books, 2003.

81. Interview with W.B. Northrop, for *The Outlook*, published November 1913.

82. *My Life*, by A.R. Wallace, Volume I, p.1, Chapman & Hall, 1905.

83. *My Life*, by A.R. Wallace, Volume I, p.196, Chapman & Hall, 1905.

84. *My Life*, by A.R. Wallace, Volume I, p.240, Chapman & Hall, 1905.

85. *My Life*, by A.R. Wallace, Volume I, p.247, Chapman & Hall, 1905.

86. *My Life*, by A.R. Wallace, Volume I, p.311, Chapman & Hall, 1905.

87. Interview with C.H. Smith for *The Book Monthly*, published May 1905.

88. Interview with C.H. Smith for *The Book Monthly*, published May 1905.

89. *My Life*, by A.R. Wallace, Volume I, p.355, Chapman & Hall, 1905.

90. *My Life*, by A.R. Wallace, Volume I, p.360, Chapman & Hall, 1905.

91. *My Life*, by A.R. Wallace, Volume I, p.360, Chapman & Hall, 1905.

92. *My Life*, by A.R. Wallace, Volume I, p.361, Chapman & Hall, 1905.

93. *My Life*, by A.R. Wallace, Volume I, p.361, Chapman & Hall, 1905.

94. *My Life*, by A.R. Wallace, Volume I, p.361, Chapman & Hall, 1905.

95. Proceedings of the Linnaean Society, Volume 3, p.55, Longmans, 1867.

96. Proceedings of the Linnaean Society, Volume 3, p.55, Longmans, 1867.

97. Proceedings of the Linnaean Society, Volume 3, p.55, Longmans, 1867.

98. *My Life*, by A.R. Wallace, Volume I, p.361, Chapman & Hall, 1905.

99. Proceedings of the Linnaean Society, Volume 3, p.61, Longmans, 1867.

100. Proceedings of the Linnaean Society, Volume 3, p.61, Longmans, 1867.

101. *My Life*, by A.R. Wallace, Volume I, p.362, Chapman & Hall, 1905.

102. *My Life*, by A.R. Wallace, Volume I, pp.362-363, Chapman & Hall, 1905.

103. *Life & Letters of C. Darwin*, Volume II, edited by F. Darwin, p.116, J. Murray, 1887.

104. *Life & Letters of C. Darwin*, Volume II, edited by F. Darwin, p.116, J. Murray, 1887.

105. *Life & Letters of C. Darwin*, Volume II, edited by F. Darwin, p.116, J. Murray, 1887.

106. *Life & Letters of C. Darwin*, Volume II, edited by F. Darwin, p.117, J. Murray, 1887.

107. *Life & Letters of C. Darwin*, Volume II, edited by F. Darwin, p.118, J. Murray, 1887.

108. *Life & Letters of C. Darwin*, Volume II, edited by F. Darwin, p.116, J. Murray, 1887.

109. *Life & Letters of C. Darwin*, Volume II, edited by F. Darwin, p.116, J. Murray, 1887.

110. *Life & Letters of C. Darwin*, Volume II, edited by F. Darwin, p.116, J. Murray, 1887.

111. *Alfred Russel Wallace: Letters & Reminiscences*, by J. Marchant, Volume I, p.129, Cassel, 1916.

112. *Alfred Russel Wallace: Letters & Reminiscences*, by J. Marchant, Volume I, p.129, Cassel, 1916.

113. *My Life*, by A.R. Wallace, Volume I, p.363, Chapman & Hall, 1905.

114. *My Life*, by A.R. Wallace, Volume I, p.359, Chapman & Hall, 1905.

115. *The Autobiography of Charles Darwin*, p.59, by C. Darwin, Icon Books, 2003.

116. *The Autobiography of Charles Darwin*, p.60, by C. Darwin, Icon Books, 2003.

117. *The Autobiography of Charles Darwin*, p.61, by C. Darwin, Icon Books, 2003.

118. *The Origin of the Species*, p.54. by C. Darwin, 1985, Penguin Books, 1985.

119. *The Origin of the Species*, p.54. by C. Darwin, 1985, Penguin Books, 1985.

120. *The Origin of the Species*, p.133. by C. Darwin, 1985, Penguin Books, 1985.

121. Proceedings of the Royal Society (1862-1863), Volume XII, p.272, Taylor & Francis, 1863.

122. *The Autobiography of Charles Darwin*, p.63, by C. Darwin, Icon Books, 2003.

123. *The Correspondence of Charles Darwin*. Vol.13 1865, p.135, edited by F. Burkhardt, Cambridge University Press, 2002.

124. *The Correspondence of Charles Darwin*. Vol.13 1865, p.135, edited by F. Burkhardt, Cambridge University Press, 2002.

125. *The Correspondence of Charles Darwin*. Vol.13 1865, p.135, edited by F. Burkhardt, Cambridge University Press, 2002.

126. *The Correspondence of Charles Darwin*. Vol.13 1865, p.135, edited by F. Burkhardt, Cambridge University Press, 2002.

127. *The Correspondence of Charles Darwin*. Vol.13 1865, p.100, edited by F. Burkhardt, Cambridge University Press, 2002.

128. *The Correspondence of Charles Darwin*. Vol.13 1865, p.107, edited by F. Burkhardt, Cambridge University Press, 2002.

129. *The Descent of Man*, by C. Darwin, p.7, J. Murray, 1901.

130. *The Descent of Man*, by C. Darwin, p.11, J. Murray, 1901.

131. *The Descent of Man*, by C. Darwin, p.11, J. Murray, 1901.

132. *The Descent of Man*, by C. Darwin, p.12, J. Murray, 1901.

133. *The Descent of Man*, by C. Darwin, p.99, J. Murray, 1901.

134. *The Descent of Man*, by C. Darwin, p.99, J. Murray, 1901.

135. *The Descent of Man*, by C. Darwin, p.195, J. Murray, 1901.

136. *The Descent of Man*, by C. Darwin, p.195, J. Murray, 1901.

137. *The Descent of Man*, by C. Darwin, p.196, J. Murray, 1901.

138. *The Autobiography of Charles Darwin*, p.68, by C. Darwin, Icon Books, 2003.

139. *The Expression of Emotions in Man & Animals*, p.360, by C. Darwin. Harper Collins, 1998.

140. *The Expression of Emotions in Man & Animals*, p.360, by C. Darwin. Harper Collins, 1998.

141. *The Autobiography of Charles Darwin*, p.83, by C. Darwin, Icon Books, 2003.

142. *The Autobiography of Charles Darwin*, p.72, by C. Darwin, Icon Books, 2003.

143. *The Autobiography of Charles Darwin*, pp.86-87, by C. Darwin, Icon Books, 2003.

144. *Alfred Russel Wallace: Letters & Reminiscences*, by J. Marchant, Volume II, p.113, Cassel, 1916.

145. Interview with A. Dawson, for *The Bookman*, published January 1898.

146. Interview with W.B. Northrop, for *The Outlook*, published November 1913.

147. Interview with W.B. Northrop, for *The Outlook*, published November 1913.

148. *The Descent of Man*, by C.Darwin, p.224, J. Murray, 1901.

149. *Alfred Russel Wallace: Letters & Reminiscences*, by J. Marchant, Volume I, p.127, Cassel, 1916.

150. *My Life*, by A.R. Wallace, Volume II, p.102, Chapman & Hall, 1905.

151. *Alfred Russel Wallace: Letters & Reminiscences*, by J. Marchant, Volume II, p.103, Cassel, 1916.

152. Interview with H. Begbie for *The Daily Chronicle*, published May 1910.

153. Interview with W.B. Northrop, for *The Outlook*, published November 1913.

154. *Alfred Russel Wallace: Letters & Reminiscences*, by J. Marchant, Volume I, p.114, Cassel, 1916.

CHAPTER 6 – ERRORS OF JUDGEMENT

1. *Frankenstein*, by M. Shelley, pp.52–53, Penguin Classics 1992.
2. *Charles Babbage: Passages from the Life of a Philosopher*, by C. Babbage, p.30, IEEE Press, 1994.
3. *Charles Babbage: Passages from the Life of a Philosopher*, by C. Babbage, p.31, IEEE Press, 1994.
4. *Charles Babbage: Passages from the Life of a Philosopher*, by C. Babbage, p.89, IEEE Press, 1994.
5. *Charles Babbage: Passages from the Life of a Philosopher*, by C. Babbage, p.89, IEEE Press, 1994.
6. *Birth of the Computer*, by R. Millington, p.30, 3M Publishing, 1972.
7. *Alan M. Turing*, by S. Turing, p.11, W. Heffer & Sons, 1959.
8. *Alan M. Turing*, by S. Turing, p.11, W. Heffer & Sons, 1959.
9. *Alan M. Turing*, by S. Turing, p.12, W. Heffer & Sons, 1959.
10. *Alan M. Turing*, by S. Turing, p.13, W. Heffer & Sons, 1959.
11. *Alan M. Turing*, by S. Turing, p.17, W. Heffer & Sons, 1959.
12. *Alan M. Turing*, by S. Turing, pp.26–27, W. Heffer & Sons, 1959.
13. *Alan M. Turing*, by S. Turing, p.35, W. Heffer & Sons, 1959.
14. *Alan M. Turing*, by S. Turing, p.36, W. Heffer & Sons, 1959.
15. On Computable Numbers, by A.M. Turing, p.42, Proceedings of the London Mathematical Society, 1937.
16. On Computable Numbers, by A.M. Turing, p.42, Proceedings of the London Mathematical Society, 1937.
17. The *Times*, 16th November 1954.
18. *Alan M. Turing*, by S. Turing, p.57, W. Heffer & Sons, 1959.
19. *Alan M. Turing*, by S. Turing, p.57, W. Heffer & Sons, 1959.
20. *Chronicle of the 20th Century in Quotations*, by D. Milstead, p.83, Guinness Publishing, 1996.
21. *The Hut Six Story*, by G. Welchman, p.200, M&M Baldwin, 1997.
22. *Station X: The Code Breakers of Bletchley Park*, by M. Smith, p.60, Channel Four Books, 1998.
23. *Top Secret Ultra*, by P. Calvocoressi, p.136, M&M Baldwin, 2001.
24. *Machines Who Think*, by P. McCrouch, p.53, Freeman, 1979.
25. Biographical Memoirs of the Fellow of the Royal Society, Volume I, p.254, Royal Society, 1955.
26. *Theory of Games & Economic Behaviour*, by J. von Neumann & D. Morgenstern, p.2, Princeton, 1947.
27. *Theory of Games & Economic Behaviour*, by J. von Neumann & D. Morgenstern, p.2, Princeton, 1947.
28. *The Making of the Atomic Bomb*, by R. Rhodes, Touchstone, 1986.
29. *The Making of the Atomic Bomb*, by R. Rhodes, Touchstone, 1986.
30. *Chronicle of the 20th Century in Quotations*, by D. Milstead, p.96, Guinness Publishing, 1996.
31. Draft Report for EDVAC, by J. von Nuemann, Moore School, April 1945.
32. *Intelligent Machinery* by A.M. Turing, NPL, 1946.
33. *Intelligent Machinery* by A.M. Turing, NPL, 1946.
34. *Alan M. Turing*, by S. Turing, p.56, W. Heffer & Sons, 1959.
35. *A.M. Turing's ACE Report & Other Papers*, edited by R.W. Doran, p.20, MIT Press, 1986.

36. *A.M. Turing's ACE Report & Other Papers*, edited by R.W. Doran, p.20, MIT Press, 1986.
37. *A.M. Turing's ACE Report & Other Papers*, edited by R.W. Doran, p.20, MIT Press, 1986.
38. 'On the Automatic Computing Engine' London Mathmatical Society, 20th February 1947.
39. 'On the Automatic Computing Engine' London Mathmatical Society, 20th February 1947.
40. 'On the Automatic Computing Engine' London Mathmatical Society, 20th February 1947.
41. 'On the Automatic Computing Engine' London Mathmatical Society, 20th February 1947.
42. 'On the Automatic Computing Engine' London Mathmatical Society, 20th February 1947.
43. 'On the Automatic Computing Engine' London Mathmatical Society, 20th February 1947.
44. 'On the Automatic Computing Engine' London Mathmatical Society, 20th February 1947.
45. Computng Machinery & Intelligence, *Mind*, October 1950.
46. Computng Machinery & Intelligence, *Mind*, October 1950.
47. *The Essential Turing*, edited by J. Copeland, p.393, Clarendon Press, 2004.
48. *The Essential Turing*, edited by J. Copeland, p.393, Clarendon Press, 2004.
49. 'On the Automatic Computing Engine' London Mathmatical Society, 20th February 1947.
50. *Alan M. Turing*, by S. Turing, p.91, W. Heffer & Sons, 1959.
51. *The Computer & the Brain*, by J. von Neumann, preface, Yale University Press, 1957.
52. *The Computer & the Brain*, by J. von Neumann, preface, Yale University Press, 1957.
53. Early Computers at Manchester University, by F. Williams, Radio & Electronic Engineer, Volume XD, 1975.
54. Computing Machinery & Intelligence, Mind, October 1950.
55. Philosophical Transactions of the Royal Society, Volume 237, Page pp.37-72, 1951.
56. *The Computer & the Brain*, by J. von Neumann, preface, Yale University Press, 1957.
57. *The Computer & the Brain*, by J. von Neumann, preface, Yale University Press, 1957.
58. *The Computer & the Brain*, by J. von Neumann, preface, Yale University Press, 1957.
59. John von Neumann: As Seen by his Brother, by N. von Neumann, p.3, Library of Congress, 1987.
60. *The Computer & the Brain*, by J. von Neumann, preface, Yale University Press, 1957.
61. The *Times*, 10th February, 1957.
62. *The Computer & the Brain*, by J. von Neumann, p.65, Yale University Press, 1957.

CHAPTER 7 – SMALL WONDERS

1. Interview with Jack Kilby, conducted by Gosta Ekspog, World Scientific Publishing, 2002.
2. Nobel Laureates Autobiography, Elsevier Publishing, 2000.
3. Proceedings of the IEEE, Volume 89, Number 1, p.109-110, January 2000.
4. Proceedings of the IEEE, Volume 89, Number 1, p.110, January 2000.
5. Interview with BBC World Service, October 2000.

6. *Nobel Lectures*, p.475, Elsevier Publishing, 2000.

7. *Nobel Lectures*, p.476, Elsevier Publishing, 2000.

8. *Nobel Lectures*, p.478, Elsevier Publishing, 2000.

9. *Nobel Lectures*, p.476, Elsevier Publishing, 2000.

10. Nobel Laureates Autobiography, Elsevier Publishing, 2000.

11. *Nobel Lectures*, p.479, Elsevier Publishing, 2000.

12. *Nobel Lectures*, p.484, Elsevier Publishing, 2000.

13. Proceedings of the IEEE, Volume 89, Number 1, p.110, January 2000.

14. Proceedings of the IEEE, Volume 89, Number 1, p.110, January 2000.

15. Interview with BBC World Service, October 2000.

16. *Nobel Lectures*, p.480, Elsevier Publishing, 2000.

17. *Nobel Lectures*, p.481, Elsevier Publishing, 2000.

18. *Nobel Lectures*, p.479, Elsevier Publishing, 2000.

19. *Nobel Lectures*, p.481, Elsevier Publishing, 2000.

20. *Nobel Lectures*, p.482, Elsevier Publishing, 2000.

21. Interview with Jack Kilby, conducted by Gosta Ekspog, World Scientific Publishing, 2002

22. Proceedings of the IEEE, Volume 89, Number 1, p.111, January 2000.

23. Proceedings of the IEEE, Volume 89, Number 1, p.110, January 2000.

24. U.S. Patent 3138743 'Miniaturised Electronic Circuits', 24th December, 1963.

25. *Nobel Lectures*, p.482, Elsevier Publishing, 2000.

26. Interview with Jack Kilby, conducted by Mike Green, Electronic Product News, July 2005.

27. *Twentieth Century in Quotation*, edited by David Milsted, p.125, Guinness Publishing, 1996.

28. Interview with Jack Kilby, conducted by Gosta Ekspog, World Scientific Publishing, 2002.

29. Interview with Jack Kilby, conducted by Gosta Ekspog, World Scientific Publishing, 2002.

30. Interview with Gordon Moore, Silicon Genesis, 3rd March, 1995.

31. Interview with Arthur Rock, Silicon Genesis, 12th November, 2002.

32. *Miniature & Microminiature Electronics*, by G.W.A Dummer & J.W. Granville, p.263, Pitman, 1961.

33. *Miniature & Microminiature Electronics*, by G.W.A Dummer & J.W. Granville, p.263, Pitman, 1961.

34. Proceedings of the IEEE, Volume 89, Number 1, p.110, January 2000.

35. Interview with Jack Kilby, conducted by Mike Green, Electronic Product News, July 2005.

36. Interview with Jack Kilby, conducted by Mike Green, Electronic Product News, July 2005.

37. Nobel Laureates Autobiography, Elsevier Publishing, 2000.

38. Texas Instruments Press Release, September 1997.

39. Interview with Jack Kilby, conducted by Mike Green, *Electronic Product News*, July 2005.

40. Nobel Laureates Autobiography, Elsevier Publishing, 2000.

41. Interview with Jack Kilby, conducted by Mike Green, *Electronic Product News*, July 2005.

42. Interview with Jack Kilby, conducted by Gosta Ekspog, World Scientific Publishing, 2002.

43. Interview with Gordon Moore, *Silicon Genesis*, 3ʳᵈ March, 1995.
44. Interview with Gordon Moore, *Silicon Genesis*, 3ʳᵈ March, 1995.
45. Interview with Jack Kilby, conducted by Mike Green, *Electronic Product News*, July 2005.
46. Interview with Jack Kilby, conducted by Mike Green, *Electronic Product News*, July 2005.

CONCLUSION

1. *The Wordsworth Dictionary of Quotations*, edited by C. Robertson, Wordsworth Reference, 1997.
2. *The Wordsworth Dictionary of Quotations*, edited by C. Robertson, Wordsworth Reference, 1997.
3. The Diary & Sundry Observations of Thomas Alva Edison, edited by D. Runes, p.179, Philosophical Library, 1948.
4. *My Inventions*, by N. Tesla, p.27, edited by B. Johnston, Hart Bros, 1982.
5. *The Prince*, by Machiavelli, pp.17-18, Pengiun Books, 1999.
6. *Sir Joseph Swan* F.R.S. by M.E. & K.R. Swan, p.189, Oriel, 1968.
7. *Sir Joseph Swan* F.R.S. by M.E. & K.R. Swan, p.189, Oriel, 1968.
8. *Sir Joseph Swan* F.R.S. by M.E. & K.R. Swan, p.189, Oriel, 1968.
9. *The Time Machine*, by H.G. Wells, epilogue, Everyman, 1995.

Suggested Reading

This book covers the lives of sixteen quite remarkable people. It can clearly only offer a brief description of their exploits and achievements. However there are many books which give far more comprehensive portrayals of particular individuals discussed. Here are a few I highly recommend, which proved indispensable in my research for this book.

With regard to Alexander Graham **Bell** I suggest *Sounds out of Silence* by James Mackay, though there are various others which are worth a look.

There are many great works that discuss the discoveries of Charles **Darwin**, as you would expect. These include Cyril Aydon's *Charles Darwin* and *Darwin* by Adrian Desmond and James Moore. For a more indepth understanding of his work, there are of course his own writings, such as *Origin of the Species*, *Descent of Man* and his *Beagle Diaries*, as well as his autobiography.

Likewise there are no shortage of publications on the life and times of Thomas **Edison**. *Edison: A Life of Invention* by Paul Israel, and the anti-biography *The Invented Self* by David Nye are among the best.

To learn about the exploits of Elisha **Gray**, check out Burton H. Baker's *The Gray Matter*.

Robert **Hooke** has become a very fashionable subject in recent years (just three short centuries after he died). I can strongly recommend reading Stephen Inwood's *The Man Who Knew Too Much*, as well as Lisa Jardine's *Curious Life of Robert Hooke*, and a compilation of essays by a number of historians entitled *London's Leonardo*.

To find out more about Jack **Kilby** then Robert Cringley's *Accidental Empires*, and Jeffrey Zygmont's *Microchip* are entertaining and informative reads.

For information on Jean-Baptiste **Lamarck** pick up a copy of *The Age of Lamarck* by Pietro Corsi or *Lamarck: The Mythical Precursor* by Madeline Barthelemy-Madaule.

Guglielmo **Marconi's** story is told in Gavin Weightman's *Signor Marconi's Magic Box* by looking at some of the different ways that his radio communication system changed the shape of the early twentieth century; from catching Dr Crippen to rescuing passengers from the *Titanic*. You may also want to consider *My Beloved Marconi*, by his wife Maria Marconi, *Marconi's Transatlantic Leap* by Gordon Bussey, and *Guglielmo Marconi* by David Gunston.

The unfortunate tale of Antonio **Meucci** is faithfully recorded in both *Antonio Meucci: The Inventor & his Times* by Basilio Catania, and *Antonio Meucci: Inventor of the Telephone* by Giovanni Schiavo.

Isaac **Newton** is another character who has unsurprisingly warranted a great deal of literary attention. I personally found Richard Westfall's *Never at Rest*, James Gleick's *Isaac Newton*, and John Fauvel's *Let Newton Be* all very well written. *Making of a Genius* by Patricia Fara is also very interesting, though it tends to go off at too many tangents for my liking. The father and son team of S. and D. Clark look at how Newton suppressed the discoveries of John Flamstead and Stephen Gray in their book *Newton's Tyranny*, and Betty Dobbs concentrates on Newton's dispute with Leibniz and his dealings in the shady world of alchemy in *The Janus Faces of Genius*. Neither of these last two deals specifically with the rivalry with Hooke, but they certainly paint an interesting picture of Newton and his rather unusual character.

Tim Jackson's *Inside Intel*, Zygmont's *Microchip*, and Cringley's *Accidental Empires* all discuss the life of Robert **Noyce** in some detail.

Many of the works on Nikola **Tesla** concentrate too heavily on the outlandish claims of his old age, however Margaret Cheney's *Tesla: Man Out of Time*, and Jill Jones' superb *Empires of Light* are certainly worth getting hold of. His autobiography *My Inventions* is also very informative, if a little self-indulgent.

Alan **Turing** life is thoroughly dissected in *Alan Turing: The Enigma* by Andrew Hodges. You might also consider Jon Agar's *Turing & the Universal Machine*, and the comprehensive anthology of his work entitled *The Essential Turing*.

Joseph **Swan's** autobiography, or the subsequent biography written by his sons make good reading. Both publications have the title *Sir Joseph Swan FRS*.

The best description of the life and times of John **von Neumann** that I managed to find was William Aspray's *John von Neumann & the Origins of Modern Computing*. It is also well worth looking at the many books that he wrote himself on the subjects of physics, mathematics, economics and computing.

To learn more about Alfred Russel **Wallace** try to get your hands on his autobiography *My Life*, as well as *The Forgotten Naturalist* by John Wilson, *In Darwin's Shadow* by Michael Shermer, and *Darwin's Moon* by Annabel Williams-Ellis.

There are also a number of articles I have written which can be found on the websites of several science, electronics and telecommunications magazines. My interview with Jack Kilby and my original article on Alan Turing are placed on the *Electronics News* website (www.electronicsnews.com). In addition, short editorials I penned on Meucci, Edison, Dummer, Shockley and Marconi have been published in *Electronic Product News* (www.epn-online.com).

Bibliography

BABBAGE, CHARLES

Charles Babbage: Pioneer of the Computer, by A. Hyman, Oxford University Press, 1983.
Charles Babbage: Passages from the Life of a Philosopher, by C. Babbage, IEEE Press, 1994.

BELL, ALEXANDER GRAHAM

Sounds Out of Silence, by J. Mackay, Mainstream Publishing, 1997.
Alexander Graham Bell, by E. Grovesnor & M. Weston, Abrahams, 1977.
The Bell Telephone, by G. Prescott, Appleton, 1884.

DARWIN, CHARLES

Darwin & Henslow, Letters 1831-1860, edited by N. Barlow, Murray, 1967.
The Autobiography of Charles Darwin, p.137, by C. Darwin, Icon Books, 2003.
Charles Darwin, by C. Aydon, Constable & Robinson, 2002.
The Origin of the Species, by C. Darwin, Penguin Books, 1985.
Voyage of the Beagle, by C. Darwin, Penguin Books, 1989.
The Voyage of Charles Darwin: His Autobiographical Writings, by C. Ralling, Ariel Books, 1982.
The Autobiography of Charles Darwin, by C. Darwin, Icon Books, 2003.
Beagle Diary, by C. Darwin, edited by R. Darwin-Keynes, Cambridge University Press, 1988.
The Correspondence of Charles Darwin. Vol.13 1865, edited by F. Burkhardt, Cambridge University Press, 2002.
The Correspondence of Charles Darwin. Vol.6, 1865, edited by F. Burkhardt, Cambridge University Press, 1990.
The Descent of Man, by C. Darwin, J. Murray, 1901.
Landmarks in Science, by R. Downs, Libraries Unlimited, 1982.
Life on Earth, by R. Attenborough, Collins, 1979.
Biology, by N. Campbell, Benjamin Cummings Publishing, 1990.
Vestiges of the Natural History of Creation & Other Evolutionary Writings, by R. Chambers, University of Chicago Press, 1994.

DARWIN, ERASMUS

Zoonomia, by E. Darwin,Vol.I, Byrne & Jones, 1794.
Erasmus Darwin, by D. King-Hele, De La Mere, 1999.

DUMMER, GEOFFREY

Miniature, and Microminiature Electronics, by G.W.A. Dummer & J.W. Granville, Pitman,
    1961.
Microminiaturisation, by G.W.A. Dummer, Pergamon Press, 1962.
Solid Circuits & Microminiaturisation, by G.W.A. Dummer, Pergamon Press, 1963.

ECKERT, PRESPER & MAUCHLY, JOHN

ENIAC, by S. McCartney, Berkley Books, 1999.
The Computer Comes of Age, by R. Moreau, MIT Press, 1984.
A History of Computing in the Twentieth Century, edited by N. Metropolis, J. Howlett and G.
    Rota, Academic Press, 1980.
Pioneers of Computing, by G. Ashurst, Frederick Muller Ltd, 1983.

EDISON, THOMAS ALVA

The Scientist, the Madman, the Thief & Their Lightbulb, by Keith Tutt, Pocket Books 2001.
Edison: The Man Who Made the Future, by R. Clark, MacDonald & James, 1977.
Empires of Light, by J. Jones, Random House, 2003.
Edison: His Life, His Work, His Genius, by W. Simonds, Bobbs-Merrill, 1934.
Edison & the Electric Chair, by M. Essig, Sutton Publishing, 2003.
Edison & The Business of Innovation, by A.J. Millard, John Hopkins Press, 1990.
Edison: A Life of Invention, by P. Israel, Wiley, 1998.
The Papers of Thomas A. Edison, by P. Israel, K. Nier, & L. Carcat, John Hopkins University
    Press, 1998.
The Invented Self, by D. Nye, Odense University Press, 1983.
The Diary & Sundry Observations of Thomas Alva Edison, edited by D. Runes,
    Philosophical Library, 1948.
Edison, by J. Josephson, Eyre & Sputtiswoode, 1961.

GRAY, ELISHA

The Gray Matter: The Forgotten Story of the Telephone, by B.H. Baker, Telepress, 2000.
Beginnings of Telephony, by F. Rhodes, Harper & Bros, 1929.
The Telephone: The First Hundred Years, Harper & Row, 1976.
The History of the Telephone, by A.C. McClurg, Chicago Press, 1911.
The Telephone Book, by H.C. Boettinger, Riverwood Publishing, 1977.
The Telephone Patent Conspiracy of 1876, by A.E. Evenson, McFarland, 2000.
American National Biography, Vol.IX, Oxford University Press, 2005.

HOOKE, ROBERT

The Curious Life of Robert Hooke, by L. Jardine, Harper Collins, 2003.

The Man Who Knew Too Much, by S. Inwood, Pan Books, 2003.

The Illustrated Pepys, edited by R. Latham, Penguin Books, 2000.

Micrographia, by R. Hooke, Royal Society, 1665.

Robert Hooke, by M. Espinasse, William Heinemann, 1956.

A More Beautiful City – Robert Hooke & the Rebuilding of London, by M. Cooper, Sutton
    Publishing, 2003.

London's Leonardo – The Life & Work of Robert Hooke, L. Jardine, J. Bennet, M. Cooper & M.
    Hunter, Oxford University, 2003.

Life of Robert Hooke, by R. Waller, London, 1705.

Aubrey's Brief Lives, by J. Aubrey, edited by D. Lawson, Seckler & Warburg, 1949.

The Diary of Robert Hooke (1672-80), edited by H. Robinson, Taylor & Francis, 1935.

Robert Hooke & the Royal Society, by R. Nichols, Book Guild, 1999.

The Diaries of Robert Hooke, by R. Nichols, Book Guild, 1994.

The Correspondence of Henry Oldenburg, Vol. II, (1663-1665), Hall & Hall, 1966.

Robert Hooke – New Studies, edited by M. Hunter & S. Schaffner, Boydell, 1989.

Lectures & Discourses on Earthquakes & Subterraneous Eruptions, by R. Hooke, Arno Press,
    1978.

Council Minutes of the Royal Society (1663-1682), Vol. I, Royal Society, 1682.

Philosophical Transactions of the Royal Society (1674-1678), Volume IX-X Royal Society, 1963.

The History of the Royal Society, by T. Birch, Volume I, 1756.

The History of the Royal Society, by T. Birch, Volume III, 1756.

The History of the Royal Society, by T. Birch, Volume IV, 1756.

The Scientific Achievements of Robert Hooke (Part 1), by P. Purgilese, Harvard University
    Press, 1982.

The Scientific Achievements of Robert Hooke (Part 2), by P. Purgilese, Harvard University
    Press, 1982.

History of the Royal Society, by T. Thompson, Royal Society, 1812.

Robert Hooke's Contributions to Mechanics, by F. Centore, Martinus Nijhoff, 1970.

Classified Papers of Robert Hooke (1660-1740), Volume XX, Royal Society, 1740.

Isaac Newton's Papers & Letters on Natural Philosophy, edited by B. Cohen, Harvard
    University Press, 1978.

Hooke's Micrographia (1665-1965), by A. Hall, p.26, University of London Press, 1966.

Extracts from Hooke's Micrographia, by R. Hooke, p.55, Oxford, 1926.

Early Science in Oxford, Volume VIII – The Cutler Lectures of Robert Hooke, edited by R.
    Gunther, Oxford, 1931.

KILBY, JACK

Microchip, by Jeffrey Zygmont, Perseus Publishing, 2003.

Accidental Empires, by R. Cringley, Penquin Books, 1996.

Proceedings of the IEEE, Volume 89, January 2000.

Interview with Jack Kilby, BBC World Service, October 2000.

Nobel Lectures, Elsevier Publishing, 2000.

Nobel Laureates Autobiography, Elsevier Publishing, 2000.

Interview with Gosta Ekspog, World Scientific Publishing, 2002.

LAMARCK, JEAN-BAPTISTE

Landmarks in Science, by R. Downs, Libraries Unlimited, 1984.
Lamarck: The Mythical Precursor, by M. Barthelemy-Madaule, MIT Press, 1982.
Zoological Philosophy, by J. Lamarck, translated by H.Elliot, University of Chicago Press, 1984.
Larmarck: His Life & Work, by A.S. Packard, Longmans, 1901.
The Age of Lamarck, by P. Corsi, University of Press, 1988.
The Spirit of System, by R. Burkhardt, Havard University Press, 1977.
Epigenetic Inheritance, by E. Jablonka & M. Lamb, Oxford University Press, 1995.
Late Eighteenth Century European Scientists, edited by R. Colby, Pegamon Press, 1966.
Lamarck, by L.J. Jordanova, Oxford University Press, 1984.
The Lamarck Manuscripts at Harvard, edited by W. Morton-Wheeler, & T. Barbour, 1933.
Nouvelle Archives du Museum Histoire Naturelle, Masson 1930.

MARCONI, GUGLIELMO

Tuned In: The life of Guglielmo Marconi, by M. Green, Hyperelectronics, 20 July 2001.
Nobel Lectures, Elsevier Publishing, 1967.
Signor Marconi's Magic Box, by Gavin Weightman, Harper Collins, 2003.
The Scientist, the Madman, the Thief, & their Light Bulb, by Keith Tutt, Pocket Books 2001.
Marconi: Master of Space, by B. Jacot & D. Collier, Hutchinson & Co., 1935.
Marconi, by W. Polly, Constable, 1972.
Guglielmo Marconi, by D. Gunston, 1970.
Proving the Practicality of Wireless Telegraphy, by G. Marconi, McLure, 1926.
My Father, Marconi, by D. Marconi, Frederick Muller, 1962.
Developments in Wireless Telegraphy, by G. Marconi, Birmingham & Midland Institute, 1921.
My Beloved Marconi, by M. Marconi, Dante University Press, 1999.
Marconi's Transatlantic Leap, by G. Bussey, Cambridge University Press, 2000.
Marconi & His Wireless Stations in Wales, by H. Williams, Gwasg Carreg, 1999.
Marconi & the Discovery of Wireless, by L. Reads, Faber & Faber, 1963.
Marconi on the Isle of Wight, by T. Wander, TRW, 2000.

MEUCCI, ANTONIO

The Story of Antonio Meucci, by M. Green, Hyperelectronics, 15 May, 2001.
Antonio Meucci: The Inventor & His Times, by B. Catania, Seat, 1996.
3rd International Symposium on Telecommunication History, edited by R. Azer, Denton, 1995.
Antonio Meucci (1808-1889), by M. Nesse & F. Nicotre, Italy-Italy Corp, 1989.
Antonio Meucci: Inventor of the Telephone, by G. Shiavo, Vigo Press, 1958.

NEWTON, SIR ISAAC

Newton: Texts, Background, Commentaries, edited by B. Cohen, & R. Westfall, Norton, 1995.
Isaac Newton, by J. Gleick, Harper Perennial, 2003.
Newton – The Making of Genius, by P. Fara, Picador, 2002.

Newton His Friends & His Foes, by B.R. Hall, Variorum, 1993.

Correspondence of Sir Isaac Newton, edited by J. Edleston, Frank Cass & Co., 1969.

The Optical Papers of Isaac Newton, Volume I (1670-1672), edited by A. Shapiro, Cambridge University Press, 1984.

The Principia, by I. Newton, translated by I. Cohen & A. Whitman, University of California Press, 1999.

The Janus Face of Genius, by B. Dobbs, Cambridge University Press, 1991.

Isaac Newton's Natural Philosophy, edited by J. Buchwald & I. Cohen, MIT Press, 2000.

Newton's Tyranny, by P. Clark, Freeman, 2001.

Brief Lives: Sir Isaac Newton, by E. Andrade, Collins, 1954.

Isaac Newton's Papers & Letters on Natural Philosophy, edited by B. Cohen, Harvard University Press, 1978.

Never at Rest: A Biography of Isaac Newton, by R. Westfall, Cambridge University Press, 1995.

NOYCE, BOB

'The Tinkerings of Bob Noyce', by Tom Wolfe, *Esquire Magazine*, December 1983.

Inside Intel, by T. Jackson, Plume, 1997.

Microchip, by Jeffrey Zygmont, Perseus Publishing, 2003.

Accidental Empires, by R. Cringley, Penquin Books, 1996.

Silicon Valley: Past Present Future, by Mike Green, Reed Business Information, August 2003.

SWAN, SIR JOSEPH

Sir Joseph Wilson Swan F.R.S., by J.W. Swan, Ernst Benn, 1929.

Encyclopaedia Britannica, Volume XI, 1993.

The Millennium History of the North East, by D. Simpson, Leighton, 1999.

A Short History of the Electric Light, by F. Andrews, Andrews, 2003.

Sir Joseph Swan F.R.S., by M.E. & K.R. Swan, Oriel, 1968.

Sir Joseph Swan & the Invention of the Incandecsent Lamp, by K. Swan, Longmans, 1948.

A Brilliant Ray, by F. Haveron, Godalming, 1981.

TESLA, NIKOLA

The Electrical Experimenter, September, 1917.

The Scientist, the Madman, the Thief, & their Light Bulb, by Keith Tutt, Pocket Books 2001.

Tesla: Man Out of Time, by M. Cheney, Touchstone, 2001.

Tesla: The Modern Sorcerer, by D.B. Stewart, North Atlantic Books, 1999.

Lightning in His Hand, by W. Draper, Omni Publications, 1964.

Priority in the Invention of Radio – Tesla vs. Marconi, by L. Anderson, Antique Wireless Association, 1980.

In Search of Nikola Tesla, by F. Pleat, Ashgrove Press, 1998.

Nikola Tesla: A Spark of Genius, by C. Dommermuth, Lerner Publications, 1994.

My Inventions, by N. Tesla, edited by B. Johnston, Hart Bros, 1982.

Wizard: The Life & Times of Nikola Tesla, by M. Seifer, Birch Lane Press, 1996.

Strange Brains & Genius: the Secret lives of Eccentric Scientists & Madmen, by C. Pickover, Plenum, 1998.

Empires of Light, by J. Jones, Random House, 2003.
Nikola Tesla: Lectures, Patents, Articles, edited by V. Popovic, Tesla Museum Publishing, 2004.
My Inventions, by N. Tesla, p.8, Skolska Knjiga, 1977.
Complete Patents of Nikola Tesla, edited by J. Genn, Barres & Noble, 1994.
Encyclopaedia Britannica, Volume XI, 1993.

TURING, ALAN

'Code Breakers Remembered', by M. Green, *Electronic News*, published May 2003.
Alan Turing: The Enigma, by A. Hodges, Vintage, 1992.
Alan M. Turing, by S. Turing, W. Heffer & Sons, 1959.
Philosophical Transactions of The Royal Society, Volume 237, Royal Society, 1951.
On Computable Number with an Application to the Entscheidungsproblem, London
    Mathematical Society, 1937.
Biographical Memoirs of the Fellow of the Royal Society, Volume I, Royal Society, 1955.
Station X: The Code Breakers of Bletchley Park, by M. Smith, Channel Four Books, 1998.
Top Secret Ultra, by P. Calvocoressi, M&M Baldwin, 2001.
The Emperor's Code, by M. Smith, Bantham Press, 2000.
The Hut Six Story, by G. Welchman, M&M Baldwin, 1997.
The Turing Omnibus, by A.K. Dewdney, Computer Science Press, 1989.
The Universal Turing Machine, edited by R. Herlen, Oxford University Press, 1988.
The Essential Turing, edited by J. Copeland, Clarendon Press, 2004.
Turing & the Universal Machine, J. Agar, Icon Books, 2001.

VON NEUMANN, JOHN

John von Neumann & the Origins of Modern Computing, by W. Aspray, MIT Press, 1990.
John von Neumann: Collected Works, edited by A.H. Taub, 1903-1957, Pergamon Press, 1963.
The Computer & the Brain, by J. von Neumann, Yale University Press, 1958.
The Computer Comes of Age, by R. Moreau, MIT Press, 1984.
A History of Computing in the 20th Century, edited by N. Metropolis, J. Howlett & G. Rota,
    Academic Press, 1980.
Pioneers of Computing, by G. Ashurst, Frederick Muller Ltd, 1983.
The Making of the Atomic Bomb, by R. Rhodes, Touchstone, 1986.
The Computer & the Brain, by J. von Neumann, p.65, Yale University Press, 1957.
John von Neumann & the Origins of Modern Computing, by W. Apray, MIT Press, 1990.
John von Neumann: As Seen by His Brother, by N. von Neumann, Library of Congress, 1987.
The Theory of Games & Economic Behaviour, by J. von Neumann & D. Morgenstern,
    Princeton University Press, 1947.

WALLACE, ALFRED RUSSEL

In Darwin's Shadow, by M. Shermer, Oxford 2002.
The Forgotten Naturalist: In Search of Alfred Russel Wallace, by J. Wilson, Arcadia, 2000.
My Life, by A.R. Wallace, Volume I, Chapman & Hall, 1905.
My Life, by A.R. Wallace, Volume II, Chapman & Hall, 1905.
Proceedings of the Linnaean Society, Volume 3, Longmans, 1867.

Biologist Philosospher: A Study of the Life & Writings of Alfred Russel Wallace, by W. George, Aberland-Schuman, 1964.

Interview with C.H. Smith for *Book Monthly*, published May 1905.

Interview with E.H. Rann for *Pall Mall* magazine, published May 1909.

Interview with S. Tooley, for *The Humanitarian*, published February 1894.

Interview with W.B. Northrop, for *The Outlook*, published November 1913.

Interview with A. Dawson, for *The Bookman*, published January 1898.

Interview with A. Dawson, for *The Christian Commonwealth*, published January 1903.

Interview with H. Begbie for *The Daily Chronicle*, published May 1910.

Alfred Russel Wallace: Letters & Reminiscences, by J. Marchant, Volume I, Cassel, 1916.

Alfred Russel Wallace: Letters & Reminiscences, by J. Marchant, Volume II, Cassel, 1916.

Alfred Russel Wallace: Biologist & Social & Reformer, by H. Clements, Hutchinson, 1983.

Darwin's Moon, by A. Williams-Ellis, Blackie, 1966.

Just Before the Origin, by J. Langdon-Brook, Coulmbia University Press, 1984.

Malay Archipelago, by A.R. Wallace, Macmillan & Co., 1880.

Island Life, by A.R. Wallace, Macmillan & Co., 1880.

Wallace & Natural Selection, by H. Lewis-McKinney, Yale University Press, 1972.

The Revolt on Democracy, by A.R. Wallace, Cassell, 1913.

Alfred Russel Wallace: A Life, by P. Raby, Chato & Windus, 2001.

The Evolution Debate (1813-1870), edited by D. Knight, Volume IX, Routelge, 2003.

Footsteps in the Forest, by S. Knapp, Natural History Museum, 2000.

Index

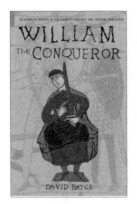

William the Conqueror
DAVID BATES

'As expertly woven as the Bayeux Tapestry'
BBC History Magazine
'No one is better qualified than David Bates to interpret King William's character' **History**

£12.99   0 7524 2960 4

Elizabeth I
FORTUNE'S BASTARD?

RICHARD REX

£9.99   978 07524 4176 4

The Cat Orchestra & the Elephant Butler
A STRANGE HISTORY OF AMAZING ANIMALS

JAN BONDESON

'Delightful' **Ali Smith**
'Animal magic' **TLS**
'Compelling' **The Daily Mail**

£20   0 7524 3934 0

The End of the Third Reich
DEFEAT, DENAZIFICATION & NUREMBERG,
JANUARY 1944 – NOVEMBER 1946

TOBY THACKER

'Absorbing...judicious... well-written'
BBC History Magazine

£20   0 7524 3939 1

TEMPUS – REVEALING HISTORY

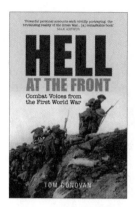

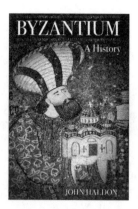

Hell at the Front
COMBAT VOICES FROM THE FIRST WORLD WAR

TOM DONOVAN

'Fifty powerful personal accounts each vividly portraying the brutalising reality of the Great War... a remarkable book' *Max Arthur*

£12.99   0 7524 3940 5

Pirates
JOEL BAER

£9.99   978 07524 4298 3

Byzantium
A HISTORY

JOHN HALDON

'A triumph of synthesis'
Speculum: A Journal of Medieval Studies

£12.99   0 7524 3472 1

The Story of England
TOM BEAUMONT JAMES

FOREWORD BY *Mick Aston*

'Skilfully blends history and archaeology... a fine, no nonsense introduction' *History Today*

£12.99   0 7524 2578 1

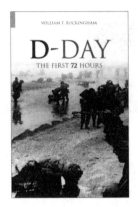

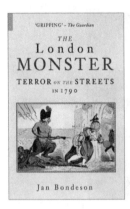

D-Day The First 72 Hours
WILLIAM F. BUCKINGHAM

'A compelling narrative' *The Observer*
A *BBC History Magazine* Book of the Year 2004

£9.99   0 7524 2842 X

The London Monster
Terror on the Streets in 1790
JAN BONDESON

'Gripping' *The Guardian*
'Excellent... monster-mania brought a reign of terror to the ill-lit streets of the capital'
The Independent

£9.99   0 7524 3327 X

London
A Historical Companion
KENNETH PANTON

'A readable and reliable work of reference that deserves a place on every Londoner's bookshelf'
Stephen Inwood

£20   0 7524 3434 9

M: MI5's First Spymaster
ANDREW COOK

'Serious spook history' *Andrew Roberts*
'Groundbreaking' *The Sunday Telegraph*
'Brilliantly researched' *Dame Stella Rimington*

£9.99   978 07524 3949 9t

TEMPUS – REVEALING HISTORY

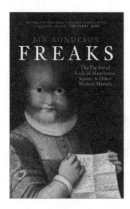

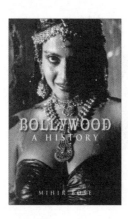

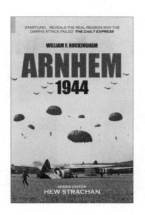

Freaks
JAN BONDESON

'Reveals how these tragic individuals triumphed over their terrible adversity' *The Daily Mail*
'Well written and superbly illustrated'
The Financial Times

£9.99   0 7524 3662 7

Bollywood
MIHIR BOSE

'Pure entertainment' *The Observer*
'Insightful and often hilarious' *The Sunday Times*
'Gripping' *The Daily Telegraph*

£9.99   978 07524 4382 9

King Arthur
CHRISTOPHER HIBBERT

'A pearl of biographers' *New Statesman*

£12.99   978 07524 3933 4

Arnhem
William Buckingham

'Reveals the reason why the daring attack failed'
The Daily Express

£10.99   0 7524 3187 0

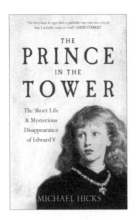
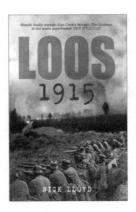

Cleopatra
PATRICIA SOUTHERN

'In the absence of Cleopatra's memoirs Patricia Southern's commendably balanced biography will do very well'
The Sunday Telegraph

£9.99   978 07524 4336 2

The Prince In The Tower
MICHAEL HICKS

'The first time in ages that a publisher has sent me a book I actually want to read' **David Starkey**

£9.99   978 07524 4386 7

The Battle of Hastings 1066
M. K. LAWSON

'A **BBC History Magazine** book of the year 2003

'The definitive book on this famous battle'
The Journal of Military History

£12.99   978 07524 4177 1

Loos 1915
NICK LLOYD

'A revealing new account based on meticulous documentary research' **Corelli Barnett**

'Should finally consign Alan Clark's Farrago, The Donkeys, to the waste paperbasket'
Hew Strachan

'Plugs a yawning gap in the existing literature... this book will set the agenda for debate of the battle for years to come' **Gary Sheffield**

£25   0 7524 3937 5

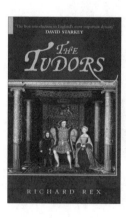

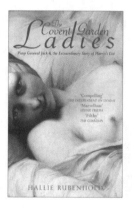

Quacks Fakers and Charlatans in Medicine
ROY PORTER

'A delightful book' *The Daily Telegraph*
'Hugely entertaining' *BBC History Magazine*

£12.99   0 7524 2590 0

The Tudors
RICHARD REX

'Up-to-date, readable and reliable. The best
introduction to England's most important
dynasty' *David Starkey*
'Vivid, entertaining... quite simply the best short
introduction' *Eamon Duffy*
'Told with enviable narrative skill... a delight for
any reader' *THES*

£9.99   0 7524 3333 4

The Kings & Queens of England
MARK ORMROD

'Of the numerous books on the kings and
queens of England, this is the best'
Alison Weir

£9.99   0 7524 2598 6

The Covent Garden Ladies
Pimp General Jack & the Extraordinary Story of Harris's List
HALLIE RUBENHOLD

'Sex toys, porn... forget Ann Summers, Miss
Love was at it 250 years ago' *The Times*
'Compelling' *The Independent on Sunday*
'Marvellous' *Leonie Frieda*
'Filthy' *The Guardian*

£9.99   0 7524 3739 9

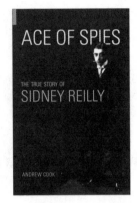

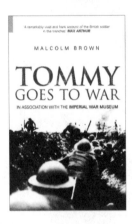

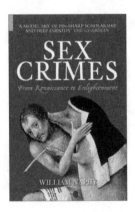

Okinawa 1945
GEORGE FEIFER

'A great book... Feifer's account of the three
sides and their experiences far surpasses most
books about war'
Stephen Ambrose

£17.99   0 7524 3324 5

Tommy Goes To War
MALCOLM BROWN

'A remarkably vivid and frank account of the
British soldier in the trenches'
Max Arthur

'The fury, fear, mud, blood, boredom and
bravery that made up life on the Western Front
are vividly presented and illustrated'
The Sunday Telegraph

£12.99   0 7524 2980 4

Ace of Spies The True Story of Sidney Reilly
ANDREW COOK

'The most definitive biography of the spying
ace yet written... both a compelling narrative
and a myth-shattering *tour de force*'
Simon Sebag Montefiore

'The absolute last word on the subject' *Nigel West*

'Makes poor 007 look like a bit of a wuss'
The Mail on Sunday

£12.99   0 7524 2959 0

Sex Crimes
From Renaissance to Enlightenment
W.M. NAPHY

'Wonderfully scandalous' *Diarmaid MacCulloch*

'A model of pin-sharp scholarship' *The Guardian*

£10.99   0 7524 2977 9

TEMPUS REVEALING HISTORY

Witchcraft
A HISTORY

P.G. MAXWELL-STUART

'Combines scholarly rigour with literary flair'
The Independent on Sunday
'Excellently illustrated *Witchcraft* is an intelligent
exploration that leaves us eager to know more'
History Today

£9.99   0 7524 2966 3

The Third Reich
A CONCISE HISTORY

MARTIN KITCHEN

'Written by one of the most admired schoalrs of
twentieth-century German history'
History Today

£9.99   978 07524 4354 6

Wizards
A HISTORY

P.G. MAXWELL-STUART

'This is a fascinating, well-researched and lucid
book' *Malcolm Gaskill*
'An excellent pioneering work and a fascinating
and entertaining book' *Ronald Hutton*

£9.99   978 07524 4127 6

Caesar
PATRICIA SOUTHERN

'Her style is delightfully approachable: lean and
lucid, witty and pacy' *Antiquity*

£9.99   978 07524 4394 2

If you are interested in purchasing other books published by Tempus,
or in case you have difficulty finding any Tempus books in your local bookshop,
you can also place orders directly through our website

www.tempus-publishing.com